T0310571

NEUROMETHODS ☐ 6

Peptides

NEUROMETHODS

Program Editors: Alan A. Boulton and Glen B. Baker

NEUROMETHODS

Program Editors: Alan A. Boulton and Glen B. Baker

Peptides

Edited by

Alan A. Boulton

University of Saskatchewan,

Glen B. Baker

University of Alberta,

and

Q. J. Pittman

University of Calgary

Humana Press • Clifton, New Jersey

Library of Congress Cataloging-in-Publication Data
Peptides:

(Neuromethods ; no. 6. Series I, Neurochemistry)
Includes bibliographies and index.
1. Neuropeptides—Analysis.   2. Neurochemistry—
Technique.   I. Boulton, A. A. (Alan A.)   II. Baker,
Glen B., 1947–   .   III. Pittmann, Q. J. (Quentin J.)
IV. Series: Neuromethods ; 6.   V. Series: Neuromethods.
Series I, Neurochemistry. [DNLM: 1. Neurochemistry—
methods.   2. Peptides—analysis.   W1 NE337G v.6 /
QU 68 P4241]
QP552.N39P47   1987   599'.0188   87-4064
ISBN 0-89603-105-5

© 1987 The Humana Press Inc.
Crescent Manor
PO Box 2148
Clifton, NJ 07015

All rights reserved

No part of this book may be reproduced, stored in a retrieval
system, or transmitted in any form or by any means, electronic,
mechanical, photocopying, microfilming, recording, or otherwise
without written permission from the Publisher.

Printed in the United States of America

Foreword

Techniques in the neurosciences are evolving rapidly. There are currently very few volumes dedicated to the methodology employed by neuroscientists, and those that are available often seem either out of date or limited in scope. This series is about the methods most widely used by modern-day neuroscientists and is written by their colleagues who are practicing experts.

Volume 1 will be useful to all neuroscientists since it concerns those procedures used routinely across the widest range of subdisciplines. Collecting these general techniques together in a single volume strikes us not only as a service, but will no doubt prove of exceptional utilitarian value as well. Volumes 2 and 3 describe current procedures for the analyses of amines and their metabolites and of amino acids, respectively. These collections will clearly be of value to all neuroscientists working in or contemplating research in these fields. Similar reasons exist for Volume 4 on receptor binding techniques, since experimental details are provided for many types of ligand-receptor binding, including chapters on general principles, drug discovery and development, and a most useful appendix on computer programs for Scatchard, nonlinear, and competitive displacement analyses. Volume 5 provides procedures for the assessment of enzymes involved in neurotransmitter synthesis and catabolism.

Volumes in the NEUROMETHODS series will be useful to neurochemists, -pharmacologists, -physiologists, -anatomists, psychopharmacologists, psychiatrists, neurologists, and chemists (organic, analytical, pharmaceutical, medicinal); in fact, everyone involved in the neurosciences, both basic and clinical.

Preface to the Series

When the President of Humana Press first suggested that a series on methods in the neurosciences might be useful, one of us (AAB) was quite skeptical; only after discussions with GBB and some searching both of memory and library shelves did it seem that perhaps the publisher was right. Although some excellent methods books have recently appeared, notably in neuroanatomy, it is a fact that there is a dearth in this particular field, a fact attested to by the alacrity and enthusiasm with which most of the contributors to this series accepted our invitations and suggested additional topics and areas. After a somewhat hesitant start, essentially in the neurochemistry section, the series has grown and will encompass neurochemistry, neuropsychiatry, neurology, neuropathology, neurogenetics, neuroethology, molecular neurobiology, animal models of nervous disease, and no doubt many more "neuros." Although we have tried to include adequate methodological detail and in many cases detailed protocols, we have also tried to include wherever possible a short introductory review of the methods and/or related substances, comparisons with other methods, and the relationship of the substances being analyzed to neurological and psychiatric disorders. Recognizing our own limitations, we have invited a guest editor to join with us on most volumes in order to ensure complete coverage of the field and to add their specialized knowledge and competencies. We anticipate that this series will fill a gap; we can only hope that it will be filled appropriately and with the right amount of expertise with respect to each method, substance or group of substances, and area treated.

Alan A. Boulton
Glen B. Baker

Preface

Neuropeptides have been in the forefront of neuropharmacology for some ten years, and the numbers of peptides found in neural tissue are far in excess of those of other known transmitter systems. Progress in identifying the roles of neural peptides in brain has come from a variety of experimental approaches, many of which are similar to those that have proven fruitful in studies of other classical transmitter systems. Therefore, rather than present a detailed methodological recipe for each individual technique, the authors, all of whom are experts in their respective fields, have been requested to evaluate their particular methodology in terms of its contribution to our understanding of neuropeptides. They have highlighted the advantages and pitfalls, and have illustrated the data that could be obtained using such techniques so that the nonexpert can develop an appreciation for the various approaches.

Perhaps one of the most exciting conceptual advances in our appreciation of neuropeptides as neurotransmitters has been the discovery that prohormones are processed into a multitude of biologically and/or structurally related fragments. Crine and Boileau have reviewed this field in the first chapter, indicating how investigations into posttranslation modifications of peptides have provided insight into their biological activities. These structurally related peptide fragments have often frustrated efforts to quantitate and measure them, but with the use of highly specific and sensitive antibodies, this task has been made easier. Benoit et al. have discussed their strategies for producing such antibodies and developing radioimmunoassays. With the use of such antibodies, it has been possible both to localize peptides to specific neuronal tracts in brain and to measure their release from brain. The former topic, that of immunohistochemical localization, is discussed by van Leeuwen, and in vivo and in vitro studies on peptide release have been critically examined by Bayón.

Given the numbers of related peptide fragments, it is not surprising that there appears to be an equally abundant diversity of receptors. To provide us with information about the molecular nature of these various receptor surfaces, St-Pierre has discussed the use of peptide analogs and structure activities. Dorsa and Baskin have then shown how peptides or peptide analogs can be utilized to characterize the type and distribution of receptors throughout neural tissue.

The remaining chapters of the book examine the postsynaptic consequences of peptide binding to receptors. Magistretti examines the biochemical consequences of peptide action with particular emphasis on cyclic AMP and glycogen hydrolysis measurements. With the use of intact behaving animals, it has been possible to provide information about peptide actions on behavior as well as on various integrative autonomic functions. Thus, Dunn and Berridge have critically examined the use of various behavioral tests to evaluate the actions of peptides on behavior, and Naylor et al. have discussed the issues relating to pharmacological investigations of peptidergic pathways involved in autonomic control in conscious animals.

The last three chapters discuss the use of electrophysiological approaches to evaluate peptide actions. Ferguson and Renaud discuss the use of in vivo electrophysiological techniques to evaluate peptide actions and specific neural circuits, whereas Pittman et al. discuss the use of in vitro preparations to permit detailed membrane analysis of peptide actions. Finally, in the last chapter, Lukowiak and Murphy discuss their use of molluscan model systems to provide information about peptide action relevant not only to the behavior and physiology of the mollusc, but also possibly to that of higher vertebrates, including mammals.

It is hoped that this book will provide the interested reader with a better appreciation of the various approaches and their relative strengths and weaknesses in the study of peptide actions and peptidergic neurons.

Q. J. Pittman

Contents

PEPTIDES: *Strategies for Antobody Production and Radioimmunoassays*
Robert Benoit, Nicholas Ling, Paul Brazeau, Solange Lavielle, and Roger Guillemin

IMMUNOCYTOCHEMICAL TECHNIQUES IN PEPTIDE LOCALIZATION: *Possibilities and Pitfalls*
F. W. van Leeuwen

IN VIVO AND IN VITRO STUDIES ON PEPTIDE RELEASE
Alejandro Bayón

FUNCTIONAL IDENTIFICATION OF PEPTIDE RECEPTORS
Serge A. St-Pierre

IDENTIFICATION OF NEUROPEPTIDE RECEPTORS
Daniel M. Dorsa and Denis G. Baskin

BIOCHEMICAL APPROACHES TO THE STUDY OF PEPTIDE ACTIONS
Pierre J. Magistretti

BEHAVIORAL TESTS: *Their Interpretation and
Significance in the Study of Peptide Action*
Adrian J. Dunn and Craig W. Berridge

INTEGRATIVE PHYSIOLOGICAL STUDIES OF PEPTIDES IN
THE CENTRAL NERVOUS SYSTEM
A. M. Naylor, W. D. Ruwe, and W. L. Veale

IN VIVO ELECTROPHYSIOLOGICAL TECHNIQUES IN THE
STUDY OF PEPTIDERGIC NEURONS AND ACTIONS
Alastair V. Ferguson and Leo P. Renaud

IN VITRO PREPARATIONS FOR ELECTROPHYSIOLOGICAL
STUDY OF PEPTIDE NEURONS AND ACTIONS
Q. J. Pittman, B. A. MacVicar, and W. F. Colmers

MOLLUSCAN MODEL SYSTEMS FOR THE STUDY OF
NEUROPEPTIDES
Ken Lukowiak and A. Don Murphy

Contributors

GLEN B. BAKER • *Neurochemical Research Unit, Department of Psychiatry, University of Alberta, Edmonton, Alberta, Canada*

DENIS G. BASKIN • *Departments of Medicine and Biological Structure, University of Washington School of Medicine, and General Medical Research, Veterans Administration Medical Center, Seattle, Washington*

ALEJANDRO BAYÓN • *Department of Molecular Biology, Instituto de Investigaciones Biomedicas, Universidad Nacional Autonoma de Mexico and Analytical Neurochemistry Unit, Instituto Mexicano de Psiquiatria, Mexico City, Mexico*

ROBERT BENOIT • *Neuroendocrinology Laboratories, The Salk Institute for Biological Studies, San Diego, California and The Montréal General Hospital Research Institute, McGill University, Montreal, Quebec, Canada*

CRAIG W. BERRIDGE • *Department of Neuroscience, University of Florida College of Medicine, Gainesville, Florida*

GUY BOILEAU • *Department de Biochimie, Université de Montréal, Montreal, Quebec, Canada*

ALAN A. BOULTON • *Psychiatric Research Division, University of Saskatchewan, Saskatoon, Saskatchewan, Canada*

PAUL BRAZEAU • *Neuroendocrinology Laboratories, The Salk Institute for Biological Studies, San Diego, California*

W. F. COLMERS • *Neuroscience Research Group, University of Calgary, Calgary, Alberta, Canada*

PHILIPPE CRINE • *Department de Biochimie, Université de Montréal, Montreal, Quebec*

DANIEL M. DORSA • *Geriatric Research Education and Clinical Center, Veterans Administration Medical Center and Departments of Medicine and Pharmacology, University of Washington School of Medicine, Seattle, Washington*

ADRIAN J. DUNN • *Department of Neuroscience, University of Florida College of Medicine, Gainesville, Florida*

ALASTAIR V. FERGUSON • *Department of Physiology, Queen's University, Kingston, Ontario, Canada*

ROGER GUILLEMIN • *Neuroendocrinology Laboratories, The Salk Institute for Biological Studies, San Diego, California*

SOLANGE LAVIELLE • *Neuroendocrinology Laboratories, The Salk Institute for Biological Studies, San Diego, California*

NICHOLAS LING • *Neuroendocrinology Laboratories, The Salk Institute for Biological Studies, San Diego, California*

KEN LUKOWIAK • *Department of Medical Physiology, The University of Calgary, Calgary, Alberta, Canada*

B. A. MACVICAR • *Neuroscience Research Group, University of Calgary, Calgary, Alberta, Canada*

PIERRE J. MAGISTRETTI • *Department de Pharmacologie, Centre Medical Universitaire, Geneva, Switzerland*

A. DON MURPHY • *Department of Medical Physiology, The University of Calgary, Alberta, Canada*

A. M. NAYLOR • *Department of Medical Physiology, The University of Calgary, Calgary, Alberta, Canada*

Q. J. PITTMAN • *Neuroscience Research Group, University of Calgary, Calgary, Alberta, Canada*

LEO P. RENAUD • *Neurosciences Unit, Montreal General Hospital and McGill University, Montreal, Canada*

W. D. RUWE • *Department of Medical Physiology, The University of Calgary, Calgary, Alberta, Canada*

SERGE A. ST-PIERRE • *INRS-Santé, Montreal, Quebec, Canada*

F. W. VAN LEEUWEN • *Netherlands Institute for Brain Research, Amsterdam, The Netherlands*

W. L. VEALE • *Department of Medical Physiology, The University of Calgary, Calgary, Alberta, Canada*

Posttranslational Processing of Peptide Precursors to Fragments

Philippe Crine and Guy Boileau

1. Introduction

In theory only one excitatory and one inhibitory transmitter should be sufficient to operate the nervous system. Therefore the recent discovery that, besides an already large number of well-established classical neurotransmitters, many small peptides are also capable of converting neural signals into physiological responses, has evoked both interest and doubt. There seem to be about two dozen peptide neurotransmitter candidates and the number is increasing rapidly. Peptides that may serve as neurotransmitters or neuromodulators have been found in both the central nervous system and in a wide variety of peripheral organs. These include angiotensin II, members of the gastrin/cholecystokinin family, as well as large peptide hormones, such as prolactin, growth hormone, insulin, glucagon, and many peptides derived from proopiomelanocortin (POMC), the common precursor to ACTH and β-endorphin in the pituitary. In peripheral tissues some of these peptides have been shown or were already known to be released in the circulation, where they could act as hormones. The cells responsible for the synthesis of these neuropeptides have been assigned to Pearse's category of APUD (amine precursor uptake and decarboxylation) cells (Pearse, 1969). APUD cells constitute a system often referred to as the diffuse neuroendocrine system (Pearse and Takor, 1979). For this reason these peptides are often called neuroendocrine peptides.

The first neurons that have been shown to produce and release peptides are those that originate in the supraoptic and paraventricular nuclei of the hypothalamus. These neurons release the octapeptides oxytocin and vasopressin into the blood through their nerve endings located in the posterior pituitary (Du Vigneaud, 1956). Then came the discovery of three hypothalamic "hormones," thyrotropin releasing hormone (TRH), luteinizing hormone releasing hormone (LHRH), and growth hormone release-inhibiting hormone (somatostatin) (Schally et al., 1973; Guillemin, 1978) synthesized in parvocellular tuberoinfundibular neurons (Szentagothai et al., 1962). Immunohistochemical studies have now shown that all these peptides are fairly widely distributed outside the hypothalamus.

Current information concerning the structural, biochemical, and physiological properties of peptide neurons is very limited. Nevertheless it now seems well established that replenishment of neuropeptides to nerve endings is very different from that of classical neurotransmitters (Hokfelt et al., 1980). In the case of noradrenaline, for instance, several efficient mechanisms have been shown to maintain fairly constant levels of neurotransmitters in nerve endings: (1) *de novo* enzymatic synthesis in the nerve ending themselves, (2) reuptake from the synaptic space through an active membrane mechanism, (3) supply of amine (or their precursors) in storage vesicles from the cell body via axonal transport. Peptides are synthesized exclusively on the ribosomes of the cell body, most probably in the form of a larger precursor molecule. Posttranslation maturation (and therefore generation of bioactive species) occurs during axonal transport. Thus, in the case of neuropeptides, there is no possible local synthesis in the nerve endings. Besides, no reuptake mechanism has been shown to exist for neuropeptides. It is therefore most probable that every single peptide molecule released must be replaced by axonal transport. These considerations emphasize the importance of biochemical studies aimed at solving the mode of synthesis and the mechanism by which neuropeptides are transported to nerve endings and processed.

2. Biosynthesis of Neuropeptides: The Concept of Polyprotein Precursors

Most of the neuropeptides have been discovered serendipitously with no systematic search. For instance, cholecystokinin, vasoactive intestinal polypeptide, and substance P were first known as intestinal hormones (Said and Mutt, 1970; Ivy et al., 1929; von Euler and Gaddum, 1931) and later recognized as brain constituents. Some of them, however, such as neurotensin (Carraway and Leeman, 1973) were first discovered in brain extracts. Very few have been discovered through the systematic screening of tissue fractions with the aid of a bioassay. Classical examples of peptides found with this kind of systematic search include the enkephalins (Hughes et al., 1975) and the pituitary hormone releasing or inhibiting factors (Guillemin, 1978). When a homogeneous material is obtained, its primary structure can be established by amino acid sequencing techniques. The determination of the amino acid sequence of the neuropeptide, however, tells us very little about its mode of synthesis and regulation. Most of what we know about the mode of synthesis of neuropeptides has come from classical biochemical studies on the synthesis of polypeptide hormones. These studies have been performed by metabolic incorporation of radioactive amino acids into proteins, using tissue explants or cell culture systems. This approach is, however, seldom applicable to neuropeptides for two reasons: (1) neuropeptides are usually synthesized at low levels in brain tissue, making detection of radiolabeled proteins very difficult; and (2) brain tissue does not survive very long in culture.

The major impetus to the field of neuropeptide synthesis came with the advent of recombinant DNA technology. These techniques allowed for the construction and isolation of cDNA copies of mRNA responsible for the synthesis of neuropeptides. The sequence of cDNAs led in turn to the establishment of the complete primary structure of the protein. It was then realized that all these small neuroendocrine peptides are originally synthesized in the form of large polypeptide precursors. Moreover, it was also found that these precursors contain, within their sequence the structure of several biologically active peptides and that each of these peptides has to be excised from the precursor to exert its biological activity. Such polypeptides encoding several neuroendocrine precursors within their sequence are now known as polyproteins (Douglass, et al., 1984). Proteolytic processing of these polyproteins therefore allows the central nervous system to elaborate the variety of neuropeptides necessary to control complex neuronal functions.

In this chapter, we will review the various strategies that are most commonly used for isolating neuroendocrine peptides, for establishing the primary structure of their precursor and for studying the processing mechanism by which they are cleaved into various bioactive fragments.

3. Extraction of Neuropeptides From Tissues and Body Fluids

Peptides are extremely susceptible to proteolytic enzyme degradation. Therefore, the risk of purifying a peptide that has been generated artifactually during the extraction process is high. The most efficient way of inhibiting peptidase activity during the isolation is to employ an extraction medium with a very low pH (i.e., 1*M* HCl) at 4°C. Bennett et al. (1981a) have introduced a purification method in which this initial low pH extraction step is followed by reverse-phase liquid chromatography for the simultaneous desalting and deproteinization of the samples. This methodology represents a new general approach for the extraction and isolation of peptides. Briefly, the method includes a deproteinization step on octadecylsilylsilica (ODS-silica) cartridges (C_{18} Sep-Pak, Waters Instruments). Small peptides are able to penetrate the pores of ODS-silica, whereas proteins are usually excluded. Peptides bound to the ODS-silica cartridge can be recovered by eluting with 80% acetonitrile. After extraction and deproteinization, peptides can be readily separated by reserve-phase high performance liquid chromatography (HPLC). An "ion-pairing" reagent is usually included in the solvent to increase the resolution (Hearn and Hancock, 1979). By performing reverse-phase HPLC steps using successively trifluoroacetic acid and heptafluorobutyric acid as ion-pairing reagents, Bennett et al. (1982, 1983) have isolated and identified a variety of peptides from pituitary tissue extracts. This methodology has several advantages over more traditional procedures in which HPLC was almost invariably used in the final purification step following conventional ion exchange and gel filtration procedures. This method is simple and rapid and ensures the recovery of unmodified peptides with high yields. High yields are also ensured by direct reloading of HPLC columns between each chromatographic step and by eliminating solvent evaporation. For these reasons it has now become the method of choice for neuropeptide isolation.

4. Establishment of the Structure of Polyprotein Precursors

4.1. Peptide Approach

Once a bioactive peptide has been isolated, its primary structure is usually established by automatic Edman degradation using the spinning cup sequenator. The recently introduced gas-phase sequenator now allows for extended amino terminal analysis of proteins at the picomolar level (Hunkapiller and Hood, 1980). The next step in analyzing the precursor molecule is to prepare specific antibodies that can be used to (1) detect and assay quantitatively the concentration of peptide in tissue extracts, (2) concentrate and purify peptides from dilute solutions and mixtures, (3) study the various molecular forms immunologically related to the peptide and representing biosynthetic precursor or processing intermediates, and (4) localize peptides in tissue slices.

4.1.1. Synthetic Peptides as Immunogens

Neuropeptides are usually present in very low concentrations in tissue extracts. Therefore, synthetic peptides matching the established sequence of the natural product have most conveniently served as antigens. Since substances of molecular weights of less than 5000 are not ordinarily very antigenic, small peptides are covalently linked to carriers in order to increase their antigenicity. The protein carriers include the serum albumin of various species, hemocyanin, ovalbumin, thyroglobulin, and fibrinogen.

Among the numerous coupling strategies that have been employed to prepare immunogenic peptide–protein conjugates, the carbodiimide method is very popular because it is carried out under mild conditions and the reaction is relatively fast (Bauminger and Wilchek, 1980). The conjugation of two compounds by the carbodiimide method requires the presence of an amino and a carboxyl group. In most cases the amino groups involved in the reaction are lysine residues of the protein carrier, whereas the carboxyl groups are contributed by the hapten. When the native hapten molecule lacks such groups, insertion of carboxyl groups through succinylation of an amino group can be easily performed. In the case of LHRH, which has blocked N- and C-terminals, a carboxyl group was introduced by attaching p-diazophenylacetic acid, which results in the formation of a mixture of azohistidyl and azotyrosyl derivatives (Koch et al., 1973). ACTH and bradykinin are two other examples of peptides that have been coupled to carrier proteins through the carbodiimide method (McGuire et al., 1965; Goodfriend et al., 1964).

Gluteraldehyde is another reagent that has been used successfully for conjugating small peptides to carrier proteins. The chemistry of the reaction of gluteraldehyde with proteins has not been elucidated definitively. It is likely that several reactions occur, giving rise to a number of products. Gluteraldehyde has been employed for coupling ACTH and glucagon to larger proteins (Reichlin et al., 1968; Frohman et al., 1970).

4.1.2. Preparation of Antisera

4.1.2.1. POLYCLONAL ANTISERA.   This technique, which is widely used, will not be discussed here in great detail since a number of reviews on this subject have appeared over the past 10 years dealing either with the theory of the method (Maurer and Callahan, 1980) or with more practical aspects (Hurn and Chantler, 1980; Sofroniew et al., 1983).

It is remarkable, however, that antisera that perform well in radioimmunoassays are not always suitable for immunoprecipitation of metabolically labeled antigen. The most common problem is the nonspecific precipitation of background radioactive peptides unrelated to the one of interest. Numerous methods have been used to reduce nonspecific background in immunoprecipitation techniques (Kessler, 1981). Sometimes it is useful to purify the antiserum on an affinity column made of sepharose-linked antigen.

4.1.2.2. MONOCLONAL ANTIBODIES.   In the case of the opioid peptides, immunoassays are complicated by the presence in the same tissue, of several, related, modified forms, each with distinct biological activity, that share common antigenic determinants. Most of these peptides cross-react when using conventional antisera (Mains et al., 1977; Zakarian and Smyth, 1982). The use of monoclonal antibodies should theoretically overcome some of these disadvantages. Moreover, the use of the hybridoma technique also provides a powerful method for generating virtually unlimited amounts of specific antibodies against peptide hormones or neuroendocrine peptides. Several articles have recently been published about the theory and practice of antibody production by hybridoma (Galfre and Milstein, 1981; Goding, 1980; Fazekas de St. Groth and Scheidegger, 1980).

Thorpe et al. (1983) have produced a monoclonal antibody for β-lipotropin that binds to the N-terminal (γ-lipotropin) portion of the molecule. This antibody can be used to detect β-lipotropin as well as proopiomelanocortin using a radiobinding assay or immunoblotting techniques. Purification of these peptides has also been readily achieved by affinity chromatography using the monoclonal antibody covalently bound to sepharose-4B. Cuello et al. (1984) have produced monoclonal antibodies against Leu- and Met-enkephalins. Two monoclonal antibodies were derived from mice immunized with either [Leu⁵]enkephalin conjugated to bovine serum albumin or [Met⁵]enkephalin-keyhole limpet hemocyanin conjugates. These monoclonal antibodies did not discriminate between Leu- and Met-enkephalin in either radioimmunoassay or immunocytochemistry. The Leu-enkephalin antibody displayed considerable cross-reactivity with peptides with C-terminal end modifications, but as the number of amino acids at the COOH end increased, the activity decreased to undetectable values for β-endorphin and dynorphin. These monoclonal antibodies did not perform in radioimmunoassay with sufficient sensitivity for their application in quantitative studies of small discrete areas of the brain. This is probably because of the poor affinity of monoclonal antibodies compared to polyclonal antisera. In another study, Meo et al. (1983) have also prepared and characterized a monoclonal antibody to the Tyr-Gly-Gly-Phe sequence common to β-endorphin and enkephalins. Competition experiments in radioimmunoassays and immunohistochemistry showed that the antibody failed to bind the β-endorphin precursor, β-lipotropin, and did not discriminate among opioid peptides that share the same *N*-terminal sequence, but have different COOH-terminal extensions. Interestingly, the antibody recognition of opioid peptides is abolished by those molecular changes that affect their receptor binding competence.

An interesting development in the field of monoclonal antibody techniques came with the advent of in vitro immunization protocols. This technique was first reported by Luben and Mohler (1980). A review on the subject has recently been published (Reading, 1982). The process utilizes a murine allogenic thymocyte culture to generate lymphokines that stimulate lymphocytes in the presence of antigen to promote antibody production. The main advantage of the technique is that it permits a reduction in the amount of antigen required and enables enhanced recovery of specific antigen-activated clones. These features certainly render this technique very attractive as a way of producing antibodies to neuropeptides usually present in brain tissues in amounts of only picomoles per gram. The in vitro immunization method therefore bypasses the difficulty of preparing sufficient quantities of high purity materials necessary to produce antibodies by standard in vivo immunization procedures. Luben et al. (1982) have reported the production of monoclonal antibodies specific for rat hypothalamic growth hormone-releasing factor (GHRF) by in vitro immunization of mouse spleen cells with less than 1 nmol of material containing approximately 1% of GHRF. Hybridoma supernantants were screened for anti-GHRF activity by use of a pituitary culture assay system. This procedure may be regarded as of general applicability for producing monoclonal antibodies against peptides available only in femtomolar quantities, even in the early steps of the purification. These antibodies can in turn be used in immunochemical procedures for large-scale purification and isolation of these peptides.

4.1.3. Use of Antibodies To Search for Polyprotein Precursors

Endogenous levels of neuropeptides in brain tissue are in the submicrogram range and too low to detect directly by UV absorbance. However, fractionation of tissue samples by reverse-phase HPLC or other chromatographic methods can be followed by determination of the amount of peptide of interest in each fraction with an appropriate radioimmunassay (Carraway and Leeman, 1976; Verhoef et al., 1980). If the antiserum recognizes poorly carboxy or amino extensions of the neuropeptides (the precursor itself or processing intermediates), these larger molecular species can be proteolytically cleaved before the immunoassay step. Mild trypsin digestion is usually most appropriate, since neuropeptide precursors are known to contain pairs of basic amino acids at the sites they are cleaved during the in vivo processing mechanism. Following trypsin digestion, the intermediate forms of the bioactive peptides can be detected by the antibody (Russel et al., 1979).

Although the use of antibodies greatly facilitates the search for neuropeptide precursors, cross-reactivity with other unknown proteins is always possible (Drager et al., 1983). The existence of a polyprotein precursor must therefore be confirmed by at least two additional criteria: (1) the immunoreactive large molecular weight material must share a common peptide sequence with the smaller neuropeptide and (2) pulse and pulse-chase studies must be conducted to prove the posttranslational cleavage of the precursor into the smaller neuropeptides.

4.1.4. Criteria for Identifying Polyprotein Precursors

4.1.4.1. PULSE AND PULSE CHASE STUDIES. In order to define a biosynthetic pathway leading from a polyprotein precursor to its different products, one needs a way in which to introduce radioactive amino acids into a tissue actively synthesizing the peptides. One also has to elaborate a strategy to determine the fate of the newly synthesized molecules as a function of time. A pulse-chase protocol is a convenient scheme for defining this type of relationship. Identical cell populations are incubated in culture medium containing a labeled amino acid for a brief period of time (15–30 min). One sample (pulse) is then extracted and analyzed by immunoprecipitation and sodium dodecylsulfate polyacrylamide gel electrophoresis (SDS-PAGE). The second sample is incubated for an additional period of time in medium containing only unlabeled amino acid (chase). During this "chase," the labeled precursor synthesized during the pulse is converted to labeled products. Moreover, since labeled amino acids are present only during the pulse period, any molecules newly synthesized during the chase incubation period will not be labeled. When analyzed by SDS-PAGE, the immunoprecipitates from the "pulse" sample should reveal a single radioactive peak corresponding to the initial precursor. As the period of chase increases, the amount of radioactivity recovered under the precursor peak decreases while the end products of the maturation process progressively accumulate. Processing intermediates appear as transient species at positions corresponding to molecular weights comprised between the original precursor and the final maturation products. This procedure depends on the availability of a good antiserum capable of immunoprecipitating the various large molecular weight species containing the sequence of interest. The fate of the other regions of the original precursor that might also yield other biologically active peptides can best be studied by using another antiserum directed against a synthetic peptide corresponding to a segment of precursor sequence in this region. This requires, of course, that the complete sequence of the precursor has previously been determined by cDNA analysis.

Studies of this sort imply that viable cells can be maintained long enough under in vitro incubation conditions in order to allow for the labeling of peptides with a sufficient specific activity.

Alternatively, radioactive amino acids can also be injected into various regions of the rat brain, using stereotaxically implanted cannula. Using this procedure, pulse and pulse-chase experiments have been conducted in vivo to follow the synthesis and processing of the vasopressin- and oxytocin-neurophysin precursors in the hypothalamo-neurohypophysial system (Brownstein et al., 1980). A similar protocol was also used by Torrens et al. (1982) to study substance P biosynthesis in the rat striatum and its transport into the substantia nigra.

4.1.4.2. Peptide Mapping of Radiolabeled Polypeptides by HPLC. HPLC has become the method of choice for the characterization of polyprotein precursors and of the various maturation products produced after complete processing of the polyprotein precursors. The usefulness of this method has been illustrated by the analysis of various fragments cleaved from the proopiomelanocortin precursor in the pars intermediate of the rat pituitary (Crine et al., 1980a,b). The method requires minute amounts of material (typically 10,000–20,000 cpm) and is nondestructive. Tryptic peptides isolated after HPLC separation can be recovered for further microsequencing studies or for the analysis of various posttranslational modifications (see below).

4.2. Nucleic Acid Approach

Development of methodologies to study mRNA populations in cells or tissues has led to the identification and characterization of peptide precursors that could not be reached by the classical protein chemistry technology. Two main techniques have been used: the cell-free translation of purified mRNA and the generation of cDNA libraries. We will now briefly explain how information concerning the structure and maturation of a protein precursor can be obtained using the nucleic acid approach.

4.2.1. Translation of mRNA in Cell-Free Systems

The nucleic acid methodology relies essentially on the isolation of intact mRNA from cultured cells or tissues. Several protocols have been published for the preparation of total cellular RNA (Chirgwin et al., 1979; Favaloro et al., 1980; Feramisco et al., 1982). Since most eukaryotic mRNAs contain a poly(A) sequence at their 3' end, they can be easily isolated from total RNA by affinity chromatography on oligo(dT) cellulose (Aviv and Leder, 1972).

In cell-free translation, the poly(A) mRNA is used as a template to direct the synthesis of proteins. Several cell-free systems can be used to achieve in vitro translation of a population of mRNA. However, the wheat germ extract (Roberts and Paterson, 1973) and the rabbit reticulocyte lysate (Pelham and Jackson, 1976) are the most widely used. Both systems are highly efficient and do not proteolytically cleave or process the newly synthesized proteins. In the case of rabbit reticulocyte lysate, it is necessary to degrade endogenous globin mRNAs before translation of exogenous mRNAs. This degradation is achieved by incubation of the lysate with micrococcal nuclease, which can subsequently be totally inhibited by chelating Ca^{2+} with EGTA. Radiolabeled proteins can be synthesized by the addition of radioactive amino acids in the cell-free translation systems and the newly synthesized proteins can be analyzed by SDS-PAGE. If antibodies against the peptide of interest are available, the precursor can be purified from the total translation products by immunoprecipitation and directly identified by SDS-PAGE. The radioactive proteins in the polyacrylamide gel are detected by autoradiography (Bonner and Laskey, 1974).

The precursor proteins synthesized in cell-free systems possess a signal sequence at their N-terminus that is used for translocation in the rough endoplasmic reticulum (RER) (Blobel and Dobberstein, 1975a,b). This signal sequence is removed from the precursor when a preparation of microsomial membranes with signal peptidase activity is added to the cell-free translation system (Katz et al., 1977).

Cell-free translation of mRNAs can provide limited but nevertheless very important information concerning protein precursors to neuropeptides. First it can be used to determine the size of primary translation products. For instance, the POMC (Nakanishi et al., 1976; Roberts and Herbert, 1977a,b; Boileau et al., 1983), enkephalin (Sabol et al., 1983), vasopressin (Schmale and Richter, 1981), and oxytocin (Schmale and Richter, 1980) precursors were identified by cell-free translation of mRNAs. In some instances multiple precursors were identified (Shields, et al., 1981; Lund et al., 1981; Shields, 1980; Goodman et al., 1980). Second, the structure of the radiolabeled precursor can be studied further by peptide mapping and sequencing by automated Edman degradation (Boileau et al., 1983; Policastro et al., 1981).

4.2.2. Recombinant DNA Technology

Recombinant DNA technology is used to obtain information about the structure and the maturation of neuropeptide precursors. The information obtained by this technology is usually more complete than that obtained by cell-free translation of mRNAs, and it is often the only way to fully characterize neuropeptide precursors present in low levels in neurons.

Recombinant DNA technology uses poly(A) RNA isolated as described previously to prepare a complementary DNA (cDNA) library. The general procedure to prepare a cDNA library is outlined in Fig. 1. In the first step, the poly(A) RNA is used as a

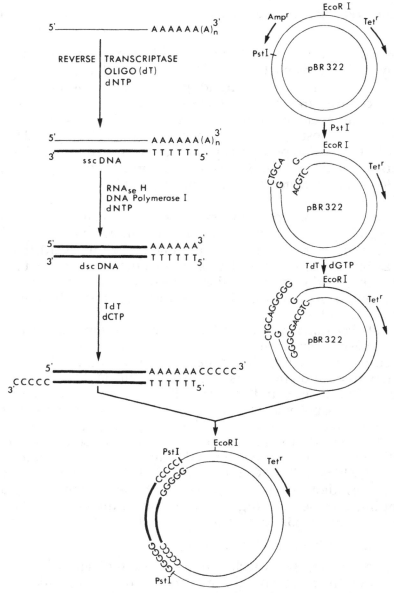

Fig. 1.   General procedure to prepare a cDNA library. The technique depicted to synthesize the dscDNA from the sscDNA molecules is that described by Gubbler and Hoffman (1983). dNTP refers to desoxy nucleotide triphosphate and pBR322 is a bacterial plasmid often used in the cloning procedure. PstI and EcoRI are recognition sites for the corresponding restriction endonucleases. Amp^r and Tet^r indicate the phenotypes carried by the plasmid. TdT refers to the enzyme terminal desoxy nucleotidyltransferase used to add homopolymer tails to the 3' ends of DNA molecules.

template to synthesize a complementary single-stranded cDNA (sscDNA). The enzyme DNA polymerase-RNA dependent (reverse transcriptase) catalyses the reaction that is primed by hybridizing an oligo(dT) polymer to the poly(A) tail of mRNAs (details of the procedure described here have been published by Maniatis et al., 1982a). DNA polymerase is then used to convert the single-stranded molecules into their double-stranded form. There are many variations to this protocol using different polymerases or different ways of priming the reaction.

Once the cDNA has been synthesized, it must be inserted into a bacterial vector. The plasmids that are generally used as vectors are small circular DNA molecules that replicate autonomously in the bacterial cell. These plasmids also carry one or two antibiotic-resistance genes that are generally used to identify bacterial colonies carrying the recombinant molecules. Insertion of the cDNA molecules in plasmids is generally achieved by annealing dCMP-tailed cDNA to dGMP-tailed plasmid. The resulting chimeric molecules are introduced in bacteria that are then spread on agar containing the appropriate antibiotic as a selective agent. Since the resistance phenotype is carried by the plasmid, bacterial cells that have not acquired a copy of plasmid will not grow. Collections of these transformed bacterial cultures are often referred to as cDNA libraries.

Once the cDNA library has been constructed, the next step is to identify clones containing the peptide precursor of interest. Several screening techniques are available. The choice of the procedure is dictated by the abundance of the mRNA in the cellular pool and by the amount of information already known about the corresponding protein, i.e., cell-free translation identification of the precursor, immunological identification, or partial amino acid sequence.

For relatively abundant peptide precursors that have already been identified by cell-free translation of their mRNA and immunoprecipitation, screening can be performed by hybridizing cDNAs purified from randomly picked clones to the population of mRNA originally used to generate the library (Maniatis et al., 1982b). The mRNA species that specifically hybridize to the cDNAs are recovered and translated in vitro. The translation products are then immunoprecipitated and analyzed by SDS-PAGE. This procedure has allowed for the identification of cDNA clones for pre-prosomatostatin (Goodman et al., 1980).

For less abundant mRNAs, the cDNA library is usually

screened with synthetic oligonucleotides matching the amino acid sequence of the biologically active peptide (Noyes et al., 1979; Mevarech et al., 1979). Because of the degeneracy of the genetic code, many nucleic acid sequences can code for a unique amino acid sequence. Usually a pool of oligonucleotides is synthesized that represents all possible combinations. However, to keep the number of combinations to a minimum, it is best to look for stretches of five or six amino acids coded for by one or two co-dons only. The oligonucleotide probes are then labeled with [^{32}P] at their 5' end and hybridized to a copy of the cDNA library, prepared by growing the bacterial clones on nitrocellulose paper and im-mobilizing the DNAs on the support (Hanahan and Meselson, 1980). The clones with sequences complementary to the synthetic probe are detected by autoradiography.

Probes of 14–17 nucleotides allow for the use of stringent hybridization conditions. These conditions are necessary to avoid the isolation of cDNA clones with sequences only partially homologous to the probe. Screening of the positive clones with a probe derived from a different amino acid sequence is sometimes useful for rejecting "false positives." When the mRNA of interest is sufficiently abundant in the poly(A) mRNA population, it can be labeled with [^{32}P] and used to hybridize the cDNA library. Using this procedure, Nakanishi et al. (1978) have identified cDNA clones corresponding to the POMC precursor.

Oligonucleotide pools have been used to prime the synthesis of cDNA from mRNA populations. The highly radioactive single-stranded cDNA thus generated can be used as a hybridization probe (Taylor et al., 1981; Boel et al., 1983) or directly for nucleotide sequencing. It is also possible to prepare cDNA libraries enriched for the cDNA clone of interest by priming the synthesis of cDNA with oligonucleotide pools (Noda et al., 1982; Kakidani et al., 1982). Such libraries can be screened with oligonucleotides derived from a different region of the protein.

Once a "positive" clone has been found, its nucleotide se-quence is determined (Maxam and Gilbert, 1980; Sanger et al., 1977), and the amino acid sequence and structural organization of the polyprotein precursor can be established. From this amino acid sequence, potential sites of proteolytic cleavage of the precursor can be identified since they usually occur at pairs of basic amino acids. Similary, the site of some posttranslational modifications such as amidation and glycosylation can also be determined (see below).

More important, a cDNA clone can be used as a hybridization probe for detecting and quantifying the corresponding mRNA in different tissues and studying the mechanisms regulating expression of the neuropeptide gene. This possibility is best illustrated by the study of the regulation of the POMC gene expression in the anterior lobe of the rat pituitary anterior lobe (Birnberg et al., 1983). Hybridization probes also revealed an interesting mechanism by which the cells can diversify their peptide production. Studies of the expressing of the calcitonin gene (Rosenfeld et al., 1980; Amara et al., 1982) and the preprotachykinin gene (Nawa et al., 1984) showed a tissue-specific alternate mode of splicing of the primary transcript of these genes, leading to different mRNAs. Expression of the calcitonin gene in thyroid "C" cells produces calcitonin, whereas expression of the same gene in the brain results in the synthesis of a different peptide called calcitonin-gene-related peptide (CGRP). From the preprotachykinin gene, the brain produces two mRNAs. One of them contains the sequence of substance P only and the other the sequence of both substance P and substance K. The intestines and thyroid generate only the latter. Finally, cDNAs can be used as probes to isolate genomic DNA clones that contain the corresponding gene. Such clones can give valuable information about the structure and expression of the gene.

5. POMC as a Model for the Other Neuroendocrine Peptides

Because of low amounts of neuropeptides in neurons and the difficulties in isolating homogeneous and viable neuron populations, most of the data accumulated so far on neuropeptide biosynthesis have been obtained by studying, in peripheral organs, peptide or hormone models that are also found in the central nervous system. Proopiomelanocortin, the common precursor to ACTH and β-endorphin, is such a peptide that is synthesized both in a peripheral organ, the pituitary gland, and in neurons of the hypothalamus. Most of what is known about polyprotein processing comes from studies on POMC biosynthesis in the pituitary gland. These studies were facilitated by the availability of two cell types that were first shown to incorporate radioactive amino acids into the peptides of interest with high efficiency: (1) the AtT-20 tumor cell line, which derives from a mouse anterior pituitary tumor (Eipper and Mains, 1980) and (2) primary cultures of the intermediate lobe of the pituitary, in which radioactive amino acid incorporation into β-endorphin was first reported (Crine et al., 1977a,b).

The first studies on the biosynthesis of ACTH in the AtT-20 tumor cell line (Eipper and Mains, 1975; Mains and Eipper, 1976) revealed the existence of a large glycoprotein precursor of approximately 31,000 daltons. Pulse-chase studies showed that the initial precursor is first cleaved into intermediate forms of ACTH (21,000–26,000 daltons) and finally into the 39-amino-acid native ACTH peptide. Peptide maps of the initial precursor synthesized in AtT-20 cells (Mains et al., 1977; Roberts and Herbert, 1977a,b) or in the intermediate lobe (Crine et al., 1978) showed that it also contains the sequence of β-endorphin. By establishing the nucleotide sequence of the cDNA cloned in *Escherichia coli*, Nakanishi and coworkers (1979) were able to predict for the first time the detailed structure of the whole precursor: β-LPH constitutes the carboxy-terminal segment of the precursor. The ACTH sequence is in the middle, next to an amino terminal peptide of approximately 100 amino acids (according to the species). All three segments of the precursor are separated from each other by pairs of basic amino acids. A conserved melanotropin-like sequence flanked by pairs of basic amino acids is found within each of the three main segments: α-MSH in the ACTH segment, β-MSH in the γ-LPH molecule, and γ-MSH in the cryptic N-terminal peptide. This type of structure was later found in all the other polyprotein precursors (*see* Fig. 2 for the structure of POMC and a few examples of other polyprotein precursors). From the primary structure of the POMC molecule, it is therefore impossible to predict which peptide will be produced by the intracellular processing mechanism. This problem was solved by performing detailed pulse and pulse-chase studies on the intermediate lobe of the pituitary (Crine et al., 1980 a; Mains and Eipper, 1979), as well as on AtT-20 cell cultures (Roberts et al., 1978; Eipper and Mains, 1980). These studies showed that the precursor is first cleaved between the β-LPH and ACTH sequence within the first 30 min after its synthesis (Fig. 3). This cleavage releases β-LPH and a molecule called "big ACTH," consisting of the N-terminal peptide linked to the ACTH sequence. β-LPH is then cleaved at the pair of basic amino acids lys_{59}-arg_{60} sequence to form β-endorphin (Crine et al., 1979) and γ-LPH (β-LPH 1-58) (Seidah et al., 1978). This process is rapid in intermediate lobe cells and much slower in the AtT-20 tumor cell line. For this reason, these tumor cells are believed to mimick closely the corticotropic

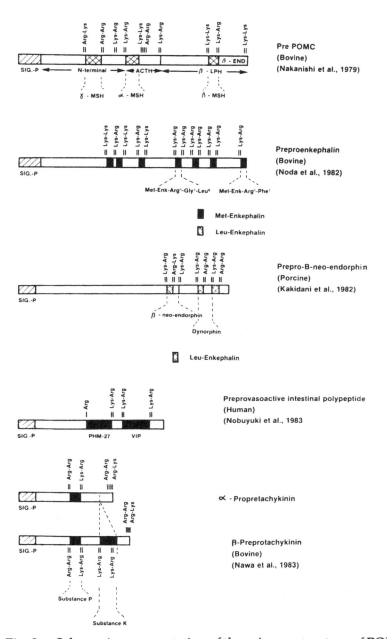

Fig. 2. Schematic representation of the primary structure of POMC and several other polyprotein precursors (SIG-P, signal peptide).

17

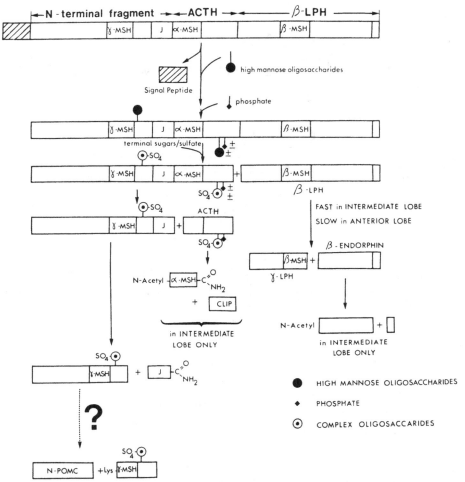

Fig. 3. Posttranslational processing of POMC in the pituitary. Vertical bars in the structure of POMC represent possible cleavage sites by either the signal peptidase or by the "processing enzyme" (dibasic amino acid pairs).

cell of the anterior pituitary from which β-LPH can be recovered with a good yield (Liotta et al., 1978). At the same time, the "big ACTH," which exists only as a transient processing intermediate, is further cleaved into native ACTH, which constitutes the final product of the maturation mechanism in the AtT-20 cell line, as well as in the normal corticotropic cells of the anterior pituitary (Eipper and Mains, 1980). In the intermediate lobe cells, α-MSH

(*N*-acetyl ACTH 1-13 valinamide) and CLIP (corticotropic-like intermediate lobe peptide) are produced instead of ACTH (Crine et al., 1979, 1980a; Bennett et al., 1982).

Processing of the *N*-terminal portion of POMC is less well understood. Crine et al. (1980a,b) have described the production, during pulse-chase studies, of 17,000–19,000 dalton peptides that were shown to contain the same *N*-terminal sequence as the original precursor. A similar observation by Boileau et al. (1982), who have purified *N*-terminal POMC peptides from the porcine pituitaries, further suggests that microheterogeneity of the peptides released from this part of the precursor molecule might be related to the facultative cleavage of a sequence 19–31 amino acids long (according to the species) adjacent (on the *N*-terminal side) of the ACTH molecule and called the joining peptide (Seidah et al., 1981). Smaller peptides resulting from the cleavage of additional pairs of basic amino acid residues have also been proposed to occur. A lysine γ_3-MSH has been detected in intermediate lobe extracts of bovine and rat pituitaries (Browne et al., 1981a,b; Esch et al., 1981; Hammond et al., 1982). An amidated form of γ_2-MSH could also be extracted from the intermediate lobe of bovine pituitary (Shibasaki et al., 1980), but not from the anterior pituitary.

6. Posttranslational Modifications of Polyprotein Precursors

During their journey from the rough endoplasmic reticulum, where they are synthesized on membrane bound polysomes, to their final destination in the neurosecretory granules of nerve endings, polyprotein precursors are submitted to various posttranslational modifications. In most cases no precise role has been assigned to these modifications. However if the various enzymes responsible for each modification can be localized in a particular cell compartment (i.e., Golgi apparatus, granules, and so on), these events can constitute landmarks for studying the intracellular transport of proteins.

6.1. Removal of Signal Peptides

As shown for all secretory proteins, polyprotein precursors contain at their *N*-terminus a hybrophobic sequence of 25–30 amino acids responsible for the vectorial transport of secretory proteins across the membrane of the endoplasmic reticulum (Walter et al., 1984). This sequence, called signal peptide, is cleaved from the precursor during the elongation step. The precise site of cleavage cannot be predicted accurately from the primary structure of the precursor as determined by the cDNA sequence. Only in the case of POMC has the cleavage site been determined by comparing the complete sequence of pre-POMC deduced from the nucleotide sequence with the partial amino acid sequence obtained by microsequencing POMC molecules labeled by incubating whole cells with various radioactive amino acids (Herbert et al., 1980). The size of the signal peptides of protein precursors can also be determined when translation products generated in rabbit reticulocyte lysate are compared to products synthesized in the same system supplemented with microsomial membranes. Several reviews have appeared recently (Kreil, 1981; Docherty and Steiner, 1982; Michaelis and Beckwith, 1982) on the function and processing of signal sequences, and this subject will not be considered further here.

6.2. Glycosylation

6.2.1. Glycosylation of Asparagine Residues

Glycosylation of asparagine residues in glycoproteins normally begins in the rough endoplasmic reticulum with the transfer of a precursor oligosaccharide, from the carrier lipid, dolichol phosphate, to nascent polypeptide chains (Struck and Lennarz, 1980; Hubbard and Ivatt, 1981). The extent to which this precursor is processed determines the final structure of the asparagine-linked oligosaccharides on mature glycoproteins (Kornfeld and Kornfeld, 1980). The first step in processing is the removal of the three glucose residues, which occurs shortly after completion of the polypeptide chain, whereas the protein is still in the rough endoplasmic reticulum (Hubbard and Ivatt, 1981; Grinna and Robbins, 1979). The initial removal of glucose residues is then followed by the removal of a variable number of mannose residues. The processing of oligosaccharide side chains that remain on "high mannose" species stops at this stage, whereas the synthesis of "complex" oligosaccharides involves the further removal of all but three of the mannose residues and the sequential addition of N-acetylglucosamine, galactose, sialic acid, and fucose (Hubbard and Ivatt, 1981; Kornfeld and Kornfeld, 1980). These steps take place in the Golgi apparatus.

It is now well established that the glycosylated asparagine residues are always part of the sequence Asn-X-Ser/Thr. However, by examination of the amino acid sequences of various glycoproteins, it was realized that the occurrence of this sequence is a necessary, but not sufficient, condition for the attachment of N-glycosidically linked carbohydrate chains to protein in vivo (Welply et al., 1983). When a potential site of glycosylation is found in the sequence of a polyprotein precursor, glycosylation of the peptide must be verified by incorporating radioactive sugars during in vitro incubations of dispersed cells. In the past ten years, new tools have also been developed for studying the biosynthesis of glycoproteins. One of them is tunicamycin, an antibiotic that blocks the synthesis of the dolichol intermediate, which serves as a donor for transferring preformed oligosaccharides into asparagine residues (Takatsuki et al., 1975; Tkacz and Lampen, 1975). POMC molecules synthesized in cells treated with tunicamycin have a higher mobility on SDS-PAGE than those in control cells, reflecting the absence of carbohydrate side chains (Crine et al., 1979; Herbert et al., 1980). Pulse and pulse-chase studies in tunicamycin-treated cells demonstrated, however, that glycosylation is not an essential step for correct cleavage or secretion of POMC or its products (Budarf and Herbert, 1982). These results are somewhat at variance with another study, in which Loh and Gainer (1979) have reported that unglycosylated POMC synthesized in tunicamycin-treated toad intermediate lobe pituitary is unstable and poorly processed to its products.

Various glycosidases have also been purified from several sources and are now being used extensively for structural studies of oligosaccharides. Endo-β-N-acetylglucosaminidase H (endo H) can be used to cleave specifically high mannose oligosaccharide side chains from glycoproteins (Tarentino and Maley, 1974). Sensitivity to endo H digestion is very often taken as an operational criteria for documenting the addition of terminal sugars on carbohydrate side chains in glycoproteins. Other glycosidases such as the almond emulsin N-glycosidase can cleave indiscriminately high mannose and complex oligosaccharide off the peptide backbone (Plummer and Tarentino, 1981).

Glycosylation has been shown to affect the biological activities of some peptide hormones via the in vivo half-life (Ashwell and Harfold, 1982), although the extent of this effect varies greatly among species.

6.2.2. Glycosylation of Serine and Threonine Residues

The process involving addition of N-acetylgalactosamine to form the key linkage found in oligosaccharide chains O-linked to serine or threonine residues has not been studied as extensively as the N-glycosylation process. Most findings indicate that it is a posttranslational process occurring in the Golgi complex. In vivo kinetic studies of the N- and O-glycosylation of human chorionic gonadotropin (hCG) are consistent only with posttranslational addition of the O-linked chains (Hanover et al., 1982). In vitro studies have also shown that the enzyme activity is enriched in smooth membranes (Golgi complex plus smooth ER), rather than in the rough ER (Hanover et al., 1980). Completion of O-linked chains after attachment of the N-acetylgalactose unit consists of the addition of other sugars, such as sialic acid and galactose, that are transferred one at a time from nucleotide sugar donors.

6.3. Phosphorylation

When the heterogeneity of CLIP and ACTH in the rat pituitary was studied by reverse-phase high-performance liquid chromatography, the existence of variant forms of ACTH and CLIP became apparent (Browne et al., 1981a,b; Bennett et al., 1981a). Some of these peptides were identified as forms of ACTH and CLIP in which asparagine 29 is glycosylated. In the intermediate lobe, additional forms of CLIP were also isolated in which residue 31 was O-phosphoserine (Browne et al., 1981a,b). In the cell of the anterior lobe, two forms of rat ACTH (1-39) were also found (Bennett et al., 1981b); one of them could be labeled with radioactive inorganic phosphate during in vitro incubations. Incorporation of [^{32}P]-phosphate was also demonstrated for CLIP in the intermediate lobe and for the precursor of POMC in both the anterior and intermediate lobes. These results are consistent with the hypothesis that, within the rat pituitary, phosphorylation occurs initially at the level of the precursor (Eipper and Mains, 1982). Similar observations were also done in the human pituitary, in which approximately 30% of ACTH was found in its phosphorylated forms (Bennett et al., 1983).

The biological activities of phosphorylated and nonphosphorylated forms of ACTH isolated from adult human pituitaries were studied using the human fetal adrenal cell bioassay. Both forms were found equipotent in terms of dehydroepiandrosterone and cortisol production (Bennett et al., 1983). A role, if any, for the phosphorylation of human ACTH has yet to be determined. It is not known whether phosphorylated forms of POMC-derived peptides also occur in the brain.

6.4. Sulfation

Although sulfated carbohydrate residues in glycoproteins and sulfated amino acids in many other types of proteins are now being documented in an increasing number of reports, the exact role of this posttranslational modification is still poorly understood. Sulfated proteoglycans have been found in many secretory granules, including those of the mouse pituitary cell line AtT-20. In these cells, granules also contain sulfated forms of adrenocorticotropin (Moore et al., 1983). Sulfate has also been found in a considerable number of fast-transported proteins and glycoproteins in neurons (Stone et al., 1983). These observations suggest that sulfation may serve as a molecular determinant for the intracellular routing of secreted proteins. Sulfation of glycoprotein side chains has also been proposed to confer protection against degradation by neuraminidase as well by proteolytic enzymes (Parsons and Pierce, 1980; Mian et al., 1979). The hydroxyl group of tyrosine has also been found to be sulfated in some proteins (Huttner, 1982).

Bourbonnais and Crine (1985) have observed that [^{35}S]-sulfate could be incorporated into different glycoprotein variants of POMC. Sulfation of POMC could be inhibited by tunicamycin. Other structural studies of the sulfated tryptic peptide showed that sulfate groups preferentially enter asparagine-linked carbohydrate side chains and not amino acid residues of POMC. The biological role of sulfation in POMC is totally unknown.

Gastrin is a gut hormone that is also synthesized in hypothalamic neurons (Rehfeld, 1981). Tyrosine *O*-sulfation of gastrin was demonstrated by incorporation of [^{35}S]-sulfate in rat central mucosa (Brand et al., 1984). The same posttranslational modification has been found in cholecystokinin (Mutt and Jorpes, 1968) and Leu-enkephalin (Unsworth et al., 1982) among neuroendocrine peptides. Although tyrosine *O*-sulfation markedly increases the biological activity of gastrin and cholecystokinin (Jensen et al., 1980; Johnson 1976), it inactivates Leu-enkephalin (Unsworth et al., 1982).

6.5. Proteolytic Processing of Polyprotein Precursors

6.5.1. Role of Basic Amino Acid Pairs

The pioneering works of Steiner et al. (1967) and Chretien and Li (1967) led to the formulation of the hypothesis that biologically active peptides such as peptide hormones are cleaved from their precursors at basic amino acid pairs. Since then, this hypothesis has been extended to numerous peptides or proteins. This subject has recently been reviewed by Lazure et al. (1983). As can be seen from the sequence of most of the polyprotein precursors known to date, the basic amino acid pairs seem to act as signals for directing the release of biologically active products. It also appears that the preferred site of cleavage is very often Lys-Arg. Two other arguments confirm the importance of these pairs of basic amino acids as substrates for the cleaving enzymes: (1) variant forms of precursors with point mutations in one of these pairs have been described and found to either remain unprocessed or to be cleaved into atypical maturation products (Brennan and Carrell, 1978; Abdo et al., 1981; Gabbay et al., 1979; Robbins et al., 1981) and (2) the metabolic incorporation of analogs of lysine or arginine interferes dramatically with precursor processing (Noe, 1981; Crine and Lemieux, 1982).

6.5.2. Cleavage at Sites Other Than Pairs of Basic Amino Acids

Although most of the polyprotein precursors are cleaved at the position of pairs of basic amino acid residues, absence of a basic pair has been found in a few cases. In the propressophysin, for instance, cleavage is known to occur between the neurophysin II and the C-terminal glycopeptide, yet there is only one arginine residues at this position (Land et al., 1982). Similarly, the yeast α-factor precursor, which can be considered a prototype of polyprotein precursor, contains basic amino acid residues on only one side of the biologically active peptide (Kurjan and Herskowitz, 1982). An interesting case can be found in the sequence of pre-prorelaxin (Kakidani et al., 1982). The cleavage that separates the B from the C chain occurs after the Leu residue in the sequence Leu-Ser-Gln-Glu, even though an adjacent Arg-Lys pair is never cleaved.

6.5.3. Tissue-Specific Processing of Precursors

As briefly mentioned earlier, processing of POMC is different in the corticotropic cells of the anterior pituitary and in the pars intermedia. The most dramatic difference resides in the fate of the

ACTH portion, which is the sole end product of the maturation in the anterior lobe; it is further cleaved into α-MSH and CLIP in the pars intermedia. A marked difference also exists in the processing kinetics of β-LPH into β-endorphin: this process is slow in the anterior lobe and very rapid in the intermediate lobe. Marked differences also exist in the nature of enkephalin-containing peptides present in the adrenal medulla and in the brain (Udenfriend and Kilpatrick, 1983). However, in this case there have been no pulse-chase studies on either of these systems, and it is difficult at this stage to decide which of these peptides are actually end-products of the processing and which are intermediates. The precursors to several other neuroendocrine peptides are also processed differentially according to the tissue. Gastrin and cholecystokinin are members of a family of structurally related peptides that are synthesized in different tissues, including the pituitary and the brain. The gastrin precursor has also been found to be processed differently in the two lobes of the pituitary (Rehfeld and Larsson, 1981). Similar findings came from studies of the cholecystokinin precursor in the duodenum and nervous system (Rehfeld et al., 1980). Glucagon-related peptides are also different in pancreas and intestine (Holst, 1980; Patzelt et al., 1979).

Two hypotheses have been proposed to take into account tissue-specific processing of polyprotein precursors: (1) different sets of processing enzymes are present in each cell type and (2) precursor molecules have different structures in different tissues.

In the case of POMC, only one functional gene has been found in rat and mouse (Uhler et al., 1983). Moreover, "Northern" blots of pituitary mRNA have failed to detect different forms of POMC mRNA (Herbert et al., 1984). Similarly, only one proenkephalin gene has been detected so far (Comb et al., 1983). It is therefore unlikely that differential processing of proenkephalin and POMC polyprotein precursors is caused by tissue-specific gene expression or pre-mRNA splicing.

6.5.4. The Nature of Processing Enzymes

The search for the true processing enzymatic activities has been and still is a very difficult one (for a recent review, see Lazure et al., 1983). Important problems to be solved are the nature of the substrate used to test the activity, and the separation of the processing enzyme from the numerous intracellular protease activities mainly of lysosomal origin (Chertow, 1981). Ideally, the enzymatic activity must be able to process the polyprotein precursor into all the peptides characterized in vivo. So far, this approach could only be performed with prohormones such as proinsulin (Docherty et al., 1982) and proPTH (MacGregor et al., 1978), as well as with a few neuroendocrine peptide precursors such as proenkephalin (Lindberg et al., 1982), high molecular weight forms of somatostatin (Zingg and Patel, 1980; Morel et al., 1981), and POMC (Loh and Gainer, 1982; Chang and Loh, 1983). Purification of the processing activity should also benefit from the observation that, in the case of most polyprotein precursors, the cleavage steps are localized mainly in the secretion granules (Glembotski, 1981; Devault et al., 1984; Gumbiner and Kelly, 1981). Preliminary isolation of secretory granule fractions should therefore help future studies aimed at the complete purification and characterization of the processing enzyme.

7. Posttranslational Modifications of Neuropeptides

Once the neuropeptides have been cleaved from their polyprotein precursors, they can be further submitted to several additional posttranslational modifications.

7.1. Acetylation

Although enzymatic acetylation of proteins is a common cellular process, α-MSH and β-endorphin are the only N-acetylated neuroendocrine peptides known to date. Acetylation of the NH_2 group of α-MSH is required for melanocyte-stimulating activity (Lowry et al., 1977), whereas acetylation of β-endorphin completely inhibits opiatelike activity (Smyth et al., 1979). Recent studies by Glembotski (1982a,b) have shown that acetylation of α-MSH and β-endorphin is carried out by a specific enzyme present in the secretory granules of the intermediate lobe of the rat pituitary. The enzymatic activity uses acetyl-CoA as an acetate donor, is not membrane-bound, and has a pH optimum of 7.0. β-Endorphin acetyltransferase activity has also been found in the hypothalamus, which is known to synthesize β-endorphin (Donohue, 1983). Regions that do not synthesize β-endorphin were devoid of acetyltransferase activity. The β-endorphin acetyltransferase is therefore a highly specialized regulatory enzyme that seems to play an important role in the modulation of the opiate activity secreted from β-endorphin-synthesizing cells and neurons.

7.2. Amidation

Many of the bioactive peptides isolated from neural and endocrine tissues have an α-amide moiety at their carboxyl terminus (Mains et al., 1983; Tatemoto et al., 1982). In general, the presence of the α-amide moiety is essential for full biological potency of the peptide. The amino acid sequences of many precursors to α-amidated peptides have been deduced over the last several years. In every case, the amino acid residue that is amidated in the product peptide is followed in the precursor by a glycyl residue (Gly). The combined proteolysis/amidation signal has generally been found to consist of the sequence X-Gly-basic-basic, where the basic residues can be Lys or Arg. An enzymatic activity capable of producing an α-amidated peptide product from its glycine-extended precursor has been identified in secretory granules of anterior, intermediate, and neural pituitary. This enzymatic activity is copper dependent, requires the presence of molecular oxygen, and is stimulated by ascorbic acid (Eipper et al., 1983; Glembotski, 1984).

Other amidated neuropeptides include γ-MSH, POMC-joining peptide, corticotropin-, growth hormone-, and thyrotropin-releasing factors, oxytocin and vasopressin, substance P, cholecystokinin and gastrin pancreatic polypeptide, metorphamide, and amidorphin (*see* Glembotski, 1984, for a review).

8. Targeting of Neuropeptide in Secretory Granules

Direct evidence consistent with the existence of at least two secretory pathways for neuroendocrine peptides has been provided by Gumbiner and Kelly (1982). They have studied ACTH secretion in the AtT-20 tumor cell line, which synthesizes both POMC and the gp70 envelope glycoprotein of an endogenous murine leukaemia virus. Mature ACTH is sequestered intracellularly in secretion granules and is released slowly in the absence of secretagogs. In contrast, gp70 appears on the cell surface and is released more rapidly ($t_{1/2} = 30$ min). Incompletely processed ACTH precursors ($M_r = 23,000$ and $30,000$) were also released with kinetics identical to gp70. These presumably represent peptides that escaped complete proteolytic processing and condensation in granules. Secretagogs that cause a rapid increase in the rate of ACTH secretion have relatively little effect on

ACTH precursor or gp70 release. Similarly, the membrane of isolated secretion granules was found to contain only an insignificant fraction of total cellular gp70. These results suggest that intact POMC and gp70 on one hand, and mature ACTH on the other hand, are transported by distinct mechanisms. Mature ACTH is probably transported by a route found only in specialized secretory cells or neurons. In contrast, gp70 and intact POMC transport may occur via a pathway that is involved in the continuous delivery of secretory products and new (or recycling) membrane components to the cell surface. These obervations therefore provide a first clue as to the existence of a mechanism responsible for sorting and targeting neuroendocrine peptides into secretion granules. Elucidation of the molecular basis of this crucial mechanism will hopefully come from more extensive studies concerning biochemical and morphological aspects of neuropeptide secretion in cell models such as the AtT-20 cell line.

Acknowledgments

Work in the authors' laboratory has been supported by grants from the Medical Research Council of Canada and by the Fonds de la Recherche en Santé du Québec. The authors wish to thank Rose-Mai Roy for her patience in typing and editing this manuscript. The careful work of Marcel Smit in the preparation of the figures is also acknowledged.

References

Abdo Y., Rousseau J., and Dautrevaux M. (1981) Proalbumin Lille, a new variant of human serum albumin. *FEBS Lett.* **131,** 286–288.
Amara S. G., Jonas V., Rosenfeld M. G., Ong E. S., and Evans R. M. (1982) Alternative RNA processing in calcitonin gene expression generates mRNAs encoding different polypeptide products. *Nature* **298,** 240–244.
Ashwell G. and Harford J. (1982) Carbohydrate-specific receptors of the liver. *Ann. Rev. Biochem.* **51,** 531–554.
Aviv H. and Leder P. (1972) Purification of biologically active globin messenger RNA by chromatography on oligothymidylic acid-cellulose. *Proc. Natl. Acad. Sci. USA* **69,** 1408–1412.

Bauminger S. and Wilchek M. (1980) The use of carbodiimides in the preparation of immunizing conjugates. *Meth. Enzymol.* **70**, 151–159.

Bennett H. P. J., Browne C. A., and Solomon S. (1981a) Purification of the two major forms of rat pituitary corticotropin using only reversed-phase liquid chromatography. *Biochemistry* **20**, 4530–4538.

Bennett H. P. J., Browne C. A., and Solomon S. (1981b) Biosynthesis of phosphorylated forms of corticotropin-related peptides. *Proc. Natl. Acad. Sci. USA* **78**, 4713–4717.

Bennett H. P. J., Browne C. A., and Solomon S. (1982) Characterization of eight forms of corticotropin-like intermediary lobe peptide from the rat intermediary pituitary. *J. Biol. Chem.* **257**, 10096–10102.

Bennett H. P. J., Brubaker P. L., Seger M. A., and Solomon S. (1983) Human phosphoserine 31 corticotropin 1-39. Isolation and characterization. *J. Biol. Chem.* **258**, 8108–8112.

Birnberg N. C., Lissitzky J.-C., Hinman M., and Herbert E. (1983) Glucocorticoids regulate proopiomelanocortin gene expression in vivo at the levels of transcription and secretion. *Proc. Natl. Acad. Sci. USA* **80**, 6982–6986.

Blobel G. and Dobberstein B. (1975a) Transfer of proteins across membranes. I. Presence of proteolytically processed and unprocessed nascent immunoglobin light chains on membrane-bound ribosomes of murine myeloma. *J. Cell Biol.* **67**, 835–851.

Blobel G. and Dobberstein B. (1975b) Transfer of proteins across membranes. II. Reconstitution of functional rough ribosomes from heterologous components. *J. Cell Biol.* **67**, 852–862.

Boel E., Vunst J., Norris F., Norris K., Wind A., Rehfeld J. F., and Marcker K. A. (1983) Molecular cloning of human gastrin cDNA: Evidence for evolution of gastrin by gene duplication. *Proc. Natl. Acad. Sci. USA* **80**, 2866–2869.

Boileau G., Gossard F., Seidah N. G., and Chretien M. (1983) Cell-free synthesis of porcine pro-opiomelanocortin: Two different primary gene products. *Can. J. Biochem. Cell Biol.* **61**, 333–339.

Boileau G., Lariviere M., Hsi K.-L., Seidah N. G., and Chretien M. (1982) Characterization of multiple forms of porcine anterior pituitary pro-opiomelanocortin N-terminal glycopeptide. *Biochemistry* **21**, 5341–5346.

Bonner W. M. and Laskey R. A. (1974) A film detection method for tritium-labelled proteins and nucleic acids in polyacrylamide gels. *Eur. J. Biochem.* **46**, 83–88.

Bourbonnais Y. and Crine P. (1985) Post-translational incorporation of [^{35}S]sulfate into oligosaccharide side chains of pro-opiomelanocortin in rat intermediate lobe cells. *J. Biol. Chem.* **260**, 5832–5837.

Brand S. J., Klarlund J., Schwartz T. W., and Rehfeld J. F. (1984) Biosynthesis of tyrosine 6-sulfated gastrins in rat antral mucosa. *J. Biol. Chem.* **259,** 13246–13252.

Brennan S. O. and Carrell R. W. (1978) A circulating variant of human proalbumin. *Nature* **274,** 908–909.

Browne C. A., Bennett H. P. J., and Solomon S. (1981a) Isolation and characterization of corticotropin- and melanotropin-related peptides from the neurointermediate lobe of the rat pituitary by reversed-phase liquid chromatography. *Biochemistry* **20,** 4538–4546.

Browne C. A., Bennett H. P. J., and Solomon S. (1981b) The isolation and characterization of γ_3-melanotropin from the neurointermediary lobe of the rat pituitary. *Biochem. Biophys. Res. Commun.* **100,** 336–343.

Brownstein M. J., Russell J. T., and Gainer H. (1980) Synthesis, transport and release of posterior pituitary hormones. *Science* **207,** 373–378.

Budarf M. L. and Herbert E. (1982) Effect of tunicamycin on the synthesis, processing, and secretion of pro-opiomelanocortin peptides in mouse pituitary cells. *J. Biol. Chem.* **257,** 10128–10135.

Carraway R. and Leeman S. E. (1973) The isolation of a new hypotensive peptide, neurotensin, from bovine hypothalami. *J. Biol. Chem.* **248,** 6854–6861.

Carraway R. and Leeman S. E. (1976) Radioimmunoassay for neurotensin, a hypothalamic peptide. *J. Biol. Chem.* **251,** 7035–7044.

Chang T.-L. and Loh Y. P. (1983) Characterization of proopiocortin activity in rat anterior pituitary secretory granules. *Endocrinology* **112,** 1832–1838.

Chertow B. S. (1981) The role of lysosomes and proteases in hormone secretion and degradation. *Endocrinol. Rev.* **2,** 137–173.

Chirgwin J. M., Przybyla A. E., MacDonald R. J., and Rutter W. J. (1979) Isolation of biologically active ribonucleic acid from sources enriched in ribonuclease. *Biochemistry* **18,** 5294–5299.

Chretien M. and Li C. H. (1967) Isolation, purification, and characterization of γ-lipotropic hormone from sheep pituitary glands. *Can. J. Biochem.* **45,** 1163–1174.

Comb M., Rosen H., Seeburg P., Adelman J., and Herbert E. (1983) Primary structure of the human pro-enkephalin gene. *DNA* **2,** 213–229.

Crine P. and Lemieux E. (1982) Incorporation of canavanine into rat pars intermedia proteins inhibits the maturation of pro-opiomelanocortin, the common precursor to adrenocorticotropin and β-lipotropin. *J. Biol. Chem.* **257,** 832–838.

Crine P., Benjannet S., Seidah N. G., Lis M., and Chretien M. (1977a) In vitro biosynthesis of β-endorphin in pituitary glands. *Proc. Natl. Acad. Sci. USA* **74,** 1403–1406.

Crine P., Benjannet S., Seidah N. G., Lis M., and Chretien M. (1977b) In vitro biosynthesis of β-endorphin, γ-lipotropin, and β-lipotropin by the pars intermedia of beef pituitary glands. *Proc. Natl. Acad. Sci. USA* **74**, 4276–4280.

Crine P., Gianoulakis C., Seidah N. G., Gossard F., Pezalla P. D., Lis M., and Chretien M. (1978) Biosynthesis of β-endorphin from β-lipotropin and a larger molecular weight precursor in rat pars intermedia. *Proc. Natl. Acad. Sci. USA* **75**, 4719–4723.

Crine P., Gossard F., Seidah N. G., Blanchette L., Lis M., and Chretien M. (1979) Concomitant synthesis of β-endorphin and α-melanotropin from two forms of pro-opiomelanocortin in the rat pars intermedia. *Proc. Natl. Acad. Sci. USA* **76**, 5085–5089.

Crine P., Seidah N. G., Routhier R., Gossard F., and Chretien M. (1980a) Processing of two forms of the common precursor to α-melanotropin and β-endorphin in the rat pars intermedia. Evidence for and partial characterization of new pituitary peptides. *Eur. J. Biochem.* **110**, 387–396.

Crine P., Seidah N. G., Jeannotte L., and Chretien M. (1980b) Two large glycoprotein fragments related to the NH_2-terminal part of the adrenocorticotropin-β-lipotropin precursor are the end products of the maturation process in the rat pars intermedia. *Can. J. Biochem.* **58**, 1318–1322.

Cuello A. C., Milstein C., Couture R., Wright B., Priestley J. V., and Jarvis J. (1984) Characterization and immunocytochemical application of monoclonal antibodies against enkephalins. *J. Histochem. Cytochem.* **32**, 947–957.

Devault A., Zollinger M., and Crine P. (1984) Effects of the monovalent ionophore monensin on the intracellular transport and processing of pro-opiomelanocortin in cultured intermediate lobe cells of the rat pituitary. *J. Biol. Chem.* **253**, 5146–5151.

Docherty K. and Steiner D. F. (1982) Post-translational proteolysis in polypeptide hormone biosynthesis. *Ann. Rev. Physiol.* **44**, 625–638.

Docherty K., Carroll R. J., and Steiner D. F. (1982) Conversion of proinsulin to insulin: Involvement of a 31,500 molecular weight thiol protease. *Proc. Natl. Acad. Sci. USA* **79**, 4613–4617.

Donohue T. L. (1983) Identification of endorphin acetyltransferase in rat brain and pituitary gland. *J. Biol. Chem.* **258**, 2163–2167.

Douglass J., Civelli O., and Herbert E. (1984) Polyprotein gene expression: Generation of diversity of neuroendocrine peptide. *Ann. Rev. Biochem.* **53**, 665–715.

Dräger U. C., Edwards D. L., and Kleinschmidt J. (1983) Neurofilaments contain α-melanocyte-stimulating hormone (α-MSH)-like immunoreactivity. *Proc. Natl. Acad. Sci. USA* **80**, 6408–6412.

Du Vigneaud V. (1956) Hormones of the Posterior Pituitary Gland: Oxytocin and Vasopressin, in *Harvey Lectures 1954–1955*, 1–26, Academic, New York.

Eipper B. A. and Mains R. E. (1975) High molecular weight forms of adrenocorticotropic hormone in the mouse pituitary and in a mouse pituitary tumor cell line. *Biochemistry* **14**, 3834–3836.

Eipper B. A. and Mains R. E. (1980) Structure and biosynthesis of pro-adrenocorticotropin/endorphin and related peptides. *Endocrinol. Rev.* **1**, 1–27.

Eipper B. A. and Mains R. E. (1982) Phosphorylation of pro-ACTH/endorphin-derived peptides. *J. Biol. Chem.* **257**, 4907–4915.

Eipper B. A., Mains R. E., and Glembotski C. C. (1983) Identification in pituitary tissue of a peptide α-amidation activity that acts on glycine-extended peptides and requires molecular oxygen, copper, and ascorbic acid. *Proc. Natl. Acad. Sci. USA* **80**, 5144–5148.

Esch F. S., Shibasaki T., Bohlen P., Wehrenberg W. B., and Ling N. (1981) Purification and partial characterization of γ-melanotropins from bovine pituitary. *Peptides* **2**, 485–488.

Favaloro J., Treisman R., and Kamen R. (1980) Transcription maps of polyomo virus-specific RNA: Analysis by two-dimensional nuclease S7 gel mapping. *Meth. Enzymol.* **65**, 718–749.

Fazekas de St. Groth S. and Scheidegger D. (1980) Production of monoclonal antibodies: Strategy and tactics. *J. Immunol. Meth.* **35**, 1–21.

Feramisco J. R., Smart J. E., Burridge K., Helfman D. M., and Thomas G. P. (1982) Co-existence of vinculin and a vinculin-like protein of higher molecular weight in smooth muscle. *J. Biol. Chem.* **257**, 11024–11031.

Frohman L. A., Reichlin M., and Sokal J. E. (1970) Immunologic and biologic properties of antibodies to a glucagon-serum albumin polymer. *Endocrinology* **87**, 1055–1061.

Gabbay K. H., Bergenstal R. M., Wolff J., Mako M. E., and Rubenstein A. H. (1979) Familial hyperprosinsulinemia: Partial characterization of circulating proinsulin-like material. *Proc. Natl. Acad. Sci. USA* **76**, 2881–2885.

Galfre G. and Milstein C. (1981) Preparation of monoclonal antibodies: Strategies and procedures. *Meth. Enzymol.* **73**, 3–46.

Glembotski C. C. (1981) Subcellular fractionation studies on the post-translational processing of pro-adrenocorticotropic hormone/endorphin in rat intermediate pituitary. *J. Biol. Chem.* **256**, 7433–7439.

Glembotski C. C. (1982a) Acetylation of α-melanotropin and β-endorphin in the rat intermediate pituitary. *J. Biol. Chem.* **257**, 10493–10500.

Glembotski C. C. (1982b) Characterization of the peptide acetyltransferase activity in bovine and rat intermediate pituitaries responsible for the acetylation of α-endorphin and β-MSH. *J. Biol. Chem.* **257**, 10501–10503.

Glembotski C. C. (1984) The α-amidation of a melanocyte stimulating hormone in intermediate pituitary requires ascorbic acid. *J. Biol. Chem.* **259**, 13041–13048.

Goding J. W. (1980) Antibody production by hybridomas. *J. Immunol. Meth.* **39**, 285–308.

Goodfriend T. L., Levine L., and Fasman G. D. (1964) Antibodies to bradykinin and angiotensin: A use of carbodiimides in immunology. *Science* **144**, 1344–1346.

Goodman R. H., Lund P. K., Jacobs J. W., and Habener J. F. (1980) Pre-prosomatostatins. Products of cell-free translations of messenger RNAs from anglerfish islets. *J. Biol. Chem.* **255**, 6549–6552.

Grinna L. S. and Robbins P. W. (1979) Glycoprotein biosynthesis. Rat liver microsomal glucosidases which process oligosaccharides. *J. Biol. Chem.* **254**, 8814–8818.

Gubler V. and Hoffman B. (1983) A simple and very efficient method for generating cDNA libraries. *Gene* **25**, 263–269.

Guillemin R. (1978) Peptides in the brain: The new endocrinology of the neuron. *Science* **202**, 390–402.

Gumbiner B. and Kelly R. B. (1981) Secretory granules of an anterior pituitary cell line, AtT-20, contain only mature forms of corticotropin and β-lipotropin. *Proc. Natl. Acad. Sci. USA* **78**, 318–322.

Gumbiner B. and Kelly R. B. (1982) Two distinct intracellular pathways transport secretory and membrane glycoproteins to the surface of pituitary tumor cells. *Cell* **28**, 51–59.

Hammond G. L., Chung D., and Li C. H. (1982) Isolation and characterization of γ-melanotropin, a new peptide from bovine pituitary glands. *Biochem. Biophys. Res. Commun.* **108**, 118–123.

Hanahan D. and Meselson M. (1980) Plasmid screening at high colony density. *Gene* **10**, 63–67.

Hanover J. A., Lennarz W. J., and Young J. D. (1980) Synthesis of *N*- and *O*-linked glycopeptides in oviduct membrane preparations. *J. Biol. Chem.* **255**, 6713–6716.

Hanover J. A., Elting J., Mintz G. R., and Lennarz W. J. (1982) Temporal aspects of the *N*- and *O*-glycosylation of human chorionic gonadotropin. *J. Biol. Chem.* **257**, 10172–10177.

Hearn M. T. W. and Hancock W. S. (1979) Ion pair partition reversed phase HPLC. *Trends Biochem. Sci.* **4**, N58–N62.

Herbert E., Budarf M., Phillips M., Rosa P., Policastro P., Oates E., Roberts J. L., Seidah N. G., and Chretien M. (1980) Presence of a pre-sequence (signal sequence) in the common precursor to ACTH and endorphin and the role of glycosylation in processing of the precursor and secretion of ACTH and endorphin. *Ann. NY Acad. Sci.* **343**, 79–93.

Herbert E., Gates E., Martens G., Comb M., Rosen H., and Uhler M. (1984) Generation of diversity and evolution of opioid peptides. *Cold Spring Harbor Symp. Quant. Biol.* **48**, 375–384.

Hokfelt T., Johansson O., Ljungdahl A., Lundberg J., and Schultzberg M. (1980) Peptidergic neurones. *Nature* **284**, 515–521.

Holst J. J. (1980) Evidence that glicentin contains the entire sequence of glucagon. *Biochem. J.* **187**, 337–343.

Hubbard S. C. and Ivatt R. J. (1981) Synthesis and processing of asparagine-linked oligosaccharides. *Ann. Rev. Biochem.* **50**, 555–584.

Hughes J., Smith T. W., Kosterlitz H. W., Fothergill L. A., Morgan B. A., and Morris H. R. (1975) Identification of two related pentapeptides from the brain with potent opiate agonist activity. *Nature* **258**, 577–579.

Hunkapiller M. W. and Hood L. E. (1980) New protein sequenator with increased sensitivity. *Science* **207**, 523–525.

Hurn B. A. L. and Chantler S. M. (1980) Production of reagent antibodies. *Meth. Enzymol.* **70**, 104–142.

Huttner W. B. (1982) Sulfation of tyrosine residues—a widespread modification of proteins. *Nature* (Lond.) **299**, 273–276.

Ivy A. C., Kloster G., Lueth H. C., and Drewyer G. E. (1929) On the preparation of "cholecystokinin." *Am. J. Physiol.* **91**, 336–344.

Jensen S. L., Rehfeld J. F., Holst J. J., Fahrenkrug J., Nielsen O. V., and de Muckadell O. B. S. (1980) Secretory effects of gastrins on isolated perfused porcine pancreas. *Am. J. Physiol.* **238**, E186–E192.

Johnson L. R. (1976) The trophic action of gastrointestinal hormones. *Gastroenterology* **70**, 278–288.

Kakidani H., Furutani Y., Takahashi H., Noda M., Morimoto Y., Hirose T., Asai M., Inayama S., Nakanishi S., and Numa S. (1982) Cloning and sequence analysis of cDNA for porcine β-neo-endorphin/dynorphin precursor. *Nature* **298**, 245–249.

Katz F. M., Rothman J. E., Lingappa V. R., Blobel G., and Lodish H. F. (1977) Membrane assembly in vitro: Synthesis, glycosylation, and asymmetric insertion of a transmembrane protein. *Proc. Natl. Acad. Sci. USA* **74**, 3278–3282.

Kessler S. W. (1981) Use of protein A-bearing staphylococci for the immunoprecipitation and isolation of antigens from cells. *Meth. Enzymol.* **73**, 442–459.

Koch Y., Wilchek M., Fridkin M., Chobsieng P., Zor U., and Lindner H. R. (1973) Production and characterization of an antiserum to synthetic gonadotropin-releasing hormone. *Biochem. Biophys. Res. Commun.* **55**, 616–622.

Kornfeld R. and Kornfeld S. (1980) Structure of Glycoproteins and Their Oligosaccharide Units, in *The Biochemistry of Glycoproteins and Proteoglycans* (Lennarz W. J., ed.) Plenum, New York.

Kreil G. (1981) Transfer of proteins across membranes. *Ann. Rev. Biochem.* **50**, 317–348.

Kurjan J. and Herskowitz I. (1982) Structure of a yeast pheromone gene (MF): A putative α-factor precursor contains four tandem copies of mature α factor. *Cell* **30**, 933–943.

Land H., Schutz G., Schmale H., and Richter D. (1982) Nucleotide sequence of cloned cDNA encoding bovine arginine vasopressin-neurophysin II precursor. *Nature* **295**, 299–303.

Lazure C., Seidah N. G., Pelaprat D., and Chretien M. (1983) Protease and posttranslational processing of prohormones: A review. *Can. J. Biochem. Cell Biol.* **61**, 501–515.

Lindberg I., Yang H. Y. T., and Costa E. (1982) An enkephalin-generating enzyme in bovine adrenal medulla. *Biochem. Biophys. Res. Commun.* **106**, 186–193.

Liotta A. S., Suda T., and Kreiger D. T. (1978) β-Lipotropin is the major opioid-like peptide of human pituitary and rat pars distalis: Lack of significant β-endorphin. *Proc. Natl. Acad. Sci. USA* **75**, 2950–2954.

Loh Y. P. and Gainer H. (1979) The role of the carbohydrate in the stabilization, processing, and packaging of the glucosylated adrenocorticotropin-endorphin common precursor in toad pituitaries. *Endocrinology* **105**, 474–487.

Loh Y. P. and Gainer H. (1982) Characterization of pro-opiocortin-converting activity in purified secretory granules from rat pituitary neurointermediate lobe. *Proc. Natl. Acad. Sci. USA* **79**, 108–112.

Lowry P. J., Silman R. E., Hope J., and Scott A. P. (1977) Structure and biosynthesis of peptides related to corticotropins and β-melanotropins. *Ann. NY Acad. Sci.* **297**, 49–60.

Luben R. A. and Mohler M. A. (1980) "In vitro" immunization as an adjunct to the production of hybridomas producing antibodies against the lymphokine osteoclast activating factor. *Mol. Immunol.* **17**, 635–639.

Luben R. A., Brazeau P., Bohlen R., and Guillemin R. (1982) Monoclonal antibodies to hypothalamic growth hormone-releasing factor with picomoles of antigen. *Science* **218**, 887–889.

Lund P. K., Goodman R. H., and Habener J. F. (1981) Pancreatic preproglucagons are encoded by two separate mRNAs. *J. Biol. Chem.* **256**, 6515–6518.

MacGregor R. R., Chu L. L. H., Hamilton J. W., and Cohn D. V. (1978) The Intracellular Translocation and Metabolism of Bovine Parathyroid Hormone, in *Endocrinology of Calcium Metabolism* (Copp D. H. and Talmadge R. V., eds.) Excerpta Medica, Amsterdam.

Mains R. E. and Eipper B. A. (1979) Synthesis and secretion of corticotropins, melanotropins, and endorphins by rat intermediate pituitary cells. *J. Biol. Chem.* **254,** 7885–7894.

Mains R. E. and Eipper B. A. (1976) Biosynthesis of adrenocorticotropic hormone in mouse pituitary tumor cells. *J. Biol. Chem.* **251,** 4115–4120.

Mains R. E., Eipper B. A., and Ling N. (1977) Common precursor to corticotropins and endorphins. *Proc. Natl. Acad. Sci. USA* **74,** 3014–3018.

Mains R. E., Eipper B. A., Glembotski C. C., and Dores R. M. (1983) Strategies for the biosynthesis of bioactive peptides. *Trends Neurosci.* **6,** 229–235.

Maniatis T., Fritsh E. F., and Sambrook J. (1982a) Synthesis and Cloning of cDNA, in *Molecular Cloning, A Laboratory Manual* Cold Spring Harbor Lab., New York.

Maniatis T., Fritsh E. F., and Sambrook J. (1982b) Identification of cDNA clones by Hybridization Selection, in *Molecular Cloning, A Laboratory Manual* Cold Spring Harbor Lab., New York.

Maurer P. H. and Callahan H. J. (1980) Proteins and polypeptides as antigens. *Meth. Enzymol.* **70,** 49–70.

Maxam A. M. and Gilbert W. (1980) Sequencing end-labeled DNA with base-specific chemical cleavages. *Meth. Enzymol.* **65,** 499–560.

McGuire J., McGill R., Leeman S., and Goodfriend T. L. (1965) The experimental generation of antibodies to α-melanocyte stimulating hormone and adrenocorticotropic hormone. *J. Clin. Invest.* **44,** 1672–1678.

Meo T., Gramsch C., Inan R., Hollt V., Weber E., Herz A., and Riethmuller G. (1983) Monoclonal antibody to the message sequence Tyr-Gly-Gly-Phe of opioid peptides exhibits the specificity requirements of mammalian opioid receptors. *Proc. Natl. Acad. Sci. USA* **80,** 4084–4088.

Mevarech M., Noyes B. E., and Agarwal K. L. (1979) Detection of gastrin-specific mRNA using oligodeoxynucleotide probes of defined sequence. *J. Biol. Chem.* **254,** 7472–7475.

Mian N., Anderson C. E., and Kent P. W. (1979) Neuraminidase inhibition by chemically sulfated glycopeptides. *Biochem. J.* **181,** 377–385.

Michaelis S. and Beckwith J. (1982) Mechanism of incorporation of cell envelope proteins in Escherichia coli. *Ann. Rev. Microbiol.* **36,** 435–465.

Moore H. P., Gumbiner B., and Kelly R. B. (1983) A subclass of proteins and sulfated macromolecules secreted by atT-20 (mouse pituitary tumor) cells is sorted with adrenocorticopin into dense secretory granules. *J. Cell Biol.* **97,** 810–817.

Morel A., Lauber M., and Cohen P. (1981) Selective processing of the 15,000 M_r prosomatostatin by mouse hypothalamic extracts releases the tetradecopeptide. *FEBS Lett.* **136**, 316–318.

Mutt V. and Jorpes J. E. (1968) Structure of porcine cholecystokinin-pancreozymin. *Eur. J. Biochem.* **6**, 156–162.

Nakanishi S., Taii S., Hirata Y., Matsukura S., Imura H., and Numa S. (1976) A large product of cell-free translation of messenger RNA coding for corticotropin. *Proc. Natl. Acad. Sci. USA* **73**, 4319–4323.

Nakanishi S., Inoue A., Kita T., Mura S., Chang A. C. Y., Cohen S. N., Nunberg J., and Schimke R. T. (1978) Construction of bacterial plasmids that contain the nucleotide sequence for bovine corticotropin-β-lipotropin precursor. *Proc. Natl. Acad. Sci. USA* **75**, 6021–6025.

Nakanishi S., Inoue A., Kita T., Nakamura M., Chang A. C. Y., Cohen S. N., and Numa N. (1979) Nucleotide sequence of cloned cDNA for bovine corticotropin-β-lipotropin precursor. *Nature* **278**, 423–437.

Nawa H., Kotani, H., and Nakanishi S. (1984) Tissue-specific generation of two preprotachykinin mRNAs from one gene by alternative RNA splicing. *Nature* **312**, 729–734.

Nawa H., Tadaaki H., Takashima H., Inayama S., and Nakanishi S. (1983) Nucleotide sequences of cloned cDNAs for two types of bovine brain substance P precursor. *Nature* **306**, 32–36.

Nobuyuki I., Ohata K. I., Yanaihara N., and Ohamoto H. (1983) Human preprovasoactive intestinal polypeptide contains a novel PHI-27-like peptide, PHM-27. *Nature* **304**, 547–549.

Noda M., Furutani Y., Takahashi H., Toyosato M., Hirose T., Inayama S., Nakanishi S., and Numa S. (1982) Cloning and sequence analysis of cDNA for bovine adrenal preproenkephalin. *Nature* **295**, 202–206.

Noe B. D. (1981) Inhibition of inlet prohormone to hormone conversion by incorporation of arginine and lysine analogs. *J. Biol. Chem.* **256**, 4940–4946.

Noyes B. E., Mevarech M., Stein R., and Agarwal K. L. (1979) Detection and partial sequence analysis of gastrin mRNA by using an oligodeoxynucleotide probe. *Proc. Natl. Acad. Sci. USA* **76**, 1770–1774.

Parsons T. F. and Pierce J. G. (1980) Oligosaccharide moieties of glycoprotein hormones: Bovine lutropin resists enzymatic deglycosylation because of terminal O-sulfated N-acetylhexosamines. *Proc. Natl. Acad. Sci. USA* **77**, 7089–7093.

Patzelt C., Tager H. S., Carroll R. J., and Steiner D. F. (1979) Identification and processing of proglucagon in pancreatic islets. *Nature* **282**, 260–266.

Pearse A. G. E. (1969) The cytochemistry and ultrastructure of polypeptide hormone-producing cells of the APUD series and the embryologic, physiologic and pathologic implications of the concept 1. *J. Histochem. Cytochem.* **17**, 303–313.

Pearse A. G. E. and Takor T. T. (1979) Embryology of the diffuse neuroendocrine system and its relationship to the common peptides. *Fed. Proc.* **38**, 2288–2294.

Pelham H. R. B. and Jackson R. J. (1976) An efficient mRNA-dependant translation system from reticulocyte lysates. *Eur. J. Biochem.* **67**, 247–256.

Plummer T. H. and Tarentino A. L. (1981) Facile cleavage of complex oligosaccharides from glycopeptides by almond emulsin peptide: N-glycosidase. *J. Biol. Chem.* **256**, 10243–10248.

Policastro P., Phillips M., Oates E., Herbert E., Roberts J. L., Seidah N., and Chretien M. (1981) Evidence for a signal sequence at the N-terminus of the common precursor to adrenocorticotropin and β-lipotropin in mouse pituitary cells. *Eur. J. Biochem.* **116**, 255–259.

Reading C. L. (1982) Theory and methods for immunization in culture and monoclonal antibody production *J. Immunol. Meth.* **53**, 261–291.

Rehfeld J. F. (1981) Four basic characteristics of the gastrin-cholecystokinin. *Am. J. Physiol.* **240**, G255–G266.

Rehfeld J. F. and Larsson L.-I. (1981) Pituitary gastrins. Different processing in corticotrophs and melanotrophs. *J. Biol. Chem.* **256**, 10426–10429.

Rehfeld J. F., Larsson L.-I., Goltermann N. R., Schwartz T. W., Holst J. J., Jensen S. L., and Morley J. S. (1980) Neural regulation of pancreatic hormone secretion by the C-terminal tetrapeptide of CCK. *Nature* **284**, 33–38.

Reichlin M., Schnure J. J., and Vance V. K. (1968) Induction of antibodies to porcine ACTH in rabbits with nonsteroidogenic polymers of BSA and ACTH. *Proc. Soc. Exp. Biol. Med.* **128**, 347–350.

Robbins D. C., Blix P. M., Rubenstein A. H., Kanazawa Y., Kosaka K., and Tager H. S. (1981) A human proinsulin variant at arginine 65. *Nature* **291**, 679–681.

Roberts B. E. and Paterson B. M. (1973) Efficient translation of tobacco mosaic virus RNA and rabbit globin 9S RNA in a cell free system from commercial wheat germ. *Proc. Natl. Acad. Sci. USA* **74**, 2330–2334.

Roberts J. L. and Herbert E. (1977a) Characterization of a common precursor to corticotropin and β-lipotropin: Cell free synthesis of the precursor and identification of corticotropin peptides in the molecule. *Proc. Natl. Acad. Sci. USA* **70**, 4826–4830.

Roberts J. L. and Herbert E. (1977b) Characterization of a common precursor to corticotropin and β-lipotropin: Identification of β-lipotropin peptides and their arrangement relative to corticotropin in the precursor synthesized in a cell-free system. *Proc. Natl. Acad. Sci. USA* **74**, 5300–5304.

Roberts J. L., Phillips M., Rosa P. A., and Herbert E. (1978) Steps involved in the processing of common precursor form of adrenocorticotropin and endorphin in culture of mouse pituitary cells. *Biochemistry* **17**, 3609–3618.

Rosenfeld M. G., Amara S. G., Roos B. A., Ong E. S., and Evans R. M. (1980) Altered expression of the calcitonin gene associated with RNA polymorphism. *Nature* **290**, 63–65.

Russell J. T., Brownstein M. J., and Gainer H. (1979) Trypsin liberates an arginine vasopressin-like peptide and neurophysin from a M_r 20,000 putative common precursor. *Proc. Natl. Acad. Sci. USA* **76**, 6086–6090.

Sabol S. L., Liang C. M., Dandekar S., and Kranzler L. S. (1983) In vitro biosynthesis and processing of immunologically identified methionine-enkephalin precursor protein. *J. Biol. Chem.* **258**, 2697–2704.

Said S. I. and Mutt V. (1970) Polypeptide with broad biological activity. Isolation from small intestine. *Science* **169**, 1271–1272.

Sanger F., Nicklen S., and Coulson A. R. (1977) DNA sequencing with chain-terminating inhibitors. *Proc. Natl. Acad. Sci. USA* **74**, 5463–5467.

Schally A. V., Arimura A., and Kastin A. J. (1973) Hypothalamic regulatory hormones. *Science* **179**, 341–350.

Schmale H. and Richter D. (1980) In vitro biosynthesis and processing of composite common precursors containing amino acid sequence identified immunologically as neurophysin II/arginine vassopressin. *FEBS Lett.* **121**, 358–362.

Schmale H. and Richter D. (1981) Immunological identification of a common precursor to arginine vasopressin and neurophysin II synthesized by in vitro translation of bovine hypothalamic mRNA. *Proc. Natl. Acad. Sci. USA* **78**, 766–772.

Seidah N. G., Gianoulakis C., Crine P., Lis M., Benjannet S., Routhier R., and Chretien M. (1978) In vitro biosynthesis on chemical characterization of β-lipotropin, γ-lipotropin, and β-endorphin in rat pars intermedia. *Proc. Natl. Acad. Sci. USA* **75**, 3153–3157.

Seidah N. G., Rochemont J., Hamelin J., and Chretien M. (1981) The missing fragment of the pro-sequence of human pro-opiomelanocortin: sequence and evidence for C-terminal amidation. *Biochem. Biophys. Res. Commun.* **102**, 710–716.

Shibasaki T., Ling N., and Guillemin R. (1980) A radioimmunoassay for γ-melanotropin and evidence that the smallest pituitary γ-melanotropin is amidated at the COOH-terminus. *Biochem. Biophys. Res. Commun.* **96**, 1393–1399.

Shields D. (1980) In vitro biosynthesis of somatostatin. Evidence for two distinct preprosomatostatin molecules. *J. Biol. Chem.* **255**, 11625–11628.

Shields D., Warren T. G., Roth S. E., and Brenner M. J. (1981) Cell-free synthesis and processing of multiple precursors to glucagon. *Nature* **289**, 511–514.

Smyth D. G., Massey D. E., Zakarian S., and Finnie M. D. A. (1979) Endophins are stored in biologically active and inactive forms: Isolated of α-N-acetyl peptides. *Nature* **279**, 252–254.

Sofroniew M. V., Couture R., and Cuello C. A. (1983) Immunocytochemistry: Preparation of Antibodies and Staining Specificity, in *Handbook of Chemical Neuroanatomy* vol. 1. *Methods in Chemical Neuroanatomy* (Bjorklund A. and Hokfelt T., eds.) Elsevier, New York.

Steiner D. F., Cunningham D., Spiegelman L., and Aten A. (1967) Insulin biosynthesis: Evidence for a precursor. *Science* **157**, 697–700.

Stone G. C., Hammerschlag R., and Bobinski J. A. (1983) Fast-transported glycoproteins and nonglycosylated proteins contains sulfate. *J. Neurochem.* **41**, 1085–1089.

Struck D. K. and Lennarz W. (1980) The Function of Saccharide-Lipids in Synthesis of Glycoproteins, in *The Biochemistry of Glycoproteins and Proteoglycans* (Lennarz W., ed.) Plenum, New York.

Szentagothai J. (1962) Anatomical Considerations, in *Hypothalamic Control of the Anterior Pituitary* (Szentagothai J., Flerko B., Mess B., and Halasz B., eds.) Akademiai Kiado, Budapest.

Takatsuki A., Kohno K., and Tamura G. (1975) Inhibition of biosynthesis of polyisoprenol sugars in chick embryo microsomes by tunicamycin. *Agric. Biol. Chem.* **39**, 2089–2091.

Tarentino A. L. and Maley F. (1974) Purification and properties of an Endo-β-N-acetylglucosaminidase from *Streptomyces griseus*. *J. Biol. Chem.* **249**, 811–817.

Tatemoto C., Carlquist M., and Mutt V. (1982) Neuropeptide Y—a novel brain peptide with structural similarities to peptide YY and pancreatic polypeptide. *Nature* **296**, 659–660.

Taylor W. L., Collier K. J., Deschenes R. J., Weith H. L., and Dixon J. E. (1981) Sequence analysis of a cDNA coding for a pancreatic precursor to somatostatin. *Proc. Natl. Acad. Sci. USA* **78**, 6694–6698.

Thorpe R., Spitz L., Spitz M., and Austen B. M. (1983) Detection and affinity purification of β-endorphin precursor using a monoclonal antibodies. *FEBS Lett.* **151**, 105–110.

Tkacz J. S. and Lampen J. O. (1975) Tunicamycin inhibition of polyisoprenyl *N*-acetylglucosaminyl pyrophosphate formation in calf-liver microsomes. *Biochem. Biophys. Res. Commun.* **65**, 248–257.

Torrens Y., Michelot R., Beaujonan J. C., Glowinski J., and Bockaert J. (1982) *In vivo* biosynthesis of ^{35}S-substance P from [^{35}S]methionine in rat striatum and its transport to the substantia nigra. *J. Neurochem.* **38**, 1728–1734.

Udenfriend S. and Kilpatrick D. L. (1983) Biochemistry of the enkephalins and enkephalins-containing peptides. *Arch. Biochem. Biophys.* **221**, 309–323.

Uhler M., Herbert E., D'Eustachio P., and Ruddle F. D. (1983) The mouse genome contains two nonallelic pro-opiomelanocortin genes. *J. Biol. Chem.* **258**, 9444–9453.

Unsworth C. D., Hughes J., and Morley J. S. (1982) *O*-sulfated Leu-enkephalin in brain. *Nature* **295**, 519–522.

Verhoef J., Loeber J. G., Burbach J. P. H., Gisjen W. H., Witter A., and de Wied D. (1980) α-Endorphin, γ-endorphin and their tryptic fragments in rat pituitary and brain tissue. *Life Sci.* **26**, 851–859.

von Euler U. S. and Gaddum J. H. (1931) An unidentified repressor substance in certain tissue extracts. *J. Physiol.* **72**, 74–87.

Walter P., Gilmore R., and Blobel G. (1984) Protein translocation across the endoplasmic reticulum. *Cell* **38**, 5–8.

Welply J. K., Shenbagamurthi P., Lennarz W. J., and Naider F. (1983) Substrate recognition by oligosaccharyltransferase. Studies on glycosylation of modified Asn-*X*-Thr/ser tripeptides. *J. Biol. Chem.* **258**, 11856–11863.

Zakarian S. and Smyth D. G. (1982) β-Endorphin is processed differently in specific regions of rat pituitary and brain. *Nature* **296**, 250–252.

Zingg H. H. and Patel Y. C. (1980) Processing of somatostatin precursors: Evidence for enzymatic cleavage by hypothalamic extracts. *Biochem. Biophys. Res. Commun.* **93**, 1274–1279.

Peptides

Strategies for Antibody Production and Radioimmunoassays

Robert Benoit, Nicholas Ling, Paul Brazeau, Solange Lavielle, and Roger Guillemin

1. Introduction

The explosion of our knowledge in the neuropeptide field, particularly during the last decade, was partly the result of the development of new probes essential to the localization and quantitation of peptidergic substances. Antibodies, monoclonal or polyclonal, remain our best markers for peptidergic systems in the brain. Labeled c-DNA probes used for *in situ* hybridization studies are becoming a complement to antibodies in the identification of peptidergic neurons (Gee et al., 1983; Bloch et al., 1984a). Recent techniques such as enzyme immunoassays (EIAs) and ultrasensitive enzyme radioimmunoassays (USERIAs) may eventually replace RIAs (Trivers et al., 1983). New fluoroimmunoassays such as dissociation-enhanced lanthanide fluoroimmunoassays (DELFIA, from Wallac OY) appear very promising. At present, however, the simplest and most economical approach to measure peptides in the high attomole to low femtomole range is still by RIA.

Specificity constitutes the most important characteristic of a probe. As it becomes clear that peptides with high structural homology do not necessarily play similar physiological roles (i.e., peptide histidine methionine, vasoactive intestinal peptide, and growth-hormone-releasing hormone or vasopressin and oxytocin), the search for specific antibodies for RIA and immunohistochemistry will become even more demanding (Valiquette et al.,

1986). Since monoclonal antibodies are considered specific for only one given epitope, which usually represents three to four amino acids, polyclonal antibodies that recognize multiple epitopes may end up being at least as specific as monoclonals for detecting peptides.

In the present chapter, we will deal with production of polyclonal antibodies, preparation of tracers, and the validation and characterization of RIA systems, and means to increase their sensitivity.

2. Antibody Production

Neuropeptides are usually less than 50 amino acids long and behave like haptens when injected into animals. Although isolated cases of successful immunization with free oligopeptides have been reported (Wu and Rockey, 1969), it is generally considered that to generate high-affinity and high-titer antibodies against these molecules, conjugation to a carrier protein must be performed first. The conjugation step usually requires a coupling agent. New methods without coupling have recently been introduced, however, that could also induce high-quality antibodies.

2.1. Preparation of the Immunogen

2.1.1. Methods Involving Coupling to a Carrier

Coupling agents are divalent chemicals that can form covalent bonds between a peptide and a protein. The divalent reagents most frequently used are bis-diazotized benzidine (BDB), glutaraldehyde, and carbodiimides (Fig. 1).

2.1.1.1. BIS-DIAZOTIZED BENZIDINE.   BDB is a diazonium salt that can react with free amino groups as well as with imidazole and phenol groups. It is prepared according to the method of Likhite and Sehon (1967) for conjugating proteins to erythrocytes. Briefly, 1.25 mmol benzidine dihydrochloride (Biochemical Lab Inc. or Sigma) is dissolved in 45 mL 0.2N HCl at room temperature. The solution is then cooled in an ice bath and stirred with a bar magnet while 2.5 mmol NaNO$_2$ in 5 mL H$_2$O is added dropwise. The mixture turns yellow immediately. After addition of NaNO$_2$, the yellow solution is stirred at 0°C for 30 min. Aliquots of 25 μmol BDB/mL are then quickly frozen in liquid nitrogen and stored at −20°C for later use. Coupling of the peptide is performed according

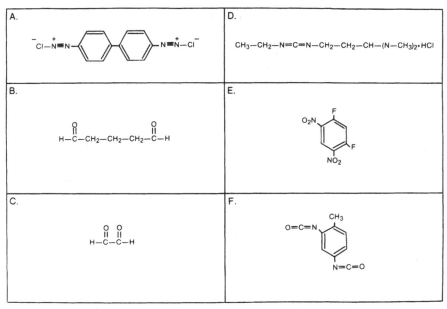

Fig. 1. Structures of (A) bis-diazotized benzidine dichloride, (B) glutaraldehyde, (C) glyoxal, (D) 1-ethyl-3-(3-dimethylaminopropyl) carbodiimide hydrochloride, (E) 1,5-difluoro-2,4-dinitrobenzene, and (F) toluene-2,4 diisocyanate.

to the procedure used by Bassiri and Utiger (1972a,b) to conjugate thyrotropin releasing hormone (TRH), with minor modification. For example, to immunize five rabbits against luteinizing hormone releasing hormone (LHRH), we dissolve 12.5 mg of the peptide in 2 mL 0.16M borate buffer, pH 9, containing 0.13M NaCl. The solution is then added to another 2.5 mL of the same buffer containing 40 mg bovine serum albumin (BSA crystalline, Pentex, Miles Lab.). The colorless mixture is stirred in an ice bath while 0.42 mL of the defrosted BDB solution is added. The solution turns ochre-colored immediately and is left stirring for 2 h. The whole mixture—including any precipitate formed—is then dialyzed against 2 L distilled water at 4°C for 12 h using a Spectrapor membrane (cut-off 12,000–14,000 mol wt, Spectrum Industries, Houston). The dialysis process is repeated six times (~12 h each time), the last two dialysis procedures being done against 2 L 150 mM NaCl. The content of the dialysis bag is then frozen in aliquots at –20°C. The titers reached after immunization with this type of conjugate are shown in Table 1.

Table 1
Mean Titers Obtained After Immunization Against Different Haptens
Using Three Methods of Preparing the Immunogens

Hapten	n^a	Method	Mean titer[b]
[D-Lys^6]-LHRH	8	Glutaraldehyde-ovalbumin	1/970,000
LHRH	3	BDB-BSA	1/700,000
[TYR^{21}]-CRF-(21-41)	6	BDB-BSA	1/257,000
[TYR^{21}]-CRF-(21-41)	4	met-BSA	1/88,000
ACTH 1-24	6	met-BSA	1/728,000
β-LPH (45-61)	2	met-BSA	1/5,000

[a]n Represents the number of immunized animals.
[b]The mean titer refers to final dilution of immune plasma obtained after the third booster injection and binding 30–40% of the iodinated hapten.

2.1.1.2. GLUTARALDEHYDE.   Glutaric dialdehyde, i.e., glutaraldehyde (Fig. 1B), conjugates mostly at the ε-amino group of lysine. Like BDB, it can also react with α-amino groups, as well as occasionally with imidazole and phenol groups (histidine, tyrosine). At pH 8, it is, like acrolein, one of the most efficient cross-linking agents (Bowes, 1963). We generated excellent antibodies against LHRH using glutaraldehyde to conjugate [D-Lys^6]-LHRH to ovalbumin (Table 1). Twenty-four milligrams of [D-Lys^6]-LHRH are dissolved in 2 mL 0.1M sodium phosphate buffer, pH 7, and the solution is added to a 2-mL solution of the same buffer containing 40 mg ovalbumin (Miles Lab.). The mixture is stirred at room temperature and 356 μL 0.1M aqueous glutaraldehyde solution (Sigma) is added. The whole solution turns pale yellow and stirring is continued for 2 h at room temperature. The mixture, including any precipitate formed, is dialyzed against 2 L H_2O for 12 h using Spectrapor membranes, as above (cut-off 12,000–14,000 mol wt). The dialysis is repeated six times; the last two dialysis procedures are done against 2 L 150 mM NaCl. The content of the dialysis bag is then aliquoted and frozen at –20°C. It contains enough conjugate to immunize five rabbits (one primary and four secondary injections).

Glyoxal (*see* Fig. 1C) is a simple bifunctional aldehyde that reacts also with α- and ε-amino groups, as well as with arginine residues (Nakaya et al., 1969). When used at pH 8, glyoxal cross-links slightly less efficiently than glutaraldehyde (Bowes, 1963).

2.1.1.3. CARBODIIMIDES. Carbodiimides are reagents that activate carboxyl groups and enable them to react with the amino groups of another peptide or protein. This reaction leads to the formation of peptide bonds between a peptide and a carrier protein. 1-Ethyl-3-(3-dimethylaminopropyl)carbodiimide HCl (ECDI-HCl) is the most frequently used coupling agent of the carbodiimide family (Fig. 1D). For immunization of five animals, 8 μmol of peptide is dissolved in 1 mL distilled water and added to 500 μL of an aqueous solution of bovine serum albumin (12 mg crystalline BSA, Pentex, Miles Lab.) in a propylene tube. The mixture is vortexed and 500 μL of an aqueous solution containing 50 mg ECDI-HCl is added. The content of the propylene tube is left on a rolling mixer in a dark, cold room at 4°C for 15 h and the pH of the mixture is kept near 6 with NaOH. The reaction mixture is then dialyzed five times against 2 L distilled water containing 0.01% mercaptoethanol (12 h each time) using a Spectrapor membrane, as above (cut-off 12,000–14,000 mol wt). The content of the dialysis bag is diluted in an equal volume of 150 mM NaCl, and 200 μL aliquots are frozen and stored at –20°C for later immunization.

2.1.1.4. DIFLUORODINITROBENZENE. Direct coupling of peptides to proteins can be achieved using halogenated dinitrobenzenes such as the aryl halide 1,5-difluoro-2,4-dinitrobenzene (DFNB, *see* Fig. 1E). Tager has successfully coupled glucagon to albumin using DFNB and produced high-titer glucagon antibodies (Tager, 1976; Tager et al., 1977).

2.1.1.5. TOLUENE-2,4-DIISOCYANATE. Shick and Singer were the first to report the protein–protein conjugate formation using toluene-2,4-diisocyanate (TC, Fig. 1F), a bivalent coupling agent that reacts with free amino groups (Shick and Singer, 1961). This coupling reagent has been used to conjugate insulin to bovine serum albumin (Mark, 1968). The reagent can yield highly homogeneous conjugates since stepwise reactions can be performed first at 0°C between position 4 of TC and the hapten, and then at 37°C between position 2 of TC and the carrier. The original procedure of Shick and Singer can be applied to peptides with minor modification. For immunization of five animals, 10 μmol peptide is dissolved in 8 mL 50 mM sodium phosphate buffer, pH 7.5, and kept at 0°C. Toluene-2,4-diisocyanate, 200 μL, is then added and the mixture is stirred vigorously at 0°C for 30 min and then centrifuged at 4°C for 30 min at 5000g. The supernatant, which contains most of the reacted diisocyanate, is removed and allowed to stand at 0°C for 1 h. The solution is then added to 8 mL borate buffer, pH 9.5, containing 40 mg bovine serum albumin (Pentex, Miles Lab.), and kept at 37°C for 1 h. The mixture is then dialyzed twice against 2 L 0.1*M* ammonium carbonate, pH 8.8, at 4°C (12 h each time), and three times against distilled H_2O using a Spectrapor membrane (cut off 12,000–14,000 mol wt). Aliquots of the conjugate are stored at –40°C.

2.1.1.6. OTHER CROSS-LINKING AGENTS. Several other cross-linking agents have been used successfully in the preparation of conjugates, namely mercurials and *N*-succinimidyl-3-(2-pyridyldithio)propionate, i.e. SPDP, for cystein-containing peptides (Wolfe and De Grado, 1985) and the maleimides, such as *N*,*N*'-(1,3-phenylene)-bis-maleimide, and *N*-ethyl benzisoxazolium fluoborate, i.e., EBIZ (Likhite and Sehon, 1967).

2.1.1.7. POLYMERIZATION OF PEPTIDES. A peptide can be made antigenic by polymerization using various coupling agents, such as carbodiimides or glutaraldehyde. Specific antibodies against glucagon have been produced by polymerization followed by injection into rabbits in Freund's adjuvant (Heding, 1969).

2.1.2. Methods Involving No Coupling Agent

2.1.2.1. METHYLATED BOVINE SERUM ALBUMIN. First introduced by Plescia and colleagues in 1964 (Plescia et al., 1964) as an approach to raise antibodies against denatured DNA, methylated bovine serum albumin (met-BSA) is now widely utilized in the production of antibodies against polypeptides (Épelbaum et al., 1977; Benoit et al., 1980; 1982a,b). Met-BSA appears to act as an adjuvant in the immune response. The advantages offered by this method include simplicity, reproducibility, and high quality of the antibodies being produced. Using the met-BSA method, we raised specific antibodies against somatostatin-14 (SS14), somatostatin-28 (SS28), adrenocorticotropin (1-24), sauvagine (17-40), neurotensin, bradykinin, arginine vasopressin, vasoactive intestinal peptide, beta-lipotropin (45-61), and [TYR[21]] ovine corticotropin-releasing factor (Benoit et al., 1980; 1982a,b; Barabe and Regoli, 1981; Morrison et al., 1983; Benoit et al., unpublished). Recently, several investigators have used this method to produce antibodies against dynorphin (1-13) (Day et al., 1982), pre-prosomatostatin (63-77) (Lechan et al., 1983), growth hormone-releasing factor (Bloch et al., 1984b), and bovine adrenal medullary peptides 12P and 22P (Baird et al., 1984). We were unsuccessful in raising antibodies against LHRH and TRH, but our attempts were limited. The amount of peptide generally used for immunization is rather large when compared to conventional conjugation methods with chemical reagents. In order to proceed to the primary immunization of five rabbits against a peptide of about 3000 daltons, 15 mg of the peptide is dissolved in 2 mL of 150 mM NaCl solution in a polypropylene tube. Three milligrams of met-BSA (Sigma) dissolved in 2 mL distilled H_2O is added to the peptide solution and the mixture is vortexed for a few seconds and kept on ice. The peptide-methylated BSA mixture is freshly prepared immediately before each immunization and a constant ratio of between 4 and 5 of peptide weight to the weight of met-BSA is maintained. For booster injections, 5 mg of peptide in 1 mL saline and 1 mg met-BSA in 1 mL water are mixed together. Usually, two or three booster injections are needed to reach a stable titer of antibodies. Table 1 shows the titers obtained after immunization of six rabbits against ACTH (1-24) and [TYR^{21}]-ovine CRF (21-41). It is also possible to use the methylated BSA method with less hapten to generate antibodies against acidic or basic peptides (Lechan et al., 1983).

2.1.2.2. ADSORPTION TO A POLYMER.   High-titer antibodies can be produced in rabbits against small peptides such as angiotensin I and II when large amounts of hapten (6 mg per animal) are emulsified with polyvinylpyrrolidone (PVP) in Freund's complete adjuvant and injected into rabbits (Worobec et al., 1972). It is assumed by Worobec and collaborators that unspecific adsorption of the haptens to the water-soluble polymer PVP would allow for prolonged release of the free peptide.

2.1.3. Method Requiring No Conventional Adjuvant

The recent interest in the production of synthetic vaccines has opened new avenues for the preparation of immunogens. An example is provided by the work of Bessler, Jung, and their collaborators (Bessler et al., 1984; Jung et al., 1986). The outer membrane of *Escherichia coli* contains a lipoprotein capable of activating B-lymphocytes. These investigators have prepared various shortened forms of this natural lipoprotein, such as tripalmitoyl-cys-ser-ser-asn-ala-OH (i.e., tripalmitoyl pentapeptide, TPP), which acts as a polyclonal activator of B-lymphocytes both in vitro and in vivo (Bessler et al., 1984). They incorporated a TPP analog at the end of various peptides produced by solid-phase synthesis and were successful in producing antibodies against these hybrid peptides without requiring Freund's adjuvant (Jung et al., 1986). Similar results were obtained using another synthetic adjuvant, *N*-

acetylmuramyl-L-alanyl-D-isoglutamine. This glycopeptide increases antibody response when simply injected in an aqueous solution with a hapten such as LHRH (Carelli et al., 1982).

2.2. Procedure for Primary Immunization and Booster Injection

There are several routes available to induce the formation of gamma-globulins in a given animal. The multitude of injection routes proposed by various investigators (im, iv, sc, id, fat pads, lymph nodes, and so on) simply reflects the lack of consensus on the "best way" to elicit the formation of high-titer and high-affinity antibodies. We propose here an approach based on our experience in immunization of rat, mouse, sheep, goat, and rabbit, against several peptides. The antibodies produced have been useful not only for RIA, but also for immunohistochemistry, immunoaffinity chromatography, and passive immunization. Sometimes the problem resides not so much in the production of antibodies, but rather in finding convenient methods to recognize these antibodies after they have been generated.

2.2.1. Selection and Preparation of Animals

Antibodies for immunoassays have been generated in a variety of animals, including primates. Those most frequently used are the rabbit, guinea pig, goat, and mouse. In our experience, unless liters of antiserum are required for commercial purposes or for experiments such as large-scale immunoaffinity chromatography, it is not necessary to use a large animal such as the goat or sheep. The guinea pig may be preferred when the supply of antigen is rather limited. A 12-wk-old rabbit of either sex remains the ideal choice for polyclonal antibody production. A strain such as the L_{18} New Zealand white rabbit, from L.I.T. Rabbitry in Montana, USA, is an excellent choice. The animal should be kept in a clean environment with adequate humidity (~50%) and minimal stress. Immediately before injection, the fur is shaved on dorsal and lateral regions and the exposed skin cleaned with 50% (vol/vol) ethyl alcohol in water. In the week preceding the primary immunization, the immune system is stimulated with 3 mL of a suspension of killed *Bordetella pertussis* (Institut Armand Frappier, Laval, Quebec, Canada) containing 6×10^{10} microorganisms and injected sc. It is advisable to immunize five to ten rabbits with a given peptide if high-affinity antibodies are required ($K_a > 10^{10}$ L/mol). If not, a group of three animals should be adequate.

2.2.2. Primary Injection

In order to immunize five rabbits against a peptide and a carrier, an aliquot containing 10–12 mg of conjugate is diluted to a 4-mL volume with sterile 150 mM NaCl. If immunization is done using met-BSA, 15 mg peptide will be required, together with 3 mg met-BSA (*vide supra*, 2.1.2.1). In either case, an identical volume of complete Freund's adjuvant (4 mL) (Difco Co., Detroit, MI) containing *Mycobacterium butyricum* is added to the solution, which is kept at 4°C in a polypropylene tube. A supplement of 50 mg killed *Mycobacterium tuberculosis* (Cuti-BCG, Institut Armand Frappier, Laval, Quebec) is added to the mixture, which is then emulsified using a polytron homogenizer (Brinkmann). The emulsification is performed in an ice bath in less than 5 min. Each animal receives 1.5 mL of the emulsion both id (100–150 sites in the dorsolateral and cervical regions using a sharp 27-gage needle) and im (one 0.2 mL injection in the thigh).

2.2.3. Booster Injections and Bleedings

Six weeks after the primary injection and every 2–4 wk thereafter, the animals are injected id with a third of the primary injection dose prepared the same way, except that incomplete Freund's adjuvant is used (Difco Co., Detroit, MI) and no bacteria are injected. In most cases, the antibody titer will reach a plateau at or even before the fourth booster injection. It is then advisable to stop or postpone the immunization until later. The animals should be bled before the primary immunization (control antiserum or plasma) and 10–14 d after each booster injection. The blood collected after the second boost is considered valuable and, from then on, 40–50 mL of blood is collected from each animal by the lateral ear vein. The vein is opened with a sharp razor blade and blood is aspirated using a vacuum pump connected to a glass apparatus (Bellco bleeding apparatus) that covers the ear and contains 0.5 mL heparin (1000 units/mL). The blood is spun at 4°C at 2000g for 30 min and the plasma is frozen after addition of sodium azide (0.02% final concentration). In certain cases, isolation of the serum fraction is preferable to the plasma. Once the testing of a small aliquot has shown that the immune plasma is valuable, it is lyophilized and kept at –20°C. Failure to do so may result in gamma-globulin damage. The titer of our ACTH antibodies, which were kept frozen for 3 yr, decreased by four–fivefold during that period.

3. Labeling of Peptides and Purification of Tracers

Most peptides can be labeled easily for RIA. The simplest approach consists of an oxidation reaction with chloramine-T, which incorporates one or more molecules of iodine (usually ^{125}I, 60 d half-life) to a peptide containing tyrosyl and/or histidyl residues (Greenwood et al., 1963) (Fig. 2A). The situation is less simple if neither of these two amino acids is present in the peptide to be labeled. In such a case, a Bolton-Hunter reagent may be useful if a free lysine or NH_2-terminus is present in the peptide. The simplest approach by far, however, is the addition of a tyrosyl residue in the peptide sequence, usually at the NH_2- or the COOH-terminus of the peptide. The tyrosine can be used not only for radioiodination, but also for conjugation of the hapten to a carrier protein using BDB *(vide supra)*.

Whenever a sensitive RIA is required, a high-specific-activity isotope such as ^{125}I is used (17 Ci/mg I). The aim is to reach the highest specific activity per peptide molecule without altering its capacity to react with the antibody. If one achieves complete separation of a monoiodinated peptide from the other components of the iodination reaction, the specific activity of the labeled peptide is close to 2000 μCi/nmol (maximum theoretical value, 2125 μCi/ nmol).

3.1. Example of Mild Radioiodination: Labeling of [TYR^{11}] SS14

Labeling [TYR^{11}] SS14 is a typical case of a small peptide containing a disulfide bond and a tyrosine residue relatively close to the antigenic determinant. The amount of reducing agent must be kept low. Two and a half micrograms of [TYR^{11}] SS14 in 25 μL 0.02M ammonium acetate, pH 4.6, is added to 25 μL 0.5M sodium phosphate, pH 7.4, in a borosilicate test tube. Six hundred microcuries Na^{125}I (in 6 μL 0.01N NaOH) is added to the solution followed by 6 μg chloramine-T in 25 μL 0.05M sodium phosphate buffer. The tube is gently shaken for 38 s, after which 19 μg sodium metabisulfite in 50 μL 0.05M sodium phosphate is added as the reducing agent, followed by 80 μL of a 10% human serum albumin solution, pH 7.4, also in the 0.05M sodium phosphate buffer.

Mild iodination of peptides can also be performed using the commercially available oxidizing agent 1,3,4,6-tetrachloro-3α,6α- diphenyl-glycol-uril (IODO-GEN, Pierce Chemicals) (*see* Fig. 2B).

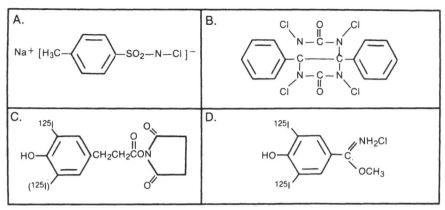

Fig. 2.   Structures of (A) chloramine-T, (B) IODO-GEN, (C) Bolton-Hunter reagent, and (D) imidoester product of Wood and collaborators (1975).

The procedure is simple and inexpensive, and allows for incorporation with minimal oxidation damage (Fraker and Speck, 1978). It does not require a reducing agent and the stability of the tracer is maintained for up to 3 mo (Salacinski et al., 1981).

3.2. Examples of Standard Radioiodination: Labeling of LHRH and ACTH (1-24)

For LHRH iodination, 5 μg of the decapeptide in 50 μL 0.1*M* sodium phosphate buffer, pH 7.4, is added to 25 μL 0.5*M* sodium phosphate buffer, pH 7.4, in a borosilicate glass tube. One and a half millicuries of Na^{125}I contained in 15 μL is added to the solution followed by 50 μg chloramine-T in 25 μL 0.05*M* sodium phosphate. The tube is shaken gently for 32 s, at which time 175 μg sodium metabisulfite in 50 μL sodium phosphate buffer is added, followed by 80 μL of a 10% human serum albumin solution in 0.05*M* sodium phosphate.

ACTH (1-24) can be iodinated in a polypropylene or a borosilicate glass tube. Five micrograms of ACTH (1-24) in 50 μL 0.2*M* ammonium acetate, pH 4.6, is added to 25 μL 0.5*M* sodium phosphate buffer, pH 7.4, followed by 1.5 mCi Na^{125}I in 15 μL 0.01*N* NaOH. Forty micrograms of chloramine-T in 20 μL 0.05*M* sodium phosphate buffer is added and, 30 s later, 87 μg sodium metabisulfite in 25 μL 0.05*M* sodium phosphate buffer is also added, followed by 80 μL of a 10% human serum albumin in the same phosphate buffer, pH 7.4.

Table 2
Parameters for Radioiodination and Tracer Purification on Carboxy-
methyl Cellulose for Several Neuropeptides

Peptide[a]	CT, μg	Na$_2$S$_2$O$_5$ μg	Duration of oxidation, s	Molarity of elution buffer, NH$_4$OAc
ACTH 1-24	20–30	100–125	30–38	0.9
Rat ACTH 1-39	20–30	100–125	30–38	0.9
Human ACTH 1-39	20–30	100–125	30–38	0.3
BAM 22P	20–30	100–125	30–38	0.8
[TYR^{21}] CRF(21–41)	20–30	100–125	30–38	0.2
hpGRF (1-44)-NH$_2$	20–30	100–125	30–38	0.5
LHRH	20–30	100–125	30–38	0.2
Neurotensin	20–30	100–125	30–38	0.2
TYR-Sauvagine (17–40)	20–30	100–125	30–38	0.1
[TYR^1] SS14	6	19	38	0.2
[TYR^{11}] SS14	6	19	38	0.2
SS28-TYR	6	19	38	0.55
TYR-SS28(1-12)	20	100	38	0.07
VIP	25	125	32	0.55

[a]The amount of peptide used is 2–5 μg.

Standard radioiodinations often lead to formation of diiodinated peptides and damage resulting from radiolysis. Stability
of tracers will be improved by using moderate amounts of oxidizing and reducing agents such as those suggested in Table 2 for
various peptides.

In situations in which a methionine residue is present in the
antigenic determinant region of a peptide that would be oxidized to
the methionine sulfoxide by low doses of an oxidizing agent
(Rehfeld, 1978), or in cases in which the antigenic determinant
contains a tyrosine, or in situations in which the peptide available
for iodination does not contain a tyrosine or a histidine, one may
have to use a tritiated peptide. However, the specific activity of
tritiated peptides remains low compared to the iodinated peptides,

and as a result an alternative iodination reagent may be preferable. Two such reagents are commercially available: the reagent of Bolton and Hunter prepared in collaboration with Rudinger and Ruegg (Bolton and Hunter, 1973; Rudinger and Ruegg, 1973) and the reagent prepared by Wood and collaborators (Wood et al., 1975). The Bolton-Hunter reagent (New England Nuclear, ICN, or Amersham), is mono- or diiodinated 3-(4-hydroxyphenyl)propionic acid *N*-hydroxysuccinimide ester that reacts with the free amino groups of a peptide (Fig. 2C). Instead of a direct substitution of ^{125}I into a tyrosyl or histidyl residue, as in the chloramine-T or lactoperoxydase procedure, the peptide is conjugated to the ^{125}I-containing acylating agent. The acylation reaction is allowed to proceed for about 20 min at 0°C and terminated by adding acetic acid and glycine. Separation of cold from labeled peptide must then be performed by gel filtration or another type of chromatography. Wood's reagent is an ^{125}I imidoester, methyl *p*-hydroxybenzimidate HCl, which also reacts with free amino groups of peptides (Fig. 2D). High-specific-activity cholecystokinin octapeptide tracers have been prepared for RIA using this reagent (Praissman et al., 1982), which is also available with a specific activity of ~4000 μCi/nmol from Amersham. Two other reagents related to those of Wood and Bolton-Hunter that could also be used in RIAs are ^{125}I-labeled diazotized aniline (Hayes and Goldstein, 1975) and *t*-butyloxycarbonyl-L-[^{125}I]-iodotyrosine *N*-hydroxysuccinimide ester (Assoian et al., 1980).

3.3. Purification of Labeled Peptides

Ideally the material obtained after iodination should be rapidly purified by reverse-phase or ion-exchange chromatography using a liquid chromatograph (Vale et al., 1983). However, the cost of the reverse-phase and ion-exchange columns and contamination of the injector with radioactive material at several million counts per min may constitute the major limiting factors. Certain investigators rely on C_{18} cartridges for separation of iodine from labeled peptides (e.g., SEP-PAK, Waters Assoc., Milford, MA) or even on Quso G-32 glass in the case of ACTH (Philadelphia Quartz Co., Philadelphia, PA). These "fast procedures", however, do not allow one to separate the labeled from the unlabeled peptide. Efficient separation can be achieved by taking advantage of the fact that an iodinated peptide is more hydrophobic than the noniodinated molecule. A simple gel filtration column containing Sephadex G-25 (Pharmacia, 1 cm × 25 cm) and elution with 0.2*M* acetic acid will easily separate the cold peptide and sodium iodide from the labeled peptide that elutes *after* the salt. This approach has proven to be useful for purification of [arg[8]] vasopressin and SS14 tracers (Roth et al., 1966; Patel and Reichlin, 1978).

We routinely purify all basic peptides that we label using CM-52 cation-exchange chromatography (Arimura et al., 1973). The procedure is simple and inexpensive and the resolution is adequate when isocratic conditions are used. The separation is performed on a carboxymethyl-cellulose column (0.5 × 15 cm, CM-52, Whatman, England) developed at 4°C with ammonium acetate, pH 4.6 (Fig. 3). Table 2 contains a list of the different molarities of ammonium acetate buffer to be used isocratically for elution of several radioiodinated peptides. Na[125]I not incorporated into peptides elutes first, followed by the peptides of the less positively charged species, to the more basic ones. The fraction eluting after Na[125]I that shows the highest binding to the antibody is selected for RIA. The selected fraction is diluted 1/3 in buffer T [T, buffer for tracer: 20 mM sodium phosphate buffer, pH 7.4, containing 0.1% HSA (w/v), 0.1% gelatin (w/v), 0.1% Triton X-100 (v/v), 0.1% sodium azide (w/v), and 150 mM NaCl] and quickly frozen in liquid nitrogen. When kept at –20°C, the labeled peptide is stable for about 2 mo. In order to purify acidic tracers by means other than reverse-phase liquid chromatography, gel permeation chromatography (*vide supra*) or anion exchangers such as a di-ethylaminoethyl cellulose (DEAE A-25, Pharmacia) are generally employed.

---→

Fig. 3. Cation-exchange chromatography on carboxy methyl-cellulose (CM-52) of material obtained after radioiodination of LHRH and ACTH (1-24) (*see* text for conditions of iodination). Total volume from 2 iodinations (220 µL) was applied on the column bed (0.5 × 15 cm), which was eluted with 0.002M ammonium acetate, pH 4.6. The first peak of radioactivity was free sodium iodide. After 4 mL of 0.002M ammonium acetate had passed the column, the buffer was changed to 0.2M ammonium acetate, pH 4.6, which eluted three peaks of radioactive material. Fraction 23 contained 3.5 × 10[8] cpm and showed the highest binding to a LHRH antibody. The peak that follows (1.2 × 10[8] cpm) also bound to a LHRH antibody and might represent diiodinated LHRH. The last peak of radioactivity corresponded to iodinated ACTH (1-24), which was rapidly displaced from CMC with 1M ammonium acetate, pH 4.6. Fraction size was 0.9 mL and flow rate 10 mL/h.

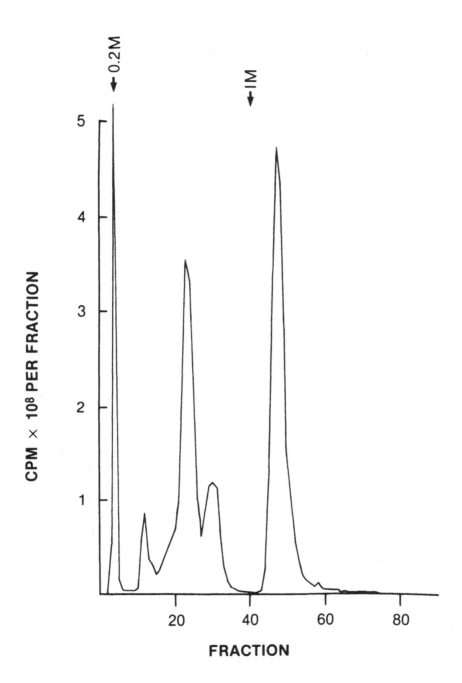

4. Validation and Characterization of RIAs

4.1. Basic Techniques and Specificity Studies

Evaluation of the antibody population of an immune plasma (or antiserum) is a tedious step in the development of a RIA. We simplify this step by carrying out all assays in a standard RIA buffer, using similar incubation conditions for all antibodies and peptides. Incubations are done at 4°C for 18–24 h in 25 mM sodium phosphate buffer containing 0.2% human serum albumin (Fraction V of Cohn, Sigma), 25 mM EDTA (disodium salt), 150 mM NaCl, 0.2M sodium acetate, and 0.1% sodium azide, and adjusted to 7.2 with NaOH and filtered (Whatman filter paper no. 541). Except in rare cases such as ACTH RIAs in which plastic tubes are used, all assays are performed in borosilicate glass tubes (12 × 75 mm) containing a total of 400 μL RIA buffer (150 μL buffer, 100 μL antibody solution, 100 μL unknown or standard, and 50 μL tracer containing ~7,000 cpm). The antigen–antibody complex is precipitated with 2 mL absolute ethanol. Once the titer is known, i.e., the highest final dilution at which the antibody will bind 30–35% of the tracer, a standard curve is generated and substances related to the hapten used during immunization are tested for cross-reactivity. This also includes fragments of the hapten. Examples are provided in Figs. 4 (upper panel) and 5, which show the results obtained after immunization against [D-Lys^6]-LHRH and porcine β-endorphin, using a conjugate in both cases. When used in the LHRH RIA system at a final dilution of $1/10^6$, LR$_1$ recognizes the C-terminus of the decapeptide but cannot recognize the precursor of LHRH. Bendo-2 (Fig. 5) does not recognize the N- nor the C-terminus of the opioid peptide and can detect the precursor β-lipotropin. Another useful approach, especially when dealing with cyclic peptides, consists of using synthetic analogs in which a single amino acid has been replaced by alanine. Table 3 shows results obtained from the characterization of four different antibodies directed against SS14 using analogs of the alanine-substituted series kindly provided by Dr. Jean Rivier, Salk Institute, La Jolla, CA. It is clear that two antibody populations (SS$_7$ and SS$_9$, produced by immunization with met-BSA and SS14) are directed against the central portion of the tetradecapeptide while the two other antibodies, S310 and S312 ([TYR^{11}] SS$_{14}$–BDB–BSA conjugate) are directed toward the NH$_2$-terminal (1–8) region of SS$_{14}$. Hence SS$_7$ and SS$_9$ both recognize SS14 and SS28 on an equimolar basis.

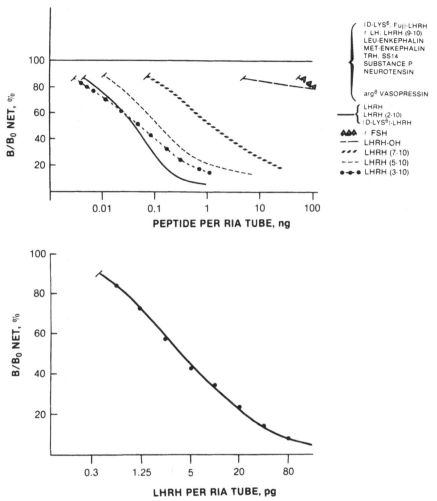

Fig. 4. Upper panel: Cross-reactivity studies performed with a LHRH antibody, LR_1, obtained after immunization of a rabbit with [D-LYS[6]]-LHRH-glutaraldehyde-ovalbumin. Final antibody dilution, 1/500,000–1/10[6]. Displacement curves indicate that residues three to five and seven to ten of the decapeptide are recognized by the antibody population. ED_{50} = 42 pg/RIA tube. A total of eight animals were immunized with this same conjugate. The antigenic determinant of antibodies obtained from animals LR_2 to LR_8 was very similar to that of LR_1. Lower panel: Results obtained with LR_1 at a final dilution of $1/3.5 \times 10^6$. Net binding = 30% of total radioactivity. Iodinated LHRH (4300 counts) is added 26 h after antibody and standards. Incubation is continued for 48 more h at 4°C followed by precipitation with 2 mL ethanol and centrifugation at 4°C (2000g for 30 min). Pellet is counted for 2 min in a LKB 1270 γ-scintillation spectrometer. ED_{50} = 3.9 pg, minimal detectable amount = 0.5 pg LHRH/tube.

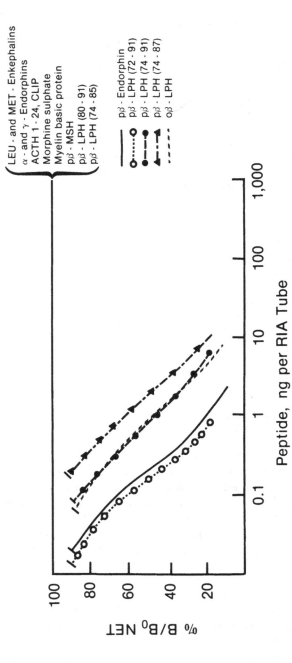

Fig. 5. Cross-reactivity studies documenting the specificity of Bendo-2, an antibody population produced in a rabbit against synthetic porcine β-endorphin coupled to bovine serum albumin using BDB. The final dilution is 1/50,000 when using iodinated β-endorphin and incubating for a total of 24 h without special measures to increase sensitivity. $ED_{50} = 190$ pg/tube. In this system, cross-reactivity with natural ovine β-lipotropin (gracious gift of Dr. Chrétien) was 57% on a molar basis.

Table 3

Antigenic Determinant Study for Four Antibody Populations Raised in Rabbits Against SS14. (Percentage Cross-Reactivity With Somatostatin Analogs)[a,b]

Peptide	SS9 anti-body, %	SS7 anti-body, %	S310 anti-body, %	S312 anti-body, %
SS14	100	100	100	100
[TYR¹] SS14	64*	54*	< 0.002	< 0.003
[ALA²] SS14	116	92	0.2	0.03
[ALA⁴] SS14	67	0.4	4.6	0.06
[ALA⁵] SS14	3.9	100	2.3	10.7
[ALA⁶] SS14	0.05	0.08	17	8.7
[ALA⁷] SS14	<0.01	<0.03	19.2	34
[ALA⁸] SS14	<0.01	0.53	37.5	34
[ALA⁹] SS14	2.2	40	100	ND[c]
[ALA¹⁰] SS14	26.9	13.5	47	53.4
[ALA¹¹] SS14	60	62	116	ND[c]
[ALA¹²] SS14	60	83	100	100

[a]Each value is in percent and represents (ED$_{50}$ analog/ED$_{50}$ SS14) × 100. Calculations are done on a weight basis. All testings are done with a ¹²⁵I[TYR¹] SS14 tracer in the case of SS9 and SS7 and with a ¹²⁵I[TYR¹¹] SS14 tracer for S310 and S312, except for the two values with asterisks, which were obtained using a ¹²⁵I[TYR¹¹] SS14 tracer.

[b]The final antibody dilutions used in the RIA performed at 4°C were 1/12,000, 1/10,000, 1/40,000, and 1/50,000 for SS7, SS9, S310, and S312, respectively. SS9 recognizes mostly amino acids 5–10 of the tetradecapeptide and SS7 recognizes amino acids 4, 6, 7, 8, and 10. S310 and S312 are directed *mostly* against amino acids 1–8 of SS14.

[c]ND indicates not done.

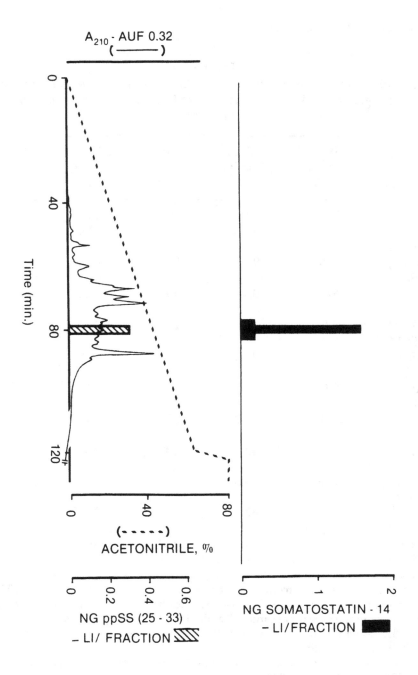

These studies are essential in order to demonstrate the specificity of any antibody, mono- or polyclonal, as well as to identify the antigenic determinant. The results of the cross-reactivity studies, as well as the recognition site determined for a given antibody population, cannot be extrapolated to other systems in which the antibody dilution, the incubation time, or the labeled peptide are substantially different. For polyclonal antibodies, one can minimize the variability among systems by using small and pure haptens and preparing homogeneous conjugates for immunization (Kopp et al., 1977).

4.2. Sources of Artifacts: Extraction Procedures

Measurement of peptides in brain extracts, plasma, or tissue perfusates needs special controls in each situation. Even if the first antibody in the RIA has been extensively characterized, cross-reactive substances may still lead to overestimation of the actual contents. Other substances unrelated to the antigenic determinant can modify substantially the antigen–antibody reaction or the precipitation of the immune complex (Table 4). In rare cases, an extract may even contain proteins or salts that increase the binding of the antigen to the antibody, leading to underestimation or false negative results.

The final demonstration that a precise quantity of a peptide is indeed present in an extract is provided by amino acid analysis and

←——————————————————————————————————————

Fig. 6.   Purification of a 10,000-dalton somatostatin from a rat brain extract obtained after G-75 gel permeation chromatography. The recovered material is purified by analytical reverse-phase liquid chromatography using a C_{18} column (Altex Ultrasphere, 4.6 × 150 mm, 5 μm particle size) and trifluoroacetic acid 0.1%/acetonitrile as the mobile phase. Virtually all the immunoreactive somatostatin-14-like material elutes in one fraction (upper panel). This same fraction contains also pre-prosomatostatin (25-33)-like material (lower panel). The evidence (molecular weight, retention time, immunoreactivity for both the N- and the C-terminus of prosomatostatin) suggests that the immunoreactive material is rat prosomatostatin. Note that the antibody directed against the N-terminal portion of the precursor underestimates considerably the amount of prosomatostatin present. One explanation could be that the N-terminus is not as exposed as the C-terminus. In order to quantitate prosomatostatin accurately in tissue extract, one needs to compare data from RIA and amino acid analysis.

Table 4

Sources of Artifacts in RIA Determinations Performed on Tissues and
Body Fluids

Degradation of labeled antigen by unextracted enzymes
Cross-reactivity with structurally related substances
Variations in binding caused by:
  Salts
  Lipids
  Excess protein
  Limited buffering capacity of buffer solutions
  "Block effect" in large assays
Interassay variability
Nonparallelism
Poor recoveries
Unsuspected heterogeneity of immunoreactive material caused by multiple molecular forms
Degradation of precursor and formation *de novo* of peptides during extraction[a]

---

[a]From Carraway, 1984.

sequencing of that peptide. Since these results are only rarely available, we rely on RIA values performed in triplicate. When the RIA has been validated, the results are generally accurate. Validation comprises the assessment of interassay variability by using internal standards. It implies the elimination of several potential sources of artifacts, some of which are listed in Table 4. In addition, it often requires gel filtration and/or reverse-phase liquid chromatography to demonstrate coelution with a standard (*see* example in Fig. 6). It should definitely include serial dilutions of extracts to demonstrate that the slopes of the standard curve and extracts are not statistically different.

The extraction step itself is most important, and the optimal extraction procedure may vary among different peptides. In general, we recommend the following procedure: (1) Freeze the tissue on dry ice immediately after dissection and store at –70°C. (2) Determine wet weight and defrost the tissue in $2M$ acetic acid (~8 mL/g wet weight) by heating at 95°C for 15 min in a solution containing pepstatin (soluble in DMSO), 20 mg/L, phenyl methyl sulfonyl fluoride (soluble in ethanol), 20 mg/L, and bacitracin, 50 mg/L. (3) Homogenize with a Polytron homogenizer (Brinkmann). (4) Centrifuge at 40,000g for 20 min at 4°C. (5) Lyophilize the supernatant and store at –20°C until assay. (6) Dissolve the lyophilized powder in 1–2 mL RIA buffer containing a pH indicator (phenol red). (7) Bring the solution to pH 7 with 1N NaOH. (8) Centrifuge at 4°C and at 12,000g for 2 min. (9) Use supernatant for RIA determination.

5. Targeting of Antibody to Specific Regions of the Peptides

Although raising antibodies against peptides is relatively simple, stimulating the immune system against selected regions of a given peptide requires more expertise. It is well known that the immune system expresses considerable variability in its antibody response in vivo. There appears to be a characteristic pattern of response, however, related to the preponderant antigenic determinants when antibodies against peptidergic haptens are generated. This could allow one to predict the major recognition sites of an antibody population. Such an approach can be useful not only for polyclonal antibody production of predetermined specificity, but also for monoclonal antibody production as a selection step before fusion.

In order to generate antibodies that will be directed against the carboxy-terminus of a peptide and that will not recognize the C-terminally extended forms of that peptide, a fragment of 6–15 amino acids corresponding to that C-terminal region is synthesized with a tyrosine added at the first position at the end of the synthesis. The hapten is then conjugated to bovine serum albumin using BDB, and the conjugate is injected as described in section 2. If, on the contrary, an antibody population directed specifically against the NH$_2$-terminus of a peptide is required, a fragment of 6–15 amino acids, corresponding to the NH$_2$-terminal region of the whole peptide, is synthesized with a tyrosine added at its carboxy-terminus. The synthetic peptide is then conjugated as above for injection. Immunizations against [TYR^{21}]ovine CRF (21-41), TYR SS28 (1-12), α-endorphin (Guillemin et al., 1977), and LHRH have led to C-terminal-specific antibodies for ovine CRF, SS28 (1-12), α-endorphin, and LHRH, respectively (Fig. 4). Examples of antibodies specific for the amino terminal region are those generated against [TYR^{11}] SS14 (Table 3) (Benoit et al., 1982b), which show negligible cross-reactivity with SS28 (<0.04%).

When the "tyrosilated hapten" is longer than 20 amino acids, the chances of eliciting antibodies specific for the N- or C-terminus of the hormone are decreased. If it is more than 25 amino acids, it will often yield antibodies of low or no affinity for the terminal region. The characteristics of the antibody populations generated after immunization against SS28-TYR and against porcine β-endorphin (which contains a tyrosine in position 1) illustrate this principle. The anti-porcine β-endorphin antibody, bendo-2 (Fig. 5), which "reads" across species, recognizes the region between residues 12 and 28 of β-endorphin. Our most specific antibody population raised against SS28 using SS28-TYR is S309. It recognizes the NH$_2$-terminus of SS28, but substitution of the first amino acid (SER) does not completely abolish the immunoreactivity, so that NH$_2$-terminal extended forms of SS28 are partially recognized (Benoit et al., 1985). Based on our experience with immunization against peptides of various sizes, the stimulation of gamma-globulin production against the central region of a peptide fragment will usually occur if the hapten to which a tyrosine is added at the C- or N-terminus for conjugation is greater than 25 amino acids long.

There are other approaches for the production of antibodies directed against the central region of a peptide. In our hands immunization against intact cyclic SS14 using methylated BSA has consistently yielded antibodies directed against the central region of the hapten, usually amino acids 4–10 (Table 3) (Benoit et al., 1980; Benoit et al., 1982a,b; Morrison et al., 1983). By extrapolation, one could conceivably generate antibodies against any selected portion of a whole peptide by simply stimulating the immune system with a hapten of ~12 amino acids corresponding to the selected region of the peptide. This hapten would need to be flanked by two cysteines to make it cyclic and injected simply with methylated BSA (section 2.1.2). Cyclization of a peptide region that is normally linear, however, could theoretically prevent antibody binding to the natural linear form because of conformational changes. In practice this does not appear to be an absolute limitation since Jackson and coworkers have recently succeeded in raising antibodies to pro-TRH using the analog CYS-LYS-ARG-GLU-HIS-PRO-GLY-LYS-ARG-CYS conjugated to thyroglobulin using glutaraldehyde (Jackson et al., 1985).

6. Procedures to Increase Sensitivity in RIAs

When a RIA is newly developed, the level of sensitivity commonly observed is about 5–10 fmol/RIA tube (i.e., minimal amount of peptide detectable per tube; empirically, 88% B/B_0 net). The challenge starts when 1 fmol or less of antigen has to be detected, as in tissue perfusion studies or for determinations in plasma. In order to reach that level of sensitivity, one usually needs an antibody with an affinity constant (K) > 1×10^{11} mol/L. In addition, the quality of the label is also important. More elaborate protocols than the ones described in section 3 may be required such as reverse-phase chromatography, prolonged isocratic elutions on ion exchangers (Marshall and Odell, 1975), or repetitive gel filtration, as described in detail for neurohypophysial hormones (Glick and Kagan, 1979). To perform the RIA under optimal conditions, a labeled peptide must be free of the corresponding cold peptide.

Delayed addition of labeled peptide until 24 h after addition of the cold antigen often increases the sensitivity by a factor of three- to fivefold. The exact mechanism of this effect is unknown, although Rodbard et al. have provided a theoretical basis for it (Rodbard et al., 1971). When the tracer is added, incubation proceeds at 4°C for another 24 or more h and at maximal dilution of antibody that is still capable of binding 30% of tracer. Instead of using 10,000–12,000 cpm/tube, only 3000–6000 cpm are added and counting is performed for 3 min with a high-efficiency gamma counter (>75%). When these measures are taken, sensitivity can increase by a factor of 10 (compare panels of Fig. 4). Last, some improvement in sensitivity has been reported, but not confirmed, after treatment of diluted antiserum with 2M NaI (Fyhrquist and Wallenius, 1975).

The power of the antigen–antibody reaction is fascinating. It is possible to produce antibodies against peptides as small as α-MSH and reach titers of $1/10^7$ (Buergisser, 1978). The best performance in terms of sensitivity can reach as low as 100–200 attomol/tube (Oyama et al., 1971; Nicholson et al., 1984). Such levels of sensitivity can be obtained with antibodies whose association constants reach 10^{12} mol/L. If it is possible to stimulate the B-lymphocytes to produce gamma-globulins with affinity constants of 10^{13} mol/L, the whole RIA standard curve will be in the attomole range (10^{-18} mol), which is close to the range of cytochemical bioassays (10^{-19} mol). This would add a new dimension to RIAs.

References

Arimura A., Sato H., Kumasaka T., Worobec R. B., Debeljuk L., Dunn J., and Schally A. V. (1973) Production of antiserum to LH-releasing hormone (LHRH) associated with gonadal atrophy in rabbits: Development of radioimmunoassays for LHRH. *Endocrinology* **93**, 1092–1103.

Assoian R. K., Blix P. M., Rubenstein A. H., and Tager H. S. (1980) Iodotyrosilation of peptides using tertiary-butyloxycarbonyl-L-[^{125}I]iodotyrosine N-hydroxysuccinimide ester. *Anal. Biochem.* **103**, 70–76.

Baird A., Klepper R., and Ling N. (1984) *In vitro* and *in vivo* evidence that the C-terminus of preproenkephalin-A circulates as an 8,500 dalton molecule. *Proc. Soc. Exp. Biol. Med.* **175**, 304–308.

Barabe J. and Regoli D. (1981) Development of a radioimmunoassay for bradykinin and des-ARG$_9$-BK. *Pharmacologist* **23** (Abst.), 127.

Bassiri R. M. and Utiger R. D. (1972a) The preparation and specificity of antibody to thyrotropin releasing hormone. *Endocrinology* **90**, 722–727.

Bassiri R. M. and Utiger R. D. (1972b) Thyrotropin-Releasing Hormone, in *Methods of Hormone Radioimmunoassay* (Jaffe B. M. and Behrman H. R., eds.), Academic, New York.

Benoit R., Ling N., Lavielle S., Brazeau P., and Guillemin R. (1980) Production of high quality antisera against brain peptides with and without coupling. *Fed. Proc.* **39**, A1166.

Benoit R., Böhlen P., Ling N., Briskin A., Esch F., Brazeau P., Ying S.-Y. and Guillemin R. (1982a) Presence of somatostatin-28 (1-12) in hypothalamus and pancreas. *Proc. Natl. Acad. Sci. USA* **79**, 917–921.

Benoit R., Ling N., Alford B., and Guillemin R. (1982b) Seven peptides derived from prosomatostatin in rat brain. *Biochem. Biophys. Res. Commun.* **107**, 944–950.

Benoit R., Böhlen P., Ling N., Esch F., Baird A., Ying S.-Y., Wehrenberg W. B., Guillemin R., Morrison V. H., Bakhit C., Koda L., and Blood F. E. (1985) Somatostatin-28 (1-12)-Like Peptides, in *Somatostatin* (Patel Y. C. and Tannenbaum G. S., eds.), Plenum, New York.

Bessler W. G., Cox M., Wiesmüller K. H., and Jung G. (1984) The mitogenic principle of escherichia coli lipoprotein: B-lymphocyte mitogenicity of the synthetic analogue palmitoyl-tetrapeptide. *Biochem. Biophys. Res. Commun.* **121**, 55–61.

Bloch B., Milner R. J., Baird A., Gubler U., Reymond C., Böhlen P., le Guelle C., and Bloom F. E. (1984a) Detection of the messenger RNA coding for preproenkephalin A in bovine adrenal by *in situ* hybridization. *Regul. Pept.* **8**, 345–354.

Bloch B., Ling N., Benoit R., Wehrenberg W. B., and Guillemin R. (1984b) Monosodium glutamate specifically depletes immunoreactive growth hormone-releasing factor (somatocrinin) in rat median eminence. *Nature* **307**, 272–273.

Bolton A. E. and Hunter W. M. (1973) The labelling of proteins to high specific radioactivities by conjugation to a ^{125}I-containing acylating agent. *Biochem. J.* **133**, 529–538.

Bowes J. H. (1963) A fundamental study on the mechanism of deterioration of leather fibers. *Br. Leather Manuf. Res. Assoc. Report.*

Buergisser E. (1978) Modell systeme fuer das studium von polypeptidhormon-receptor-wechselwirkungen (doctoral thesis). University of Zurich, Switzerland.

Carelli C., Audibert F., Gaillard J., and Chedid L. (1982) Immunological castration of male mice by a totally synthetic vaccine administered in saline. *Proc. Natl. Acad. Sci. USA* **79**, 5392–5395.

Carraway R. E. (1984) Rapid proteolytic generation of neurotensin-related peptides and biologic activity during extraction of rat and chicken gastric tissues. *J. Biol. Chem.* **259**, 10328–10334.

Day R., Denis D., Barabé J., St-Pierre S., and Lemaire S. (1982) Dynorphin in bovine adrenal medulla. *Int. J. Pept. Prot. Res.* **19**, 10–17.

Epelbaum J., Brazeau P., Tsang D., Brawer J., and Martin J. B. (1977) Subcellular distribution of radioimmunoassayable somatostatin in rat brain. *Brain Res.* **126**, 309–323.

Fraker P. J. and Speck J. C. (1978) Protein and cell membrane iodinations with a sparingly soluble chlorosamide, 1,3,4,6-tetrachloro-3α,6α-diphenylglycoluril. *Biochem. Biophys. Res. Commun.* **80**, 849–857.

Fyhrquist F. and Wallenius M. (1975) Improvement of antisera against polypeptide hormones. *Nature* **254**, 82–83.

Gee C. E., Chen C.-L. C., Roberts J. L., Thompson R., and Watson S. J. (1983) Identification of proopiomelanocortin neurones in rat hypothalamus by *in situ* cDNA-mRNA hybridization. *Nature* **306**, 374–376.

Glick S. M. and Kagan A. (1979) Vasopressin, in *Methods in Hormone Radioimmunoassay*, 2nd ed. (Jaffe B. M. and Behrman H. R., eds.), Academic, New York.

Greenwood F. C., Hunter W. M., and Glover J. S. (1963) The preparation of ^{131}I-labelled human growth hormone of high specific radioactivity. *Biochem. J.* **89**, 114–123.

Guillemin R., Ling N., and Vargo T. (1977) Radioimmunoassays for α-endorphin and β-endorphin. *Biochem. Biophys. Res. Commun.* **77**, 361–366.

Hayes C. E. and Goldstein I. J. (1975) Radioiodination of sulfhydryl-sensitive proteins. *Anal. Biochem.* **67**, 580–584.

Heding L. G. (1969) The production of glucagon antibodies in rabbits. *Horm. Metab. Res.* **1**, 87–88.

Jackson I. M. D., Wu P., and Lechan R. M. (1985) Immunohistochemical localization in the rat brain of the precursor for thyrotropin releasing hormone. *Science* **229**, 1097–1099.

Jung G., Wiesmüller K H., Metzger J., Bühring H-J, Muller C. P., Biesert L., and Bessler W. G. (1986) Enhancement of Immune Response Using B-Lymphocyte Mitogens Covalently Linked to Antigens. in *Peptides, Structure and Function*, (Deber C. M., Hruby V. J., and Kopple K. D., eds.), Pierce Chemical, Rockford, Illinois.

Kopp H. G., Eberle A., Vitins R., Lichtensteiger W., and Schwyzer R. (1977) Specific antibodies against α-melanotropin for radioimmunoassay. *Eur. J. Biochem.* **75**, 417–422.

Lechan R. M., Goodman R. H., Rosenblatt M., Reichlin S., and Habener J. F. (1983) Prosomatostatin-specific antigen in rat brain: Localization by immunocytochemical staining with an antiserum to a synthetic sequence of preprosomatostatin. *Proc. Natl. Acad. Sci. USA* **80**, 2780–2784.

Likhite V. and Sehon A. (1967) Protein–Protein Conjugation, in *Methods in Immunology and Immunochemistry* Vol. 1, (Williams C. A. and Chase M. W., eds.), Academic, New York.

Mark Y-C. (1968) Antigenicity of insulin (doctoral thesis). Department of Chemistry, McGill University, Montreal, Canada.

Marshall J. C. and Odell W. D. (1975) Preparation of biologically active ^{125}I LHRH suitable for membrane binding studies. *Proc. Soc. Exp. Biol. Med.* **149**, 351–355.

Morrison J. H., Benoit R., Magistretti P. J., and Bloom F. E. (1983) Immuno-histochemical distribution of pro-somatostatin related peptides in cerebral cortex. *Brain Res.* **262**, 344–351.

Nakaya K., Suzuki T., Taikenaka O., and Shilata K. (1969) States of amino acid residues in proteins. XIX. Modification of arginine residues in myoglobin. *Biochem. Biophys. Acta* **194**, 301–309.

Nicholson W. E., Davis D. R., Sherrel B. J., and Orth D. N. (1984) Rapid radioimmunoassay for corticotropin in unextracted human plasma. *Clin. Chem.* **30**, 259–265.

Oyama S. N., Kagan A., and Glick S. M. (1971) Radioimmunoassay of vasopressin: Application to unextracted human urine. *J. Clin. Endocrinol. Metab.* **33**, 739–744.

Patel Y. C. and Reichlin S. (1978) Somatostatin in hypothalamus, extrahypothalamic brain and peripheral tissues of the rat. *Endocrinology* **102**, 523–530.

Plescia O. J., Braun W., and Palczuk N. C. (1964) Production of antibodies to denatured deoxyribonucleic acid. *Proc. Natl. Acad. Sci. USA* **52**, 279–285.

Praissman M., Izzo R. S., and Berkowtiz J. M. (1982) Modification of the C-terminal octapeptide of cholecystokinin with a high-specific-activity iodinated imidoester: Preparation, characterization and binding to isolated pancreatic acinar cell. *Anal. Biochem.* **121**, 190–198.

Rehfeld J. F. (1978) Immunochemical studies on cholecystokinin. *J. Biol. Chem.* **253**, 4016–4021.

Rodbard D., Ruder H. J., Vaitukaitis J., and Jacobs H. S. (1971) Mathematical analysis of kinetics of radioligand assays. Improved sensitivity obtained by delayed addition of labeled ligand. *J. Clin. Endocrinol. Metab.* **33**, 344–355.

Roth J., Glick S. M., Klein L. A., and Petersen M. J. (1966) Specific antibody to vasopressin in man. *J. Clin. Endocrinol. Metab.* **26**, 671–677.

Rudinger J. and Ruegg U. (1973) Preparation of N-succinimidyl 3-(4-hydroxyphenyl) propionate. *Biochem. J.* **133**, 538–539.

Salacinski, P. R. P., McLean C., Sykes J. E. C., Clement-Jones V. V., and Lowry P. J. (1981) Iodination of proteins, glycoproteins, and peptides using a solid-phase oxidizing agent, 1,3,4,6-tetrachloro-3,6-diphenyl glycoluril (Iodogen). *Anal. Biochem.* **117**, 136–146.

Schick A. F. and Singer S. J. (1961) On the formation of covalent linkages between two protein molecules. *J. Biol. Chem.* **236**, 2477–2485.

Tager H. S. (1976) Coupling of peptides to albumin with difluorodinitrobenzene. *Anal. Biochem.* **71**, 367–375.

Tager H. S., Hohenboken M., and Markese J. (1977) High titer glucagon antisera. *Endocrinology* **100**, 367–372.

Trivers G. E., Harris C. C., Rougeot C., and Dray F. (1983) Development and Use of Ultrasensitive Enzyme Immunoassays, in *Methods in Enzymology* vol. 103, (Conn P. M., ed.), Academic, New York.

Vale W., Vaughan J., Yamamoto G., Bruhn T., Douglas C., Dalton D., Rivier C., and Rivier J. (1983) Assay of Corticotropin Releasing Factor, in *Methods in Enzymology* vol. 103, (Conn P. M., ed.), Academic, New York.

Valiquette G., Hou-Yu A., and Zimmerman E. A. (1986) Monoclonal Antibodies to Neurohypophysial Hormones, in *Monoclonal Antibodies: Basic Principles, Experimental and Clinical Application in Endocrinology* (Lipsett M., Serio M. and Forti G., eds.), Serono Symposia Series, Raven, New York, in press.

Wolfe H. R. and De Grado W. F. (1985) Convenient Method for Synthesis, Purification and Conjugation of Synthetic Peptide Antigens, in *Ab-*

stracts of the Ninth American Peptide Symposium, Toronto, Canada, p. 152.

Wood F. T., Wu M. M., and Gerhart J. C. (1975) The radioactive labeling of proteins with an iodinated amidination reagent. *Anal. Biochem.* **69,** 339–349.

Worobec R. B., Wallace J. H., and Huggins C. G. (1972) Angiotensin–antibody interaction. 1. Induction of the antibody response. *Immunochemistry* **9,** 229–238.

Wu W. H. and Rockey J. H. (1969) Antivasopressin antibody characterization of high affinity rabbit antibody with limited association constant heterogeneity. *Biochemistry* **8,** 2719–2728.

Acknowledgments

This research was supported by NIH grant AM 18811-HD 09690 and MRC grant (MA-9145) to R. Benoit. We are also grateful to Dr. Serge Saint-Pierre from the Department of Pharmacology and Physiology, C. H. U. Sherbrooke, Quebec, Canada, who provided us with part of the synthetic somatostatin-14 when we initiated these studies, and to Dr. Jean Rivier, Salk Institute, San Diego, CA, for the somatostatin analogs. We are grateful to Dr. Michel Chrétien, Clinical Research Institute, Montreal, for the ovine β-LPH. We wish to thank Mrs. Barbara Alford and Alice Moss for excellent technical assistance and Mrs. Dominique Besso for expert clerical work.

Immunocytochemical Techniques in Peptide Localization

Possibilities and Pitfalls

F. W. van Leeuwen

1. Introduction

During the past decade, immunocytochemistry (ICC) has been introduced into the neurosciences on a large scale. ICC has proved to be a powerful tool in supplementing and extending the data obtained with other techniques (e.g., electrophysiology, neuropharmacology, neurochemistry). Its strength is that it combines a high degree of specificity with the resolution of the light and electron microscope. Therefore it can be more or less regarded as "biochemistry within the tissue section," and even more than that if the amount of immunoreactivity can be quantified. ICC alone also resulted in a large number of new data that changed a number of concepts (e.g., coexistence of peptides and amines, occurrence of hitherto assumed "organ- or tissue area-unique" peptides elsewhere; Hokfelt et al., 1980; Livett, 1978). At present, a wide variety of compounds can be localized by ICC (van Leeuwen et al., 1982). Here we will focus on peptide ICC, paying special attention to the pitfalls with which every immunocytochemist will be confronted as soon as the first successful localizations have been achieved. It will be illustrated that at present the technique to prove the specificity of antipeptide sera is still in its infancy.

2. Methodology

Usually the ICC procedure starts with the fixation of the tissue followed by sectioning (with or without embedding) and incubation. These various parts will be discussed briefly since they have recently been dealt with at length in other papers (e.g., Buijs, 1982; Pool et al., 1983).

2.1. Fixation

It is impossible to give a general procedure that can be used under all circumstances; a fixative suitable for one antigen or tissue area is not necessarily suitable for another (e.g., containing a lower amount of peptide). It is necessary to keep the aim of the study in mind. If one wants to perform immunoelectron microscopy, the preservation of the ultrastructure is of the utmost importance. This may result in the choice of a fixative that, however, impairs the immunoreactivity (*see* section 2.4). When starting ICC, a series of fixations should be carried out (e.g., van Leeuwen, 1977). An example of the immunoelectron microscopical localization of enkephalin immunoreactivity will be given (section 2.4). Successful fixatives for both light and electron microscopic ICC appear to be buffered solutions of aldehydes (formaldehyde and glutaraldehyde and combinations of both) preferably administered by intracardial perfusion. The application of osmium tetroxide for immunoelectron microscopy is sometimes also possible (e.g., Li et al., 1977). Furthermore, one may change the pH of the buffer in order to enhance the penetration of the antibodies and the length of the fixation (Berod et al., 1981). The way in which the antibodies are raised may also be of importance. Peptides usually have to be coupled to a larger molecule (e.g., thyroglobulin) in order to obtain an antiserum with a high titer. If this coupling occurs with formalin or glutaraldehyde, the choice of one of these agents will influence the later optimal fixation procedure (Buijs and Pool, 1984).

2.2. Embedding and Sectioning

After the fixation and dehydration, the tissue can be embedded in a number of media (e.g., paraffin, Epon), sectioned, and stained (post-embedding staining). Post-embedding staining has the advantage that there are few penetration problems since antigens are accessible to antibodies after removal of the embedding medium. Thus, only short incubation and washing times need to

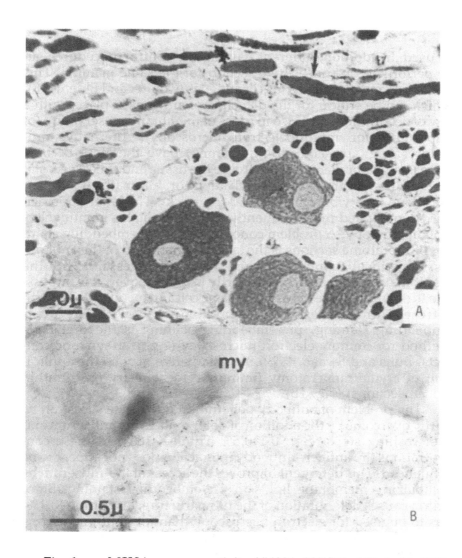

Fig. 1. α-MSH immunoreactivity (#4394, 23/4/75), (diluted 1:200) in dorsal root ganglion cells and fibers in semi-thin epon sections (van Leeuwen et al., 1979b.) (A) Arrows point to myelin sheath. (B) Immunoelectron microscopy performed on ultrathin frozen sections revealed that labeling was present on neurofilaments in axons. Labeling was performed with protein-A-gold (8 nm). Abbreviation: my, myelin.

be used. In addition, different staining and control procedures can be performed on consecutive sections of the same material. This procedure may result in a loss of antigens (Coulter and Elde, 1978), which can be critical for the detection of antigens at places with a low concentration, and also in a loss of membrane preservation (of importance for immunoelectron microscopy). Alternative procedures (e.g., pre-embedding staining) have therefore become increasingly popular. One can freeze the tissue and collect cryostat sections (2–100 μm thick) after fixation, or, alternatively, section the tissue on a vibratome (30–100 μm thick). The latter approach has been applied most frequently since it facilitates immunoelectron microscopy. A problem concerning the pre-embedding staining of Vibratome sections is the lack of penetration of the antibodies. Long incubation and washing times are necessary, with the result that the ultrastructure is partly lost. Moreover, it is impossible to obtain serial sections of the same cell and incubate them with different antibodies. However, compared with the post-embedding staining, pre-embedding staining is the preferable method for immunoelectron microscopy (e.g., for synaptic contacts, Buijs and Swaab, 1979). An alternative may be the application of cryoultramicrotomy (immunocytochemistry on ultrathin frozen sections) (Fig. 1B).

The problem of antibody penetration into tissue slices can be partly overcome by the addition of detergents to the antiserum and the washing steps (e.g., 0.1–0.5% Triton-X 100). At these concentrations the antigen–antibody interaction does not appear to be influenced. This detergent improves the access of the antiserum by solubilizing membrane lipids in pre-embedding staining. Since Triton causes deterioration of the ultrastructure, less Triton (0.1%) has to be used for electron microscopy than for light microscopy (0.5%). Because of penetration problems, a clear outside–inside gradation of reaction product can still be seen in the vibratome section. Only the outer rim (2–10 μm) of the section is stained and suitable for immunoelectron microscopy. Under special circumstances it appears to be impossible to use detergents (van Leeuwen et al., 1983, Fig. 2b). In Table 1 the advantages and disadvantages of the pre- and post-embedding staining are summarized.

2.3. Immunocytochemical Techniques

The most frequently used of ICC techniques in the neurosciences are the indirect immunofluorescence and immunoperoxidase methods (for the principles of the reactions, see Nairn, 1976;

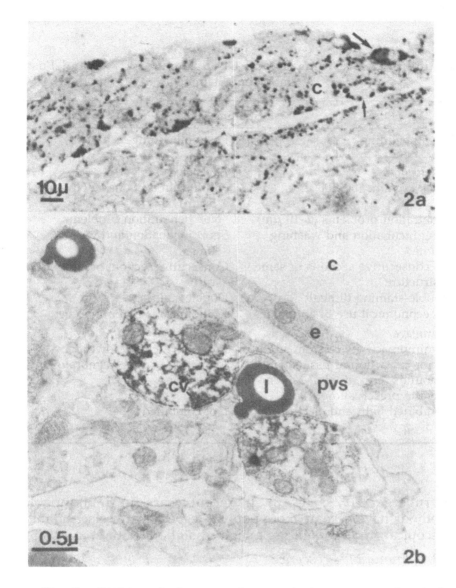

Fig. 2. Light and electron microscopic immunocytochemical demonstration of leu-enkephalin (Leu-enk) immunoreactive fibers in the rat neural lobe. (a) Semi-thin (2 μm) Epon section. Note that the beaded fibers sometimes surround structures (arrows). (b) At the electron microscopical level Leu-enk immunoreactivity is present in fibers terminating in a synaptoid fashion upon pituicytes that are easily recognized by the presence of lipid droplets (L) that are partly extracted. Abbreviations: c, capillary; cv, cluster of clear vesicles; e, endothelial cell; pvs, perivascular space.

Table 1
Listing of the Advantages and Disadvantages of Pre- and Post-embedding Staining

Staining	
Pre-embedding	Post-embedding
Disadvantage	*Advantage*
Thick sections (30–100 μm)	(Ultra) thin sections (± 60 nm– 5 μm)
Penetration problems (2–10 μm)	"No" penetration problems
Long incubation and washing times	Short incubation and washing times
No consecutive sections of same structure	Consecutive sections of same structure
Double staining difficult	Double staining
No economical use of antibodies	Economical use of antibodies
Advantage	*Disadvantage*
Immunoreactivity less impaired	Immunoreactivity lowered
Minor extraction of tissue compounds (e.g., lipids) → membrane preservation	Extraction → poor membrane preservation
Combined light and electron microscopy	Not combined

Sternberger, 1979; Fig. 3). In recent years the immunogold methods have become more popular, especially for electron microscopical studies requiring a high resolution (Roth et al., 1985).

2.3.1. Immunofluorescence

For light microscopic ICC, this method is a very good choice. The main reasons for its popularity are its high degree of resolution and the possibilities for performing double staining and quantifying the amount of fluorescence. Also, in the combination of retrograde labeling of neurons and studying their subsequent ICC characteristics, this technique appeared to be superior (Van der Kooy, 1982). Disadvantages of this method are that it is impossible to perform immunoelectron microscopy on the same section and that a loss of fluorescence (fading) occurs during longer observations.

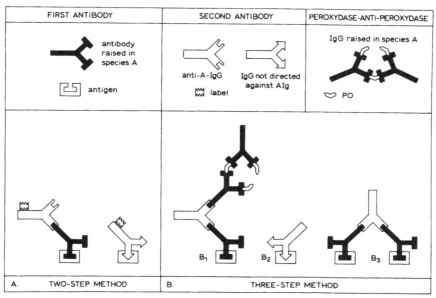

Fig. 3. Principle of some immunocytochemical staining methods. Label, e.g., fluorochrome (e.g., FITC), horseradish peroxidase (PO), colloidal gold. In the two-step method (A) the label is covalently coupled to the second antibody. Method specificity may occur by binding labeled immunoglobulins (IgGs) directly to the tissue section. Such antibodies may already be present (natural antibodies) or synthesized as antibody against an impurity in the IgG fraction of A (Sternberger and Petrali, 1977). In the three-step method (B) (the unlabeled antibody enzyme method), the second antibody forms a bridge between the first antibody and the third antibody–enzyme complex (B_1). Any second antibody binding to the tissue section will not be detected, since it is not directed against IgG of species A (B_2). If the concentrations of antigen within the tissue section and the first antiserum are too high, the two Fab fragments of the antibody become bound. This blocks the reaction with the PAP complex (B_3) (Bigbee et al., 1977).

2.3.2. Immunoenzyme Methods

The introduction of enzymes attached either chemically (e.g., Nakane and Pierce, 1967) or immunologically (Sternberger, 1979) to immunoglobulins has overcome all of the disadvantages of the immunofluorescence procedure, although the quantification of antigens is complex when this method is used (Streefkerk et al., 1975). Most often, the enzyme horseradish peroxidase is used, and the method most frequently applied in this field is the peroxidase-anti-peroxidase (PAP) method developed by Sternberger (1979). There are several reasons for preferring this method. In the first place, it is very efficient because, compared with immuno-fluorescence and indirect methods, the dilution of the first antibody can be very high. Second, the background staining appears to be minimal. One explanation of this low background is illustrated in Fig. 3. A possible disadvantage of the PAP method is that under certain circumstances a negative result can be obtained because of a high concentration of antigen in the tissue section (Bigbee et al., 1977) (Fig. 3). This problem can easily be solved by further diluting the antiserum (Vandesande, 1979). More recently the (strept) avidin-biotin techniques (ABC-techniques) have become increasingly popular. The high affinity interaction of avidin and biotin permits the performance of an ICC reaction. The efficiency of the method (dilution of the first antibody) was claimed to be better than that of the PAP-method (Hsu et al., 1981). However, using the ABC kit from Vector (Burlingame, CA, USA) and our "home-made" PAP, we found no differences in efficiency within our object of interest (vasopressin within the brain).

The indirect two-step methods are not applied often, but with correctly prepared peroxidase-conjugates (prepared by D. M. Boorsma, Free University, Amsterdam) we were able to obtain a very good immunoreactivity, whereas the background was very low. Usually it is not necessary to intensify the end-product of peroxidase stains (diamino benzidine). However, under certain circumstances silver-gold intensification resulted in a remarkable improvement of the ICC reaction (e.g., Liposits et al., 1984). Another remarkable improvement of immunoreactivity was obtained after an incubation in which the second antibody and PAP complex were applied twice. In order to reduce the background staining, somewhat longer rinse steps are useful (Ordronneau et al., 1981).

2.3.3. Immunometal Procedures

It is possible to complex colloidal gold to immunoglobulins (Gu et al., 1981) and other proteins, e.g., protein A, and, as a result, to perform post-embedding staining, especially at the electron microscopical level (e.g., Castel et al., 1986). The advantage of this method is that its resolution is much higher (5–20 nm) than that of the peroxidase methods (±80 nm; Sternberger, 1979). A disadvantage is the poor penetration of immuno- or protein-A gold complexes. Thus, only post-embedding staining seems possible, although cryoultramicrotomy may sometimes be a solution for immunoelectron microscopy (Fig. 2) (van Leeuwen et al., 1982). A disadvantage of the latter methods is that orientation during trimming (before the specimen is frozen) may be difficult. This is a serious problem for a heterogeneous tissue such as the central nervous system. On the other hand, the sections can be observed immediately after sectioning with a phase-contrast microscope.

2.4. Example of an ICC Staining: Enkephalin Immuno-reactivity in the Neural Lobe

Rossier et al. (1979) were the first to demonstrate that en-kephalin fibers innervate the neural lobe. For a number of reasons (not listed here, *see* van Leeuwen et al., 1983), this finding was very interesting. We therefore decided to reproduce the results of Ros-sier et al. with the same bleeding of anti-leu-enkephalin (Leu-enk) that they used. We were indeed able to reproduce their results. Leu-enk fibers were observed all over the neural lobe (Fig. 2a). However, it was not clear whether they terminated near the numerous capillaries or in a different way. Thus, the ultimate aim was to localize immunoelectron microscopically Leu-enk immuno-reactivity. Post-embedding immunoelectron microscopy was not a good device because of the loss of membrane structure. Therefore pre-embedding staining was achieved in 30–50 µm thick vibratome sections. Shortly before, this method had been introduced into the laboratory for immunoelectronmicroscopy on vasopressin syn-apses in the limbic system (Buijs and Swaab, 1979). The method included a perfusion fixation with 2.5% glutaraldehyde and 1% paraformaldehyde followed by sectioning and incubation with anti-vasopressin to which 0.1% Triton was added as a detergent. The PAP procedure was followed. This resulted in a clear demon-stration of vasopressin synapses. This strategy was therefore fol-lowed for the demonstration of Leu-enk in the neural lobe. Howev-er, although it appeared that this anti-Leu-enk serum in the lateral septum gave the same quality pictures as did anti-vasopressin, in the neural lobe the glutaraldehyde–paraformaldehyde mixture led to an almost complete loss of Leu-enk immunoreactivity. Moreover, the ultrastructural preservation was poor, which can be partly explained by differences in histology between the lateral septum and the neural lobe. The latter structure is heav-ily innervated by nerve fibers, has many capillaries, and displays large perivascular spaces. A mixture of 4% paraformaldehyde and 0.2% glutaraldehyde was then used, which resulted in a sufficient preservation of Leu-enk immunoreactivity. The ultrastructure was however, poor. We decided, therefore, to perform an experiment with a few variables, e.g., omission of Triton-x 100 (it is known that detergents deteriorate membranes), and, second, introduction a second fixation step (2.5% glutaraldehyde and 1% paraformaldehyde) after the PAP incubation. It appeared that a good morphology was only obtained when these two variables were incorporated into the procedure (Fig. 2b). In conclusion, peptides may be sensitive to fixatives required for the preservation of the ultrastructure. Therefore it is sometimes advisable to start with a somewhat weaker fixative followed by ICC and again by a fixation. In addition, it is clear that a procedure that is well suited for a certain brain area can be of less value or worthless for another.

3. Specificity (Antiserum Characterization and Purification)

Although it is quite easy to achieve impressive localizations of antigens by ICC, a great deal of effort is required before the method and especially the serum specificity can be proven. Work in our department has made it clear that it is difficult to satisfactorily establish serum specificity, even when relatively pure immunogens are used for immunization (e.g., Swaab et al., 1977; van Leeuwen, 1982; Pool et al., 1983). In this chapter some examples will be given of misinterpretations in the ICC literature, and an attempt will be made to supply guidelines for assessing method and serum specificity. *Method specificity* is defined as the absence of staining disturbances caused by mechanisms other than the interaction between antibodies and the antigen to be localized. *Serum specificity* means that primary antibodies must react only with that compound to which they were raised and not with other compounds.

In papers dealing with ICC there is generally no clear distinction between these two forms of specificity and, if definitions are given, small (but not essential) differences exist in the literature, especially with regard to the classification of antibodies not directed against the antigen to be localized (Vandesande, 1979; van Leeuwen, 1981; Pool et al., 1983).

4. Method Specificity

Method specificity can be demonstrated in a number of different ways. For example, the antibodies of the primary antiserum able to react with the antigen of interest can be removed effectively by using a solid phase (Sepharose 4B) containing the homologous antigen (*see* Swaab and Pool, 1975; Vandesande, 1979). Liquid phase adsorption that has been reported to be ineffective sometimes (Swaab and Pool, 1975) is nevertheless used in most of the papers. Method specificity can also be checked by omitting one of the immunoreagents in the staining sequence or by using increasing dilutions of the first antiserum (Petrusz et al., 1976). The first of these approaches meets the definition of method specificity best and should, therefore, be preferred. However, often impurities occur within immunogens, even in synthetic preparations (e.g., Mohring et al., 1982; see below, h). When antibodies against these impurities are raised, they will also be removed during adsorption with the homologous antigen, which results in a negative staining. Thus in this way the definition mentioned above is not met. Therefore, it is necessary to give a characterization of all immunoreactive compounds present within the antigen(s) used for an adsorption (see below, h). In the case of antibodies already present in the pre-immune serum and able to react with the tissue section (see below, g), even an immunological characterization (using the adsorbed antiserum) of all compounds present within a particular area is a necessity. Pool et al. (1983) avoid these difficulties by substitution of the last four words of the definition of method specificity mentioned above by "tissue."

4.1. False Positive Results

The causes of method aspecificity may be either nonimmunological (a–f) or immunological (g–i) in nature. This classification is of course arbitrary.

    a. *Pseudoperoxidase activity* is found in heme-groups in erythrocytes, for example. This activity can be eliminated by incubating the sections with methanol followed by hydrogen peroxide (Streefkerk, 1972) or by lowering the pH of the diamino benzidine solution (Vacca et al., 1978). If epoxy sections are used, it is not necessary to treat the sections (van Leeuwen et al., 1979) (Fig. 4).

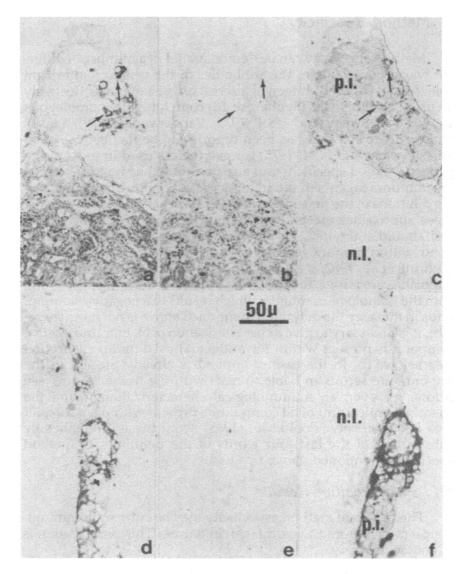

Fig. 4. Serial sections of neurointermediate lobe of Wistar (a–c) and homozygous Brattleboro rats (d–f) after incubation with anti-AVP purified with OXT-beads (a,d), anti-AVP adsorbed four times to α-MSH (b,e), and anti-α-MSH (c,f). Note that in the PI, the same cells are immunoreactive for anti-AVP and anti-α-MSH (arrows). Abbreviation: n.l., neural lobe (from van Leeuwen et al., 1979a, with permission of the publisher).

b. *Endogenous peroxidase.* Catalase present in per-
oxisomes and cytochrome c present in mitochon-
dria may be capable of reacting with the DAB-H_2O_2-
mixture, resulting in falsely positive cells found in a
number of areas in the brain (Buijs, 1978). The activity of the enzymes can be inhibited or drastically
reduced by methanol, nitroferricyanide, or phenylhydrazine (Straus, 1971, 1972). Alternatively, they
can be marked with a different chromogen at the
light microscopical level before the ICC incubation
(Robinson and Dawson, 1975). In monolayers and
explants of nervous tissue, a high background is
usually obtained, which is probably caused by enzymes capable of acting as peroxidases with the
DAB-H_2O_2 mixture. However, treatment of these
cultures with a mixture of 10% methanol and 3%
H_2O_2 in phosphate-buffered saline (G. Jirrikowski,
personal communication) results in a remarkable
decrease in background levels. The time necessary
for inhibition of these enzymes can easily be determined by observation of the cultures (as soon as
the production of oxygen bubbles diminishes, the
treatment can be terminated) (Fig. 5).

c. After aldehyde fixation, *free reactive groups* can still be
present within the section, which leads to nonimmunological binding of proteins (e.g., IgGs) and an
overall high background staining. This can be prevented by pre-incubating the section with a serum
not recognized by the immunoreagents, by reducing the remaining aldehyde groups with sodium
borohydride (Lillie and Pizzolato, 1972), or by using
buffers containing small molecules (lysine, glycine,
tris) or proteins (gelatin, serum albumin) (both with
reactive amino groups).

d. *Hydrophobic and ionic interactions or combinations of both,*
of antibodies (Fc-parts, Aarli et al., 1975), tracers, or
reaction products (e.g., DAB) with tissue components (collagen, nuclear, and ribosomal components), suggest that charged groups within proteins (as present in these organelles) are responsible
for ionic interaction (Kraehenbuhl and Jamieson,
1974). Ionic interactions can be minimized with a

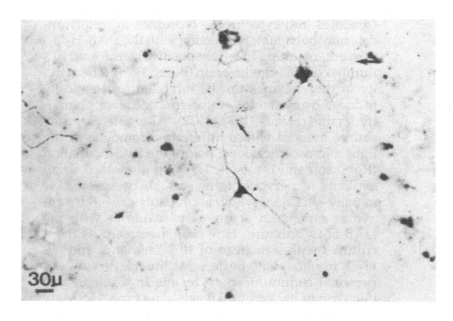

Fig. 5. Immunocytochemical localization of glutamic acid de-carboxylase immunoreactivity (antibody #10/30 obtained from Dr. J. Y. Wu, Houston, USA and diluted 1 : 400) in 2-wk old cultures of dissociated cerebral occipital cortex fixed with 4% paraformaldehyde. Note that underlying glial monolayer is barely positive. Arrows point to varicosities in the nerve fibers. If the preincubation with methanol-hydrogen peroxide had not been performed, all cells would show staining (Habets and van Leeuwen, unpublished observations).

buffer with enhanced ionic strength (Capel, 1974; Grube, 1980).

Hydrophobic interactions can result from hydrophobic parts present in antibodies and from labels with hydrophobic tissue compounds or embedding media. The addition of Triton-X 100 to the washing fluids and incubation media may decrease this interaction. Moreover, in order to minimize the influence of ionic charges, it was found that incubating the tissue section at a pH value of 8.6 results in a decrease of background levels. The rationale behind this treatment may be that for some antisera the overall charge of the IgG

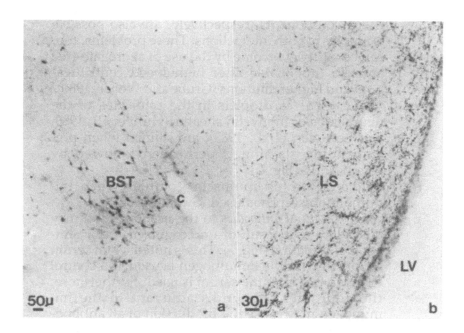

Fig. 6.   Vibratome section of the rat septal area fixed with 4% para-formaldehyde showing immunoreactivity (dilution 1:2000) for C-terminal peptide of propressophysin (c) within cell bodies of the bed nucleus of the stria terminalis (BST) (a) and fibers in the lateral septum (LS) (b) (antibody obtained from Drs. N. G. Seidah and M. Chrétien, Montreal, Canada). Without the use of O.5*M* NaCl/0.05*M* Tris (pH, 8.6) buffer, a high background staining was obtained. Abbreviations: c, capillary; LV, lateral ventricle.

population is lower at pH 8.6 than at pH 7.6. In addition, the degree of tissue swelling may be different at a higher pH. Thus if a certain antiserum gives a high background, we incubate the tissue with 0.05*M* Tris/0.5*M* NaCl/0.5% Triton-X 100, pH 8.6 (recipe of Dr. C. W. Pool). In addition, the rinsing steps and subsequent incubations can be performed with this mixture. An example of this phenomenon is given in Fig. 6.

The tendencies of peroxidase conjugates and immunoglobulin aggregates (resulting in the false appearance of antigen "hot spots") to adhere nonspecifically to hydrophobic embedding media and tissue components, respectively, are also possible examples of both interactions. These problems, too, can be partly overcome by the use of aggregate-free antisera (obtainable after high-speed centrifugation) and higher dilutions (Grube and Weber, 1980).

e. *Lipids,* present as droplets in the pituicytes of the neural lobe or in myelin sheaths (Hunt et al., 1980; van Leeuwen et al., 1983), are stained when pre-embedding staining is performed followed by osmification (Fig. 2).

f. *Autofluorescence* of lipofuscin granules when immunofluorescence is used (Swaab, 1982).

g. *Binding of natural antibodies* to the tissue section may take place as a result of unrecognized, prior antigenic stimulation (Fig. 7). This is noted when serum adsorbed with the immunogen is used as a control or when pre-immune serum is used. Furthermore, the addition of Freund's adjuvant to the immunogen may cause the production of all antibodies to be enhanced. Since the titer of these antibodies is often not very high, the contributions of such antibodies to the ICC staining (when present) can usually be reduced by the use of more diluted antisera (Sternberger, 1979). Of course, erroneous results may be obtained by the re-use of animals that have already been used for a different immunization (Clayton and Hoffmann, 1979).

h. *Contaminating antibodies.* Most antigens used in the production of antisera have to be purified from biological material (e.g., pituitary hormones, brain-specific proteins, neurotransmitter-synthesizing enzymes) and are thus never absolutely pure. Peptides that are usually obtained as synthetic preparations may contain impurities (e.g., Lembol, 1978; North et al., 1978) or be synthesized incorrectly (Brown et al., 1986). Impurities may induce the production of antibodies, which can react with unwanted compounds in the tissue section.

i. It is difficult to induce antibody formation to molecules with a molecular weight below 4 kdalton. Conjugation of proteins to these small molecules (peptides, amines, amino acids, and so on) may

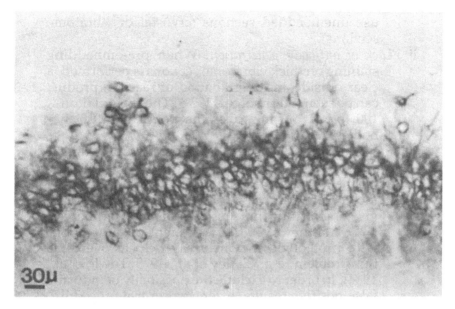

Fig. 7. Vibratome section of the rat hippocampus (dorsal part) fixed with 4% paraformaldehyde and incubated with anti-vasopressin serum (#Hilbert; 5/3/84) diluted 1:1000 showing immunoreactivity around pyramidal cells. However, the pre-immune serum of the rabbit showed this labeling pattern as well (from Caffe, unpublished results).

overcome this difficulty (Skowsky and Fisher, 1972; Geffard et al., 1984). However, in this situation, antibodies are also produced against the carrier proteins. If necessary, these antibodies can be removed by solid phase adsorption.

All the above mentioned points may give rise to false positive results.

4.2. False Negative Results

When, at the start of an ICC study, positive staining is not obtained, this does not automatically imply the absence of antigens within the tissue section. A number of causes should be considered.

i. *Modification or loss of the antigenic sites* during fixation, dehydration, embedding, and clearing procedures (Coulter and Elde, 1978; van Leeuwen, 1981; Matthews, 1981). For this reason it became popular to use unembedded sections (cryostat or vibratome sections).

ii. *Lack of antibody penetration.* When pre-embedding staining on thick vibratome sections is performed, a clear outside–inside gradation of reaction product can be seen (Sternberger, 1979). The use of Triton-X 100 (0.1–0.5%) improves the penetration of antibodies.

iii. *Steric hindrance.* This refers to the presence of antigenic sites within the tissue section that cannot be reached by the antibody. This may be caused by incomplete removal of embedding medium or the masking of antigenic sites by proteins (Sternberger, 1979; Matthews, 1981). Treatment of the sections with proteolytic enzymes may help to overcome this problem (e.g., Finley et al., 1978; Towle et al., 1984). In order to exclude the possibility of inducing false positive results, extensive control procedures should be performed (Caffe et al., 1985).

iv. *The "Bigbee" effect.* A high concentration of antigen within the tissue section and insufficient dilution of the first antibody in the unlabeled antibody enzyme method prevent the bridge function of the second antibody (Bigbee et al., 1977; Vandesande, 1979) (*see* Fig. 3).

v. *The titer of antibodies* may vary considerably during immunization (Swaab, 1975). Moreover, their detection depends on the screening method [radioimmunoassay (RIA), immunofluorescence, or PAP method]. In indirect methods, the immunoreactivity of the second antibody is also important (for details, *see* Pool et al., 1983).

4.3. Serum Specificity

A number of specificity tests have been developed in the past 10 yr. The various approaches have recently been reviewed in detail by Van der Sluis and Boer (1986). Here the most frequently used techniques will be discussed.

Criteria for a specificity test for ICC were formulated by Pool et al. (1983):

i.   The incubation procedures in the specificity test and the ICC localization should be the same. Hence, neither RIA nor immunoprecipitation techniques (immunodiffusion or immunoelectrophoresis) are suitable for an ICC specificity test. The discrepancy between model systems and immunohistochemical staining results, with regard to crossreactivity, was clearly illustrated by Milstein et al. (1983).

ii. The condition of the test antigen should be the same. The same pretreatment should be applied for the antigens in the sections and the antigens in the test system (namely, the fixation procedure may change the immunoreactivity of the antigens) (Milstein et al., 1983).

iii. The test system should include all potential antigens in the tissue.

iv. Identification of all antigens by nonimmunochemical techniques must be possible (e.g., molecular weight, isoelectric point, alone, or in combination with hormonal, pharmacological, or enzymatical activities) (e.g., Geuze et al., 1979; Bidlack et al., 1981).

v. Quantification of the (first) antibody–antigen reaction should be possible even if the primary goal is to obtain a qualitative impression of the distribution of a component; for the precise assessment of either the staining intensity of a control incubation or the success of a purification only a quantitative determination is conclusive.

To prove serum specificity (absence of cross-reaction of the first antiserum) in the ICC literature, four techniques are generally used: radioimmunoassay (RIA) (section 3.2.1), the adsorption to the homologous antigen (section 3.2.2), the adsorption to heterologous antigens (section 3.2.3), and the tissue spectrum affinity test (section 3.2.4). None of these four tests fulfills entirely the requirements for a specificity test.

4.3.1. Radioimmunoassay

With respect to the technique, it has become clear that antisera that show a negligible cross-reaction in RIA may nevertheless show a clearcut cross-reaction in ICC (Swaab et al., 1975). At the moment no explanation for this difference is available. However, one should be aware of the marked difference in the relative concentrations of antigens and antibodies used in the two techniques. Second, specificity in a RIA system only depends on the competition between labeled and unlabeled antigen for attachment to the antibody. This competition is absent in ICC because of the excess of antibodies. In ICC, specificity depends on the lack of antibodies directed against compounds other than the one of interest.

In addition, it has been shown that RIA titers can raise false hopes about the potency of the antiserum in an ICC procedure (Swaab et al., 1975). Thus, the two methods may be using different portions of the antibody population. In conclusion, RIA data (cross-reaction and titer) can be unreliable for ICC.

4.3.2. Adsorption Test (Homologous Antigen)

The use of the adsorption technique (with the homologous antigen) is often recommended as a means of determining serum specificity (e.g., Petrusz et al., 1976, 1977, 1980). Our group is forced to disagree with this view (also see below) (Swaab et al., 1977); the absence of ICC staining after adsorption to the homologous antigen only proves that all antibodies were bound to the added antigen(s). It does not, however, exclude the possibility of staining produced by unwanted or unexpected contaminating antibodies (often erroneously designated as "cross-reaction") raised by impurities within the antigen, whether used as immunogen or as adsorbant (*see* section 4). Furthermore, the possibility of cross-reaction cannot be excluded completely.

The rat hypothalamoneurohypophyseal system (HNS), in which the closely related compounds arginine-vasopressin (AVP) and oxytocin (OXT) are synthesized, provides a good illustration of this problem (Swaab and Pool, 1975; Swaab et al., 1977). Anti-AVP, solid phase adsorbed with AVP, proved negative in Wistar rat hypothalamus. However, in the HNS of the homozygous Brattleboro rat, almost deficient in AVP-synthesis, the anti-AVP was positive all over the neural lobe. This indicates that OXT fibers were stained.

The adsorption test also does not exclude binding to apparently unrelated compounds (i.e., according to their primary amino acid sequence) (Fig. 4; van Leeuwen, 1980). The next example will illustrate this. Using anti-Lysine vasopressin (LVP) serum immunoreactivity was obtained in the mouse pars intermedia (PI) (Castel, 1978). This finding was explained by assuming that the PI cells contain AVP receptors able to bind AVP molecules. These molecules would, during the subsequent ICC procedure, be recog-

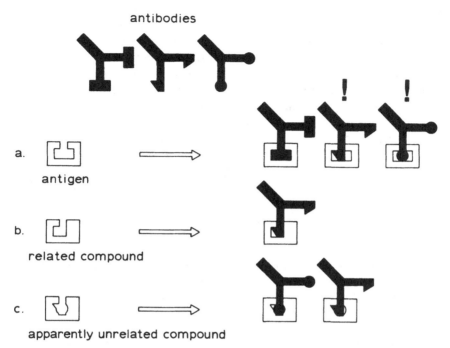

Fig. 8.   When antibodies are raised against a "pure" antigen (e.g., AVP), a population of antibodies will be obtained that all recognize epitopes on the homologous (a) antigen (e.g., AVP). However, part of the population recognizes, in addition, epitopes on heterologous compounds that are structurally related (b) (e.g., OXT) or to apparently unrelated compounds (c) according to their primary amino acid sequence (e.g., α-MSH). In ICC, this antiserum will stain both the homologous (a) and at least two heterologous (b,c) antigens, in the case of the latter two because of cross-reacting antibodies(!) In addition, it appeared that anti-AVP purified with OXT beads still showed a reaction in PI cells (Fig. 6a,d). Anti-AVP purified with α-MSH beads did not show any reaction with OXT (as observed in the neural lobe of the homozygous Brattleboro rat, lacking AVP, but not OXT) or with α-MSH in PI cells (Fig. 6b,e). Thus, that part of the population binding to AVP and OXT binds α-MSH also (c), and that part binding to AVP and α-MSH does not bind OXT (from van Leeuwen, 1980, with permission of the publisher).

nized by the anti-LVP serum. We confirmed this staining in Wistar rat PI (van Leeuwen et al., 1979a). However, staining in the PI was also shown in the homozygous Brattleboro rat (Fig. 4d). Preincubation of Brattleboro neurointermediate lobe sections with AVP did not result in an increase in staining. Thus, AVP that may be present in anti-AVP serum, with subsequent binding of this AVP to the receptor, cannot be responsible for this staining.

A remaining possibility is that contaminating antibodies (directed, for example, against fragments of AVP) are present in the anti-AVP serum. These antibodies could be reacting with α-MSH. However, after solid phase adsorption of anti-AVP to α-MSH, the reaction in the PI disappeared, whereas the reactivity in the neural lobe for AVP decreased (Fig. 4b). This points to cross-reactivity rather than to contaminating antibodies. In the latter case, the immunoreactivity in the neural lobe should be unaffected after adsorption with α-MSH (Pool et al., 1983). Therefore, this reaction can best be explained as a cross-reaction with the apparently unrelated α-MSH (Fig. 8). With a technique unrelated to ICC, it was recently shown that the intermediate lobe contains no AVP binding sites (van Leeuwen and Wolters, 1983).

In conclusion, the fact that staining was no longer found on Wistar neural lobe after solid phase adsorption to AVP (the adsorption test!) does not eliminate the possibility that antibodies directed against AVP may be able to react not only with the immunogen, but also with structurally related, or even apparently unrelated, compounds.

Another illustration of the limits of the adsorption test is the observation of α-MSH immunoreactivity throughout the brain (Swaab and Fisser, 1978) (Fig. 1A). No reaction was obtained after solid-phase adsorption of anti-α-MSH to α-MSH. However, immunoelectron microscopical studies showed that α-MSH immunoreactivity is present on neurofilaments (van Leeuwen et al., 1979b) (Fig. 1B). Drager et al. (1983), using a different anti-α-MSH serum, showed that the neurofilament M protein is α-MSH immunoreactive. It was recently reported by Shaw et al. (1985) that the N-terminus of neurofilament M protein shares two amino acids with α-MSH, which seems the most likely explanation for the results of Swaab and Fisser (1978), as well as the immunoelectron microscopical, data (van Leeuwen et al., 1979B). Indeed, using SDS gelelectrophoresis and electroblotting, the anti-α-MSH used in our studies showed a band in the 150 kdalton region at the same height as the neurofilament band (Verhaagen et al., 1986). This example shows once again the danger of using the adsorption test without additional controls.

4.3.3. Adsorption Test (Heterologous Antigen)

When it was realized that the adsorption test with the homologous antigen was of limited value, solid or liquid phase adsorptions were performed with either a low (Swaab and Pool, 1975; Lu et al., 1982) or a high number (Martin et al., 1980) of potentially immunoreactive compounds. In addition, model systems were developed to quantify the immunoreactivity. Larsson (1981) applied antigens in different concentrations to filter paper, and then fixed them. Schipper and Tilders (1983) dissolved antigens in gelatin before fixation. In both model systems, immunocytochemical staining was subsequently performed. Another possibility is coupling a number of antigens to Sepharose beads (Swaab and Pool, 1975; Childs, 1982), whereas ELISA techniques are also a possibility for testing the reactivity of antisera (Engvall and Perlmann, 1971). An important advantage of these model systems is the ability to quantify the immunoreactivity. However, the main disadvantage of all the above-mentioned techniques is the inability to predict which tissue compounds (see example of similar epitopes on α-MSH and neurofilament M) may react with the antiserum. Therefore, one has to characterize the antiserum binding to all tissue compounds. Thus, for the assessment of serum specificity it is necessary to obtain a complete spectrum of tissue antigens followed by ICC staining of this spectrum.

4.3.4. Tissue Spectrum Affinity Test

This test, in which all possible tissue antigens are present, fulfills most of the requirements as formulated by Pool et al. (1983) (see above). In these tests the antigens are extracted from the tissue homogenates, separated, and identified by nonimmunochemical techniques. For large antigens (mol wt > 10 kdalton), SDS polyacrylamide gel electrophoresis can be applied, followed by longitudinal sectioning, treatment with fixative and immunocytochemical staining (e.g., Van Raamsdonk et al., 1977). This approach meets conditions (i), (iii), and (iv), although condition (v) is realized by binding the antigens of the gel to sepharose beads. These beads can be used in an ICC reaction and the amount of (first) antibody binding can be measured by means of spectrophotometry (Pool et al., 1983). Alternatively, the antigens can be coupled to polystyrene test tubes followed by ELISA (Lutz et al., 1979). Requirement (ii) can never be fulfilled in model systems since the tissue antigen is surrounded by other compounds that may interfere with the antigen. In the test system all attempts are made to isolate each antigen from all other tissue compounds.

More recently, protein blotting methods have been introduced in which the separated proteins are transferred from the gel to a number of matrices, after which immunodetection can be performed (Towbin et al., 1979; Gershoni and Palade, 1983). A number of adaptations to this system have recently been developed (for details, *see* Van der Sluis and Boer, 1986).

Peptides, however, require a different approach because of relatively small differences in molecular weight. Small molecular compounds (peptides <10 kdalton) cannot be separated in SDS gel electrophoresis. However, on the basis of their isoelectric points, isoelectric focusing of tissue extracts can be performed. Such a procedure was developed by Van der Sluis et al. (1983) for synthetic peptides (Fig. 9). Recently, a test was developed for peptides extracted from tissue homogenates (Fig. 10). This approach is the best way to obtain an impression of the immunoreactivity of antisera directed against peptides. If it is impossible to perform such a test, an exclusion procedure (*see* section 4.3.3) of possible compounds interacting with the first antiserum is a means of determining serum specificity (Martin et al., 1980). However, as mentioned before, this procedure will never be conclusive.

4.4. Purification of the First Antibody

When the first antiserum shows affinity for more than one compound, purification of the antiserum may be necessary. In some marginal cases, dilution of the first antiserum also appeared to be a solution to the problem of reactivity with unwanted compounds (van Leeuwen and Swaab, 1977; Van Raamsdonk et al., 1977; Petrusz et al., 1980). However, this approach is very procedure-dependent (Pool et al., 1983).

Purification can be performed in two ways:

a. Elution of antibodies from a column (loaded with the antigen to be localized) subsequent to the passage of antibodies that do not show affinity (affinity-purified antibodies). The two main disadvantages of this method are: (1) cross-reactive antibodies and, usually, antibodies against contaminants are

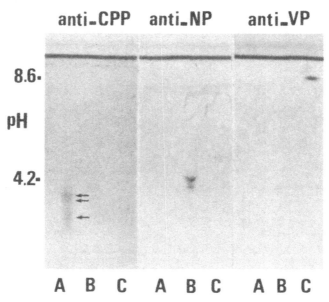

Fig. 9.   Press-immunoblot showing the immunoreactivity of anti-human C-terminal peptide of propressophysin (CPP), anti-rat NP (#RNP 7/19/73, Dr. A. G. Robinson, Pittsburgh, USA), and anti-VP (#125), with (A) human CPP, (B) rat NP A (associated with VP, Dr. B. T. Pickering, Bristol, UK), and (C) AVP (Sigma Grade VIII, 377 IU/mg). Note that the antisera only react with their homologous antigen. This indicates that these antibodies react with different parts of the vasopressin precursor. Within the smear pattern of human CPP several bands can be distinguished (arrows). The two bands shown in the NP blot most probably represent NP A (associated with VP, upper band) and NP B (associated with oxytocin, being slightly more acidic than NP A, lower band).

also retrieved; (2) only the lower avidity antibodies will be eluted, and the higher avidity antibodies will remain on the column (Sternberger, 1973, p. 166), although this does not seem to hold for all antigens (De Mey, 1983; Boorsma, personal communication). A requirement for this approach is that the antigen to be localized is the only immunoreactive compound in the adsorbance. Such "immunologically pure" antigens do not usually exist. This approach can, however, be useful as a first step in the purification.

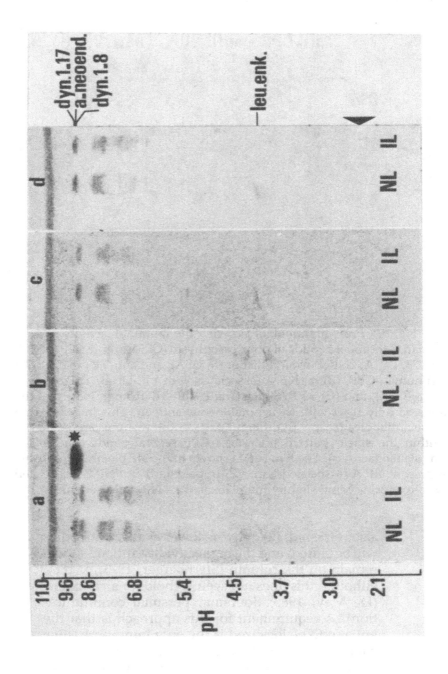

Fig. 10. Immunoblot showing isoelectric focusing patterns of acidic ethanol extracts of neural (NL) and intermediate lobe (IL) containing low-molecular-weight (mol wt < 20 kdalton) peptides. The experiment was performed by Dr. Van der Sluis. Arrowhead: site of application of extracts. The peptides were fixed by the press-blot technique (Van der Sluis et al., 1983) and the various antisera were used in dilutions of 1:1000. (a) Anti-α-neo endorphin (#58, Bleed 10). (b) Anti-dynorphin 1-17 (#54, Bleed 10). (c) anti-dynorphin 1-8 (#73, Bleed 10). (d) anti-leu-enkephalin (A206, 8/15/77). In (a), synthetic α-neo-endorphin (100 ng, Cambridge Research Biochemicals) was also focused (asterisk). Note that more than one band is stained by all antibodies and that anti-leu-enkephalin immunoreactivity is not present in the intermediate lobe (d:IL). Note also the small differences in isoelectric points between the compounds present on the basic side. Resolution can be improved by narrowing down pH intervals (e.g., 9–11). The lines (at the right) indicate the positions of the synthetic peptides. It should be noted that this technique is still in the developmental phase. Antibodies (a–c) and (d) were obtained from Dr. S. J. Watson (Ann Arbor, Michigan, USA) and Dr. R. M. Miller (Chicago, Illinois, USA), respectively (from van Leeuwen, 1986, with permission of the publisher).

b. Removal of unwanted antibody populations. A requirement for this approach is the absence of the antigen to be localized within the adsorbant. This condition is easier to achieve than the immunological purity of the adsorbant, although, in addition, the two above-mentioned disadvantages are absent. Thus, prior to purification procedures, it is necessary to check the immunoreactive compounds within each antigen used for adsorption.

Purification of the first antiserum can be achieved by liquid or solid phase adsorption. In many cases liquid phase adsorption failed to give blanks (e.g., for small peptides and in the situation of an excess of antigen; Swaab and Pool, 1975). Since immune complexes of small peptides do not easily precipitate, dissociation and subsequent binding to the tissue section can occur during incubation. The safest and quickest way, therefore, is the removal of the antibodies with a solid phase, a procedure in which sepharose beads are loaded with peptide. The procedure can be performed by repeated immersion of the beads with the antibody (e.g., Swaab and Pool, 1975). An additional advantage of using beads is that the same matrix can be used for quantification of the amount of antibodies bound, and subsequently for the assessment of the minimal amount of beads necessary for an adsorption. In addition, the success of an adsorption can be checked (see below).

4.5. Possible Steps Following Antiserum Purification

The results of the purification have to be tested again in a model system (either beads or the spectrum of the tissue antigens) and the tissue section. These experiments will reveal some more details of the properties of the remaining antibody population(s) that play a role in the ICC localization procedure (cross-reactive antibodies still present?) (for details, see Pool et al., 1983).

In addition, the results of the incubation in the model system and tissue section have to be compared, and will show whether the model system (reaction with only one tissue compound) has any predictive value (see, for example, Pool et al., 1983). The results of these two systems should be fully comparable, or very nearly so; two final tests must give negative answers to the following questions:

1. Is there any reactivity of the purified antiserum after incubation with the homologous antigen? When reaction is still obtained in the tissue section (and not in the model system), this should be attributed to lack of predictive value of the model system. This may mean that condition 2 for a specificity test (see before) is not fulfilled or that the tissue section allows a better detection than the model system. In the latter case, the sensitivity of the model system should be enhanced. The difference between the use of the adsorption test in this way and that of Petrusz et al. (1976), and most ICC papers, is that here the test is used as a control for ICC specificity as found with the model system, whereas in Petrusz' case the adsorption is the sole proof for specificity.

2. Second, the purified antiserum must give negative staining of the same tissue area lacking the antigen to be localized. A genetic mutant is almost ideal for this experiment. The homozygous Brattleboro rat, until recently assumed to be completely deficient in AVP, can be used for the demonstration of the serum specificity of the anti-AVP (Swaab and Pool, 1975; van Leeuwen and Swaab, 1977). However, recently it was shown that within the homozygous Brattleboro rat a minority of the cell bodies synthesizes the vasopressin precursor (Richards et al., 1985; van Leeuwen et al., 1986) (Fig. 11). The parvocellular suprachiasmatic nucleus (SCN), which is known to produce AVP exclusively, was used as a control for the cross-reaction of anti-OXT with AVP (Buijs et al., 1978). However, since the content of AVP in the SCN is different from that in other brain areas (Dogterom et al., 1978), a negative answer may give a false prediction about the reactivity of anti-OXT serum. Another example is the hypogonadal mouse deficient in synthesis of luteinizing hormone releasing hormone within the hypothalamus (Charlton, 1986).

Recently an interesting approach was presented by Jirikowski (1985). In order to discriminate between leu- and met-enkephalin,

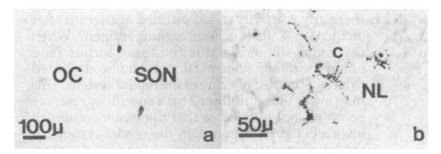

Fig. 11. Light microscopic demonstration of immunoreactivity of C-terminal peptide of propressophysin—the glycopeptide moiety in the vasopressin precursor—in cell bodies of the supraoptic nucleus (SON) and in fibers of the neural lobe (NL) of the homozygous Brattleboro rat. Abbreviations: OC, optic chiasm; c, capillary.

met-enkephalin immunoreactivity was selectively abolished by pretreatment of the sections with cyanogen bromide.

5. Concluding Remarks

In the 1970s, the first reports appeared about antibodies with "pre-defined specificity," synthesized by clonal cell lines (monoclonal antibodies) (Kohler and Milstein, 1975). The advantage of this technique is that it allows the isolation and permanent replication of desired hybrid cell clones to provide every investigator with a continuous source of antibodies (for a review on the application of monoclonal antibodies in the neurosciences, *see* Valentino et al., 1985). This overcomes the problem of individual batches of antiserum, raised against the "same" antigen in different laboratories, being variable with regard to their epitopes. On the other hand, conflicting results will force the investigators to face the possible pitfalls of their work. The hybridoma technology eliminates antibody heterogeneity. Still, the possibility of cross-reaction can never be excluded (Milstein et al., 1983). Therefore, in order to prove serum specificity of monoclonal antibodies, the same procedures have to be followed as outlined above. An important disadvantage of monoclonal antibodies is that they are selected on the basis of ICC on sections fixed in a certain way. Since the antibodies all react in the same way, small differences in the degree of fixation, which are often unavoidable, can lead to a false negative result. With

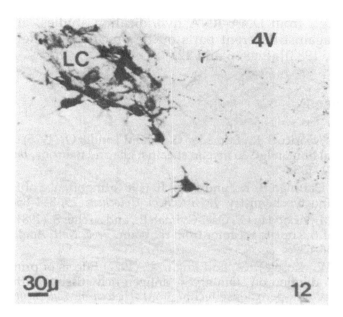

Fig. 12.   Neurophysin immunoreactivity in the rat locus coeruleus (LC) and subcoeruleus (same antibody as used in Fig. 9). Abbreviation: 4V, fourth ventricle.

polyclonal antibodies the chance that this problem occurs is rather small, because in this case another part of the antibody population will be able to react (Pool, personal communication). It has already been reported that ICC staining is improved when clones are mixed (Lin et al., 1983; Van den Oord et al., 1985).

A number of potent monoclonal antibodies has been raised against peptides (*see* Valentino et al., 1985). However, it is questionable whether monoclonal antibodies are really an improvement for peptide ICC (see above). It seems that the application of monoclonal antibodies will be more useful for the ICC of antigens that cannot be obtained in a pure form. For the reliable achievement of an ICC study, it is necessary to determine (a) the immunological purity of the antigens used for adsorptions and in test systems and (b) a complete spectrum of all immunoreactive compounds present in a particular area. In addition, quantification of the first antiserum–antigen reaction is indispensable. The manner outlined above of proving serum specificity has only very recently become practicable for peptides. In the meantime, enough possibilities are present to collect considerable circumstantial evidence [e.g., from DNA–RNA hybridization studies, antibodies directed against different parts of the precursor (Caffe and van Leeuwen, 1983)] (Figs. 6 and 12).

References

Aarli J. A., Aparico S. R., Lumsden C. E., and Tonder O. (1975) Binding of normal human IgG to myelin sheaths, glia and neurons. *Immunology* **28,** 171–185.

Berod A., Hartman B. K., and Pujol F. (1981) Importance of fixation in immunoctyochemistry. *J. Histochem. Cytochem.* **29,** 844–850.

Bidlack J. M., Abood L. G., Osei-Gyimah P., and Archer S. (1981) Purification of the opiate receptor from rat brain. *Proc. Natl. Acad. Sci. USA* **78,** 636–639.

Bigbee J. W., Kosek J. C., and Eng L. F. (1977) Effects of primary antiserum dilution on staining of "antigen-rich" tissues with the peroxidase anti-peroxidase technique. *J. Histochem. Cytochem.* **25,** 443–447.

Brown J. R., Hunter J. C., Jordan C. C., Tyers M. B., Ward P., and Whittington A. R. (1986) Problems with peptides—all that glitters is not gold. *Trends Neurosci.* **9,** 100–102.

Buijs R. M. (1978) Intra- and extrahypothalamic vasopressin and oxytocin pathways in the rat. Pathways to the limbic system, medulla oblongata and spinal cord. *Cell Tiss. Res.* **192,** 423–435.

Buijs R. M. (1982) Tissue Treatment in Immunocytochemistry, in *Second EMBO Practical Course on Immunocytochemistry* Amsterdam.

Buijs R. M. and Pool C. W. (1984) Additional Notes on Tissue Treatment in Immunocytochemistry, *Immunoctochemistry and Neurotransmitter Binding* (van Leeuwen F. W. and Sels. J., eds.) European Neuroscience Assoc., Amsterdam.

Buijs R. M. and Swaab D. F. (1979) Immunoelectron microscopical demonstration of vasopressin and oxytocin synapses in the limbic system of the rat. *Cell Tiss. Res.* **204,** 355–365.

Buijs R. M., Swaab D. F., Dogterom J., and van Leeuwen F. W. (1978) Intra- and extrahypothalamic vasopressin and oxytocin pathways in the rat. *Cell Tiss. Res.* **186,** 423–433.

Caffé A. R. and van Leeuwen F. W. (1983) Vasopressin-immunoreactive cells in the dorsomedical hypothalamic region, medial amygdaloid nucleus and locus coeruleus of the rat. *Cell Tiss. Res.* **233,** 23–37.

Caffé A. R., van Leeuwen F. W., Buijs R. M., De Vries G. J., and Geffard M. (1985) Coexistence of vasopressin, neurophysin and noradrenaline immunoreactivity in medium-sized cells of the locus coeruleus and subcoeruleus of the rat. *Brain Res.* **338,** 160–164.

Capel P. J. A. (1974) A quantitative immunofluorescent method based on the covalent coupling of protein to Sepharose beads. *J. Immunol. Meth.* **5,** 165–178.

Castel M. (1978) Immunocytochemical evidence for vasopressin receptors. *J. Histochem. Cytochem.* **26,** 581–592.

Castel M., Morris J. F., Whitnall M. H., and Sivan N (1986) Improved visualization of the immunoreactive hypothalamus-neurohypophysial system by use of immuno-gold techniques. *Cell Tiss. Res.* **243,** 193–204.

Charlton H. M. (1986) Use of Neural Transplants To Study Neuroendocrine Mechanisms, in *Frontiers in Neuroendocrinology* vol. 9, Raven, New York.

Childs G. V. (1982) Use of Immunocytochemical Techniques in Cellular Endocrinology, in *Electron Microscopy in Biology* vol. 2. (Griffith J., ed.) John Wiley, New York.

Clayton C. J. and Hoffman G. E. (1979) Immunocytochemical evidence for anti-LHRH and anti-ACTH activity in the 'F' antiserum. *Am. J. Anat.* **154,** 139–145.

Coulter H. D. and Elde R. P. (1978) Somatostatin radioimmunoassay and immunofluorescence in the rat hypothalamus: Effects of dehydration with alcohol and fixation with aldehydes and OsO4. *Anat. Rec.* **190,** 369–370.

De Mey J. (1983) Raising and Testing Antibodies for Immunocytochemistry, in *Immunocytochemistry: Applications in Pathology and Biology* (Polak J. and Van Noorden S., eds.), J. Wright, Bristol, UK.

Dogterom J., Snijdewint F. G. M., and Buijs R. M. (1978) The distribution of vasopressin and oxytocin in the rat brain. *Neurosci. Lett.* **9,** 341–346.

Drager U. C., Edwards D. L., and Kleinschmidt J. (1983) Neurofilaments contain α-melanocyte-stimulating hormone (α-MSH)-like immunoreactivity. *Proc. Natl. Acad. Sci. USA* **80,** 6408–6412.

Engvall E. and Perlmann P. (1971) Enzyme linked immunosorbent assay (ELISA). III. Quantitative assay of immunuglobulin G. *Immunochemistry* **8,** 871–880.

Finley J. C. W., Grassman G. H., Dimeo P., and Petrusz P. (1978) Somatostatin-containing neurons in the rat brain: Widespread distribution revealed by immunocytochemistry after pre-treatment with pronase 1. *Am. J. Anat.* **153,** 483–488.

Geffard M., Buijs R. M., Seguela P., Pool C. W., and LeMoal M. (1984) First demonstration of highly specific and sensitive antibodies against dopamine. *Brain Res.* **294,** 161–165.

Gershoni J. M. and Palade G. E. (1983) Protein blotting: Principles and applications. *Anal. Biochem.* **131**, 1–15.

Geuze J. J., Slot J. W., and Tokuyasu K. T. (1979) Immunocytochemical localization of amylase and chymotrypsin in the exocrine cell with special attention to the Golgi complex. *J. Cell Biol.* **82**, 697–707.

Grube D. (1980) Immunoreactivities of gastrin (G-) cells. II. Non-specific binding of immunoglobulins to G-cells by ionic interactions. *Histochemistry* **66**, 149–167.

Grube D. and Weber E (1980) Immunoreactivities of gastrin (G-) cells. *Histochemistry* **65**, 223–237.

Gu J., De Mey J., Moeremans M., and Polak J. (1981) Sequential use of the PAP and immunogold staining methods for the light microscopical double staining of tissue antigens. *Regul. Peptides.* **1**, 365–374.

Hokfelt I., Johansson O., Ljungdahl A., Lundberg J. M., and Schultzberg M. (1980) Peptidergic neurons. *Nature* (Lond.) **284**, 515–521.

Hsu S. M., Raine L., and Fanger H. (1981) The use of avidin–biotin-peroxidase complexes (ABC) in immunoperoxidase technique—A comparison between ABC and unlabeled antibody (PAP) procedures. *J. Histochem. Cytochem.* **29**, 577–580.

Hunt S. P., Kelly J. S., and Emson P. C. (1980) The electron microscopic localization of methionine enkephalin within the superficial layers (I and II) of the spinal cord. *Neuroscience* **5**, 1871–1890.

Jirikowski G. (1985) Cyanogen bromide cleavage of methionin residues as a control method for enkephalin immunocytochemistry. *Histochemistry* **83**, 93–95.

Kohler G. and Milstein C. (1975) Continuous cultures of fused cells secreting antibody of predefined specificity. *Nature* (Lond.) **256**, 495–497.

Kraehenbuhl J. P. and Jamieson J. D. (1974) Localization of intracellular antigens by immunoelectron microscopy. *Int. Rev. Exp. Pathol.* **13**, 1–53.

Larsson L. I. (1981) A novel immunocytochemical model system for specificity and sensitivity screening of antisera against multiple antigens. *J. Histochem. Cytochem.* **29**, 408–410.

Lembol H. L. (1978) The relationship of radioimmunoassay to bioassay. In vitro studies with synthetic lysine vasopressin in aqueous solution inactivated by heat. *Acta Endocrinol.* **80**, 465–473.

Li J. Y., Dubois M. P., and Dubois P. M. (1977) Somatotrophs in the human fetal anterior pituitary. An electron microscopic immunocytochemical study. *Cell Tiss. Res.* **181**, 545–552.

Lillie R. D. and P. Pizzolato (1972) Histochemical use of borohydrides as aldehyde blocking reagents. *Stain. Technol.* **47**, 13–16.

Lin C. T., Huang L. H., and Chan T. S. (1983) A comparative study of polyclonal and monoclonal antibodies for immunocytochemical localization of cytosolic aspartate aminotransferase in the rat brain. *J. Histochem. Cytochem.* **31**, 920–926.

Liposits Zs., Setalo G. Y., and Flerko B. (1984) Application of the silver-gold intensified 3,3-diaminobenzidine chromogen to the light and electron microscopic detection of the luteinizing hormone-releasing hormone system of the rat brain. *Neuroscience* **13**, 513–525.

Livett B. G. (1978) Immunohistochemical localization of nervous system-specific proteins and peptides. *Int. Rev. Cytol.* (suppl.) **7**, 53–237.

Lu C. L., Cantin M., Seidah N. G., and Chrétien M. (1982) Distribution pattern in the human pituitary and hypothalamus of a new neuropeptide, the C-terminal glycoprotein-fragment of human pro-pressophysin (CPP). *Histochemistry* **75**, 319–326.

Lutz H., Higgins J., Pederson N. C., and Theilen G. H. (1979) The demonstration of antibody specificity by a new technique. *J. Histochem. Cytochem.* **27**, 1216–1220.

Martin R., Frosch D., and Voigt K. H. (1980) Immunocytochemical evidence for melantropin- and vasopressin-like material in a cephalopod neurochemical organ. *Gen. Comp. Endocrinol.* **42**, 235–243.

Matthews J. B. (1981) Influence of clearing agent on immunohistochemical staining to paraffin embedded tissue. *J. Clin. Pathol.* **34**, 103–105.

Milstein C., Wright B., and Cuello A. C. (1983) The discrepancy between the crossreactivity of a monoclonal antibody to serotonin and its immunohistochemical specificity. *Mol. Immunol.* **20**, 113–123.

Mohring J., Bohlen P., Schoun J., Mellet M., Suss U., Schmidt M., and Pliska V. (1982) Comparison of radioimmunoassay, chemical assay (HPLC) and bioassay for arginine vasopressin in synthetic standards and posterior pituitary tissue. *Acta Endocrinol.* **99**, 371–378.

Nairn R. C. (1976) *Fluorescent Protein Tracing* 14th Ed. Churchill Livingstone, London.

Nakane P. K. and Pierce G. B. (1967) Enzyme-labeled antibodies: Preparation and application for the localization of antigens. *J. Histochem. Cytochem.* **14**, 329–332.

North W. G., La Rochelle Jr. F. T., Haldar J., Sawyer W. H., and Valtin H. (1978) Characterization of an antiserum used in radioimmunoassay for arginine-vasopressin: Implications for reference standards. *Endocrinology* **103**, 1976–1984.

Ordronneau P., Lindstrom P. B. M., and Petrusz P. (1981) Four unlabeled antibody bridge techniques: A comparison. *J. Histochem. Cytochem.* **29**, 1397–1404.

Petrusz P., Sar M., Ordronneau P., and DiMeo P. (1976) Specificity in immunocytochemical staining. *J. Histochem. Cytochem.* **24**, 1110–1115.

Petrusz P., Sar M., Ordronneau P., and DiMeo P. (1977) Reply to the letter of Swaab et al. (1977) Can specificity ever be proved in immunocytochemical staining? *J. Histochem. Cytochem.* **25,** 390–391.

Petrusz P., Ordronneau P., and Finley J. C. W. (1980) Criteria for reliability for light microscopic immunocytochemical staining. *Histochem. J.* **12,** 333–348.

Pool C. W., Buijs R. M., Swaab D. F., Boer G. J., and van Leeuwen F. W. (1983) On the Way to a Specific Immunocytochemical Localization, in *Immunocytochemistry* IBRO Handbook Series 3 (Cuello A. C., ed.) John Wiley, Chichester.

Richards S., Morris R., and Raisman G. (1985) Solitary magnocellular neurons in the homozygous Brattleboro rat have vasopressin and glycopeptide immunoreactivity. *Neuroscience* **16,** 617–623.

Robinson G. and Dawson I. (1975) Immunochemical studies of the endocrine cells of the gastrointestinal tract. II. An immunoperoxide technique for the localization of secretin containing cells in human duodenum. *J. Clin. Pathol.* **28,** 631–635.

Rossier J., Battenberg E., Pittman Q., Bayon A., Koda L., Miller R., Guillemin R., and Bloom F. E. (1979) Hypothalamic enkephalin neurones may regulate the neurohypophysis. *Nature* (Lond.) **277,** 653–655.

Roth J., Kasper M., Meitz Ph. U., and Labat F. (1985) What's new in light and electron microscopic immunocytochemistry? Application of the Protein-A-Gold techniques to routinely processed tissue. *Path. Res. Pract.* **180,** 711–717.

Schipper J. and F. J. H. Tilders (1983) A new technique for studying specificity of immunocytochemical procedures. *J. Histochem. Cytochem.* **31,** 12–18.

Shaw G., Fischer S., and Weber K. (1985) AlphaMSH and neurofilament M-protein share a continuous epitope but not extended sequences. *FEBS Lett.* **181,** 343–346.

Skowsky W. R. and Fisher D. A. (1972) The use of thyroglobulin to induce antigenicity to small molecules. *J. Lab. Clin. Med.* **80,** 134–146.

Sternberger L. A. (1973) *Enzyme Immunocytochemistry. Electron Microscopy of Enzymes, Principles and Methods* vol. 1 (M. A. Hayat, ed.) Van Nostrand Reinhold, New York.

Sternberger L. A. (1979) *Immunocytochemistry* 2nd Edn. John Wiley, New York.

Sternberger L. A., and Petrali J. P. (1977) The unlabeled antibody enzyme method. Attempted use of peroxidase-conjugated antigen as the third layer in the technique. *J. Histochem. Cytochem.* **25,** 1036–1042.

Straus W. (1971) Inhibition of peroxidase by methanol-nitroferricyanide for use in immunoperoxidase procedures. *J. Histochem. Cytochem.* **19**, 682–688.

Straus W. (1972) Phenylhydrazine as inhibitor of horseradish peroxidase for use in immunoperoxidase procedures. *J. Histochem. Cytochem.* **20**, 949–951.

Streefkerk J. G. (1972) Inhibition of erythrocyte pseudoperoxidase activity by treatment with hydrogen peroxide following methanol. *J. Histochem. Cytochem.* **20**, 829–831.

Streefkerk J. G, Van der Ploeg M., and Van Duijn P. (1975) Agarose beads as matrices for proteins in cytophotometric investigations of immunohistoperoxidase procedures. *J. Histochem. Cytochem.* **23**, 243–250.

Swaab D. F. (1982) Comments on the Validity of Immunocytochemical Methods, in *Cytochemical Methods in Neuroanatomy* (Palay S. L. and Chan-Palay V., eds.) Alan Liss, New York.

Swaab D. F. and Fisser B. (1978) Immunocytochemical localization of α-MSH-like compounds in the rat nervous system. *Neurosci. Lett.* **7**, 313–317.

Swaab D. F. and Pool C. W. (1975) Specificity of oxytocin and vasopressin immunofluorescence. *J. Endocrinol.* **66**, 263–272.

Swaab D. F., Pool C. W., and Nijveldt F (1975) Immunofluorescence of vasopressin and oxytocin in the rat hypothalamic-neurohypophysical system *J. Neural Transm.* **36**, 195–215.

Swaab D. F., Pool C. W., and van Leeuwen F. W. (1977) Can specificity ever be proved in immunocytochemical staining? *J. Histochem. Cytochem.* **25**, 388–391.

Towbin H., Staehelin T., and Gordon J. (1979) Electrophoretic transfer of proteins for polyacrylamide gels to nitrocellulose sheets, procedure and some applications. *Proc. Natl. Acad. Sci. USA* **76**, 4350–4354.

Towle A. C., Lauder J. M., and Joh T. H. (1984) Optimization of tyrosine hydroxylase immunocytochemistry in paraffin sections using pretreatment with proteolytic enzymes. *J. Histochem. Cytochem.* **32**, 766–770.

Vacca L. L., Hewett D., and Woodson G. (1978) A comparison of methods using diaminobenzidine (DAB) to localize peroxidases in erythrocytes, neutrophils and peroxidase-antiperoxidase complex. *Stain Technol.* **53**, 331–336.

Valentino K. L., Winter J., and Reichardt L. F. (1985) Applications of monoclonal antibodies to neuroscience research. *Ann. Rev. Neurosci.* **8**, 199–232.

Van den Oord J. J., De Wolf-Peeters C., Van Stapel M. J., and Desmet V. J. (1985) Improved immunohistochemical visualization of helper inducer T-cells by the simultaneous application of two non-crossblocking monoclonal antibodies. *Stain Technol.* **60**, 45–49.

Van der Kooy D. (1982) Combination of Immunocytochemistry with the Tracing of Neuronal Pathways, in *Second EMBO Practical Course on Immunocytochemistry* (Van Leeuwen F. W., ed.), pp. 65–74.

Van der Sluis P. J. and Boer G. J. (1986) The relevance of various tests to study specificity in immunocytochemical staining. *Cell Biochem. Funct.* **4**, 1–17.

Van der Sluis P., Boer G. J., and Pool C. W. (1983) Fixation and immunoperoxidase staining of oligopeptides after isoelectric focusing in thin polyacrylamide slabgels. *Anal. Biochem.* **133**, 226–232.

Vandesande F. (1979) A critical review of immunocytochemical methods for light microscopy. *J. Neurosci. Meth.* **1**, 3–23.

van Leeuwen F. W. (1977) Immunoelectron microscopical visualization of neurohypophyseal hormones: Evaluation of some tissue preparations and staining procedures. *J. Histochem. Cytochem.* **25**, 1213–1221.

van Leeuwen F. W. (1980) Immunocytochemical specificity for peptides with special reference to arginine-vasopressin and oxytocin. *J. Histochem. Cytochem.* **28**, 479–482.

van Leeuwen F. W. (1981) An introduction to the immunocytochemical localization of neuropeptides and neurotransmitters. *Acta Histochem.* (suppl.) **XXIV**, 49–77.

van Leeuwen F. W. (1982) Monospecific Localization of Neuropeptides, A Utopian Goal?, in *Techniques in Immunocytochemistry* vol. 1 (Bullock G. and Petrusz P., eds.) Academic, New York.

van Leeuwen (1986) Pitfalls in immunocytochemistry with special reference to specificity problems in the localization of neuropeptides. *Am. J. Anat.* **175**, 363–377.

van Leeuwen F. W. and Swaab D. F. (1977) Specific immunoelectronmicroscopic localization of vasopressin and oxytocin in the neurohypophysis of the rat. *Cell Tiss. Res.* **177**, 493–501.

van Leeuwen F. W. and Wolters P. (1983) Light microscopic autoradiographic localization of [^{3}H]-arginine-vasopressin binding sites in the rat brain and kidney. *Neurosci. Lett.* **41**, 61–66.

van Leeuwen F. W., De Raay C., Swaab D. F., and Fisser B. (1979a) The localization of oxytocin, vasopressin, somatostatin and luteinizing hormone releasing hormone in the rat neurohypophysis. *Cell Tiss. Res.* **202**, 189–201.

van Leeuwen F. W., Swaab D. F., De Raay C., and Fisser B. (1979b) Immunoelectron microscopical demonstration of α-melanocyte-stimulating hormone-like compounds in the rat brain. *J. Endocrinol.* **80**, 59–60P.

van Leeuwen F. W., Swaab D. F., Buijs R. M., and Sels J., eds. (1982) *Immunocytochemistry and Its Application in Brain Research*. Course manual, Second EMBO practical course, Amsterdam.

van Leeuwen F. W., Pool C. W., and Sluiter A. A. (1983) Enkephalin immunoreactivity in synaptoid elements on glial cells in the rat neural lobe. *Neuroscience* **8**, 229–241.

van Leeuwen F. W., Caffe A. R., Van der Sluis P. J., Sluiter A. A., Van der Woude T. P., Seidah N. G., and Chretien M. (1986) Propressophysin is present in neurons at multiple sites in Wistar and homozygous Brattleboro rat brain. *Brain Res.* **379**, 171–175.

Van Raamsdonk W., Pool C. W., and Heyting C. (1977) Detection of antigens and antibodies by an immuniperoxidase method applied on thin longitudinal sections of SDS polyacrylamide gels. *J. Immunol Meth.* **17**, 337–348.

Verhaagen J., Edwards P. M., Schotman P., Jennekens F. G. I., and Gispen W. H. (1986) Characterization of epitopes shown by α-melanocyte-stimulating hormone and the 150 KD neurofilament protein (NF 150): Relationship to neurotrophic sequences. *J. Neurosci. Res.*, in press.

In Vivo and In Vitro Studies on Peptide Release

Alejandro Bayón

1. Introduction

The discovery that many peptide hormones previously identified in endocrine glands and/or in the endocrine brain (Hökfelt et
al., 1980) are also present in central and peripheral neurons has
unexpectedly increased the list of potential messengers participat-
ing in neuronal communication. To this list we must add other
newly discovered peptides first detected as endogenous ligands of
brain receptors (Hughes et al., 1975) and peptides whose existence
was initially deduced from nucleotide or amino acid sequences in
genes or protein precursors studied with the powerful techniques
of modern molecular biology (Nakanishi et al., 1979; Lenoir et al.,
1985). In order to understand the role of this growing list of peptides in neurohumoral coding of brain function, we must not only
study the ways in which they interact with classical transmitter
systems and modify the behavior of their target nerve cells, but also
the ways in which they are produced, stored, and released.

Although valuable knowledge has been gathered on the synthesis and processing of neuropeptides, relatively little progress
has yet been made in understanding the mechanisms by which
these peptides are liberated in the nervous system and how these
mechanisms may differ from those controlling the release of nonpeptide transmitters or the secretion of peptide hormones. More
attention has been paid to the study of the physiological context in
which their release occurs, from the cellular to the behavioral
levels. In order to study the mechanism and regulation of the
release of neuropeptides at both the cellular or subcellular level and
in the behaving animal, we have used and modified the technical

and methodological experiences gained by studying the release of classical transmitters, to be described in the first part of this chapter. In this task, however, we have confronted new problems and limitations. Therefore, in the second part of this chapter, we present an account of some of the initial efforts to overcome these difficulties, and a sample of successful strategies and applications of these techniques in peptide neurobiology.

2. Perfusion Methods in the Study of Neuropeptide Release

All perfusion methods are based on the same working principle: the controlled infusion and withdrawal of an artificial physiological fluid that bathes organelles, cells, tissues, or organs. This procedure allows for continuous collection of the products that cells liberate under various physiological and pharmacological conditions. The rapid separation of these products from their source is the distinctive feature of perfusion techniques. Other methods used to study release processes, such as the measurement of changes in tissue content or changes in tissue–medium concentration of a particular substance during incubation, are indirect and more susceptible to variables such as changes in the synthesis, reuptake, or degradation of the releasable substances. The apparent simplicity of the experimental setups used in these methods, however, still makes them the choice of many researchers to study release from tissue fragments, small organs (ganglia, glands), and subcellular preparations. Admittedly, static incubations may be easier and even advisable when studying, for instance, cultured tissue in multiple wells, but this is not the case with most neurobiological preparations. Setting up a perfusion technique is an investment, beyond the inconveniences of alternate methods, because of its advantages: automatization, reproducibility, and the possibility of closely following the temporal course, resolution, latency, and duration of the release events (Bayón and Drucker-Colín, 1985b).

Therefore, in this section we will focus on perfusion procedures, discussing "hardware," various hydrodynamic setups, preparation of biological materials before perfusion, design and development of perfusion sessions, and analysis of the materials released. We will discuss strong points, pitfalls, and limitations of these methods, which will help in choosing among alternatives, and provide a word of caution for the interpretation of data.

2.1. Experimental Setups

The first consideration when choosing the perfusion system used to study a release process is the complexity and integrity of the biological preparation in which our scientific inquiry will be meaningful. This surely seems obvious, but it is too often overlooked: Release phenomena easily studied at the cellular or subcellular level may become impossible to analyze, or even to detect, in vivo; conversely, release events observed in the moving animal may never be detected in isolated preparations. In vitro and in vivo perfusion techniques are presented separately.

2.1.1. Perfusion In Vitro

In the study of in vitro release processes through perfusion methods, three types of preparations have been widely used: synaptosomes, tissue slices, and small organs, such as ganglia and small glands (larger organs can be efficiently perfused only through their vasculature, a procedure that is beyond the scope of this work). The widely different sizes of the elements constituting these preparations have determined the perfusion systems used. Figure 1 shows a schematic of a perfusion system for synaptosomal fractions used with various modifications in most laboratories (Raiteri et al., 1974; Redburn et al., 1975). Its main feature is the microporous filter where synaptosomes are retained and imbedded, presenting a thin layer to the perfusing fluid. At moderate flow rates (0.2–1 mL/min) this system preserves synaptosomal integrity and function and minimizes undesirable interactions between the material released and the synaptosomes (i.e., reuptake, effects on possible autoreceptors). Although this system has been successfully and widely used to study the release of classical transmitters, experiments in our laboratory on the release of peptidic materials indicate one of its intrinsic limitations: The capacity of the filters to accomodate synaptosomal elements without imparing the flow and the stability of the preparation is limited to a small amount of synaptosomal protein (about 0.5 mg in a filter of 25 mm diameter). Since neuropeptides are contained in and released from synaptosomes in much smaller quantities than other chemical messengers (usually by several orders of magnitude), the sensitivity of the assays used to detect the peptides released must match the limitation in the amount of tissue and peptides. If the quantity of peptide released and assay sensitivity cannot be matched (*see* end of section 2.2), the load of source tissue should be increased by using larger filters (which unfortunately is cumbersome and impairs the resolution of the release process) or by changing the supporting material. Recently columns packed with a mixture of synaptosomes and gel filtration support—Biogel P2—have been used to study peptide release (Klaff et al., 1982). In any case, the results obtained must be carefully pondered before stating that release does not occur.

The perfusion of tissue fragments or slices (or small organs), where the amount of perfused material is not so strictly limited, is not severely affected by the problem discussed above. Figure 2 depicts a perfusion system for tissue slices that comprises the basic characteristics of other devices used for similar purposes (Jessell, 1978; Marien et al., 1983). The tissue is placed in the chamber between grids, the mesh being small enough to retain the fragments that usually range between 100 and 500 μm. The flow is maintained by two pumping systems. This system is different from ordinary designs in that it is not a closed path; it is open to atmospheric pressure. By allowing air to enter the chamber, the withdrawal system can be set faster than the inflow, maintaining a constant pressure and speeding up the removal of the perfusing medium. Other advantages of open systems will be discussed in relation to in vivo perfusion techniques, but their use in the perfusion of tissue in vitro is particularly relevant to the study of peptide release: Peptides are exposed to degradation by proteases, and proteases are released during perfusion (Bayón et al., 1978, 1983, 1985; Greenfield and Shaw, 1982). The fast removal and inactivation of perfusates allowed by the open system is a great advantage in improving the recovery of released peptides. In this regard, it is also advantageous to reduce the time of exchange of medium in the perfusion chamber. Adjusting the size of the chamber to the amount of tissue is not always feasible; the use of inert chromatographic supports as fillings is a possible alternative (Ohno et al., 1981). These improvements in perfusion techniques will surely have relevance also to the study of the release of nonpeptide transmitters.

2.1.2. Perfusion In Vivo

The most obvious approach to studying the release of neural substances in vivo is to use superfusion systems analogous to those

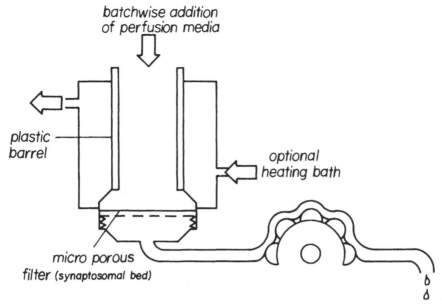

batchwise addition
of perfusion media

plastic
barrel

optional
heating bath

micro porous
filter (synaptosomal bed)

Fig. 1. Perfusion system for synaptosomal preparations. Synaptosomes are imbedded by suction on top of the filter (0.45–0.60 μm), then covered with perfusing medium. The flow rate (up to 1 mL/min) is kept constant during the experiment.

used in vitro, but instead of the tissue sample being placed inside a bath, the bath is implanted into brain tissue of a live animal. Cups, needles, concentric cannulae, dialysis bags, and fibers introduce a stream of artificial physiological fluid into the brain, allowing for the administration and sampling of chemicals into and from defined neural structures. Surface cups (mainly cortical cups) are chambers chronically adhered to the surface of the brain that contain inlet and outlet ports to allow for infusion and withdrawal of perfusing medium. Ventricular needles or cannulae are probes implanted in separate regions of the brain ventricular system that act as inlet and outlet ports to perfuse the cerebrospinal fluid (CSF)-filled cavity between them. Cups and ventricular needles have been, and still are, valuable tools for in vivo neurochemistry (*see* Myers, 1972). Unfortunately, their use is restricted to brain surfaces and cavities and does not allow for the study of most deep-brain nuclei. The perfusion cannula designed by Gaddum (1961) has found much wider use; since its inlet and outlet ports— push–pull—are concentric (see *Fig.* 3), it can be implanted just as an electrode is, and stereotaxic techniques can be used to reach

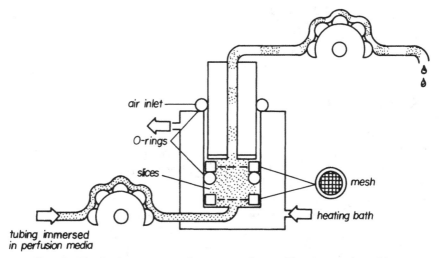

Fig. 2. Perfusion system for tissue slices. Slices are placed between grids (≥50 mesh) in the perfusion chamber (adjustable volume). The outflow is set faster than the inflow (more than two-fold).

deep neural regions (Myers, 1972). Alternative perfusion devices such as dialytrodes are also suitable for site-specific perfusions. Dialytrodes are based on the same working principles as other perfusion probes, but the diffusional exchange between the tissue and the perfusing medium is mediated by a dialysis membrane (Delgado et al., 1985; Hamberger et al., 1985). Their use is restricted, however, to accessible brain areas (*see* Fig. 4). Since the basic working principles of surface cups, ventricular needles, and dialytrodes are close to those of concentric push–pull cannulae, the latter will serve as a model for the methodology of in vivo perfusion in our discussion.

In vivo perfusion methods and systems have been the subject of extensive exploration and experimentation in an attempt to solve their main technical problems: extreme pressure changes leading to flow obstructions and tissue damage. These problems have been discussed elsewhere (Myers, 1972; Yaksh and Yamamura, 1974; Honchar, et al., 1979; Bayón and Drucker-Colín, 1985b). Briefly, pressure changes caused by tissue resistance lead to the accumulation of liquid or to erosion in the perfused region, and therefore, the success of a perfusion system lies in great part in maintaining a constant pressure in the tissue. This requirement is not fulfilled by systems in which precision delivery and withdrawal mechanisms are matched to force a constant flow through a

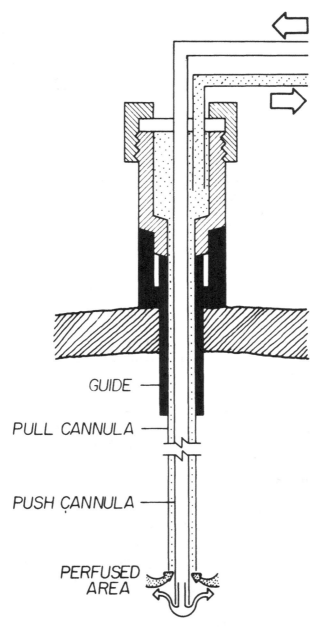

Fig. 3. In vivo perfusion push–pull cannula: concentric classical design. The inlet and outlet tubings separate in the upper chamber. The guide cannula (black) is fixed to the skull.

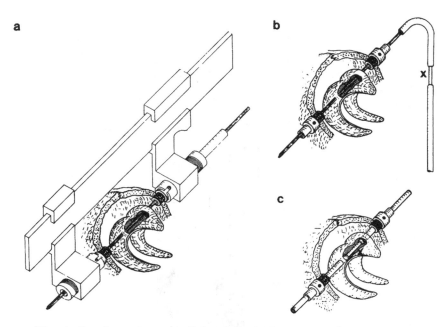

Fig. 4. Implantation of a dialytrode. (a) Insertion of screws in trepanized holes and introduction of a stainless steel guide through the anterior part of the rat hippocampus. (b) Nylon tubing glued onto guide—X, dialysis fiber. (c) Dialysis tubing pulled into position in the hippocampus (from Hamberger et al., 1985).

closed circulation path (Fig. 5A), since unbuffered pressure changes result from the changing resistance of the tissue (i.e., swelling). Systems open to the atmospheric pressure have been favored for in vivo perfusion in the last decade (Neouillón et al., 1977). Although they can maintain a steady flow only on the average, pressure changes can be buffered by the entrance of air into the system when tissue resistance changes (Fig. 5B); therefore, if an occlusion occurs, it usually subsides by itself. Obviously other perfusion systems with pressure buffering capabilities will also protect the tissue (Fig. 5C; Philippu et al., 1973); only open systems, however, allow for the rapid removal, collection, and inactivation of perfusates (*see* section 2.1.1) needed to study peptide release.

Since peptide release under ordinary experimental conditions is below the picomole range, efficient recovery of these solutes into perfusates is a determinant of success in in vivo perfusion. Diffusional exchange and extent of erosion of the tissue are com-

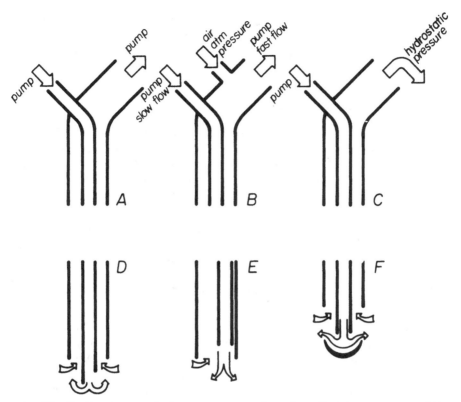

Fig. 5. In vivo perfusion systems and probes. Driving systems: (A) closed system, (B) open system, (C) single-pump open system. Probes: (D) Push cannula protruding from pull tubing, (E) shielded push cannula, (F) slotted cannula. Outer pull cannulae are usually 21–26 gage; inner push cannulae are 25–33 gage.

promised by the geometry of the tip of the perfusion probe. For probes in which the push cannula protrudes from the pull cannula (Myers, 1972; Fig. 5D), the surface of contact between the medium and the tissue can be relatively large, but the outflowing medium collides against the tissue, increasing erosion and lesion. With the use of probes in which the push cannula is shielded within the pull tubing (Philippu et al., 1973; Fig. 5E), tissue erosion is minimized at the expense of the area of the diffusion interface. An alternative device has been designed in our laboratory that helps to cope with this compromise for studying peptide release (Bayón et al., 1985; Fig. 5F): The push tubing is shielded within the pull cannula, but the interface of diffusion between the perfusion medium and the tissue is controlled by the size and number of the slots in the outer tubing. Using this probe, tissue damage has been reduced, while maintaining the same efficiency in collecting released peptides and protein materials.

2.2. Experimental Procedures

Because of their common working principles, in vivo and in vitro techniques share the same basic problems when studying release: viability of the tissue before and during perfusion, efficiency in recovering the substances released, and reliability of the methods used for their detection and measurement. The main strong points, pitfalls, and limitations of perfusion procedures have been outlined in a chapter in volume 1 of this series (Pittman et al., 1985). Here we will focus our attention on those problems particularly affecting peptide release studies.

2.2.1. Preparation of Biological Materials and Stability of the Releasable Peptides in the Tissue

Preparation of synaptosomal fractions (Gray and Whittaker, 1962; Hajós, 1975) and tissue slices (McIlwain and Rodnight, 1962) and implantation of perfusion probes into the brain (Myers, 1972; Delgado et al., 1985) necessarily produce damage to cells and tissues, which may be reflected in the stability of the stores of releasable peptides. This damage can sometimes be minimized, but frequently the best we can do is explore its effect in the release process under study and take it into account when interpreting results.

The viability of synaptosomal preparations with regard to release processes is usually tested by studying the uptake and release of exogenous, labeled transmitter (amino acids, biogenic amines) and nontransmitter (α-aminoisobutyric acid) substances (Sitges et al., 1986). Although this conventional approach has been successful so far, it is insufficient when studying peptide release. Since we do not yet know if the release mechanisms of transmitters of exogenous and endogenous origins are the same (Levi et al., 1978), the ability of synaptosomal preparations to release exogenous transmitters does not necessarily indicate that endogenous synaptosomal peptide stores are available. Furthermore, releasable peptide stores are not readily labeled; uptake mechanisms, if they indeed exist (Segawa et al., 1977; Nakata et al., 1981), do not have the capacity to label releasable stores of peptides. Since the synthesis of peptides is perikaryal, labeling them with precursor amino acids is highly inefficient and requires large amounts of radioactive material and incorporation is low and slow (Jones and Marchbanks, 1983). The use of controls testing the release of other known peptides from the preparation, as well as the measurement of the peptide stores during the course of perfusion, seems to be a first suitable approach to the problem. Testing the viability of synaptosomal preparations for peptide release studies requires additional developments; meanwhile, negative results when attempting to demonstrate the release of neuropeptides from synaptosomes have to be interpreted with caution.

The use of tissue slices in studying peptide release must also be subjected to the cautions indicated above. There is evidence that both slicing and incubation may produce a drop in the tissue content of a releasable peptide (e.g., enkephalin) (Fig. 6; Bayón et al., 1981b; Iversen et al., 1978a), whereas similar tissue processing does not affect the levels of others (e.g., somatostatin, neurotensin) (Iversen et al., 1978b). This is because enkephalin leaks from the slices during perfusion, and somatostatin does not (Iversen et al., 1978a,b; Bayón et al., 1978). The different behavior of these peptides is very likely related to the effect of slicing on the circuitry and anatomical connections of the brain regions where they were studied—the pallidum and hypothalamus, respectively. Enkephalins do not leak from the pallidum during in vivo perfusion (Bayón et al., 1981a,b).

The preparation of the tissue for in vivo perfusion can be as simple as the surgery required for the acute stereotaxic introduction of the perfusing probe into the brain of a restrained, anesthetized animal or as elaborate as the chronic implantation of a guiding cannula used to direct the insertion of the perfusing probe into the brain of an unanesthetized, freely moving animal. In both cases peptide stores may be affected. When using the first approach, besides the acute damage to the tissue (see below), the effect of the anesthetic on the release process must be explored (Cesselin et al., 1981). The second approach, the chronic implantation of a guide cannula, can follow two strategies, with different adverse consequences: In one, the guiding tube is also the pull cannula and is implanted to reach the area to be perfused (Fig. 7, right), so that at the time of perfusion no additional lesion is produced by the insertion of the inner push tubing (Ramírez, 1985); however, the chronic presence of the guide produces gliosis in the area to be perfused. This type of lesion has also been documented

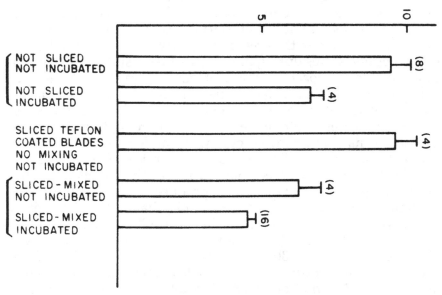

Fig. 6. Loss of enkephalin immunoreactive material in the rat globus pallidus as a consequence of slicing and incubation. Tissue slicing was performed at 200-μm intervals; incubations were for 60 min in Krebs-bicarbonate (oxygenated, 0.1% bovine serum albumin added). Enkephalin extraction and assay were as described in Iversen et al. (1978a). 1 unit - 1 ng of Leu-enkephalin immunoequivalents; bars represent the mean ± SEM, with the number of experiments in parentheses (from Bayón et al., 1981b).

with the use of chronically implanted dialytrodes (Hamberger et al., 1985). In the alternative strategy, the guide is positioned well above the region to be perfused (Fig. 7, left), so gliosis is not present in the area at the time of perfusion. The acute lesion produced by the introduction of the perfusing probe is unavoidable, however. In our experience studying the in vivo release of enkephalins (Bayón et al., 1981b), an acute lesion does not impair the resting or stimulated release processes, whereas a chronic lesion can completely block them (Fig. 8). Other investigations have shown, however, that chronic implants that reach the perfused area allow for repeated perfusion sessions in the same animal with no obvious deletereous effects on peptide release (i. e., luteinizing hormone releasing factor—LHRF) (Ramírez, 1985).

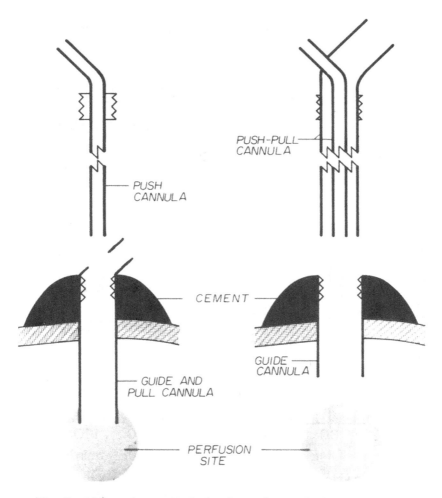

Fig. 7. Alternative cannula implants for perfusion in vivo. (Right) The guide cannula is also the pull cannula; the push tubing is inserted at the time of perfusion. (Left) The guide cannula is permanent; the perfusion probe is inserted at the time of perfusion.

This shows that the preparative strategies used for in vivo perfusion have to be tested in each particular case of brain region and peptide in question. We have discussed in section 2.1 and elsewhere (Bayón and Drucker-Colín, 1985b) several causes of tissue damage arising during in vivo perfusion: hydrodynamic malfunctions, movement of the probes, excessive flow rates, increased resistance in the perfused area. The use of lesion markers—i. e., proteins, lactic acid dehydrogenase, lysosomal proteases, and so

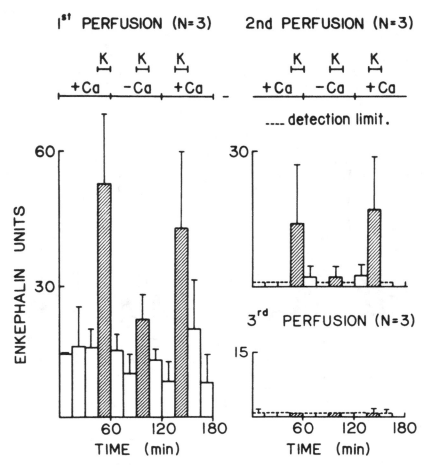

Fig. 8. In vivo release of enkephalin from the globus pallidus of the cat. Sequential push–pull cannula perfusions were performed at 1-wk intervals after the chronical implantation of the guide cannulae. In all perfusions the flow rate was held constant at 23 μL/min and 15-min fractions were collected on ice-cold 2N acetic acid. The samples were boiled, lyophilized, redissolved in water, and assayed for Leu-enkephalin (1 unit = 1 pg in the RIA). White bars correspond to resting release periods; 50 mM K⁺ medium was perfused during the periods indicated by the hatched bars. During the second hour of perfusion, Mg^{2+} was added to the medium instead of Ca^{2+}. Vertical lines represent the SEM; N is the number of bilaterally perfused cats. Enkephalin values lower than the minimum detectable amount are indicated by a discontinuous line on top of the bars (from Bayón et al., 1981b).

on—can help to detect these lesions, whatever their origin (Yaksh and Yamamura, 1974; Honchar et al., 1979; Greenfield et al., 1983; Bayón et al., 1985). With the use of appropiate perfusion probes and pumping systems, this damage can be minimized; protective agents, such as serum albumin (Iversen et al., 1978a; Bayón et al., 1978, 1981a) and low K^+ concentrations in the perfusing media, reduce tissue swelling. The viability of the tissue during a perfusion experiment must be empirically tested, however; checking the stability of the tissue response—peptide release, in this case—at different times during perfusion is necessary to validate the results obtained (Bayón et al., 1981b, 1985).

2.2.2. Efficiency of Recovery of the Released Peptides and Reliability of the Methods Used for Their Detection and Measurement

The efficiency of recovery of the peptides released into perfusates is a function of the diffusional exchange between the tissue and the perfusing medium and of the adsorption losses and degradation of the peptides in the perfusates. Some of the factors affecting this recovery have already been presented (section 2.1). Diffusional exchange depends on the interface between the tissue and the medium and on the flow rates. In vivo perfusion methods confront this problem mainly by modifying the geometry of the perfusing probe and by using moderately high flow rates. In vitro, the total surface of the sliced tissue or the synaptosomal beds is the limiting factor, since the usual flow rates are already high (about 0.5 mL/min).

Once in the perfusion medium, peptides can be lost by degradation or adsorption to the tubing (as already indicated, reuptake is not a serious concern when working with peptide release). Fast removal and fast inactivation of the perfusates upon collection—by heat or by protease inhibitors—do not seem to be enough to protect released peptides from degradation. The opioid peptides serve to exemplify this case (*see* Bayón et al., 1983): During in vitro perfusion, collection of perfusates in a boiling bath yields a low recovery of enkephalins unless a peptidase inhibitor is added to the perfusing medium (Iversen et al., 1978a; Bayón et al., 1978). Several peptidase inhibitors and mixtures of synthetic peptides have been used to protect enkephalins during perfusion (Henderson et al., 1978; *see* Bayón et al., 1983), but none can fully prevent enkephalin degradation in perfusates. This problem can be generalized to other neuropeptides since the sites in their amino acid sequences that are potential substrates of peptidases can be as varied as their peptide bonds. Thus neither the presence of peptidase inhibitors in the perfusing media nor the fast removal of peptides from the site of release—followed by protease inactivation—assure the preservation of the released peptides. The effectiveness of these protection methods has to be tested by the use of internal standards: Addition of tracer amounts of the labeled peptide to the medium to be perfused helps to monitor its degradation after release (Fig. 9; Bayón et al., 1978).

Peptide losses from adsorption to the tubings or to other materials during preparation for their assay should also be monitored with internal standards (Bayón et al., 1978, 1983; Cesselin et al., 1981; Franco-Bourland and Fernstrom, 1981). Protective methods should also be used, especially when dealing with picogram amounts of material. During perfusion, the presence of serum albumin and the high ionic strength of the Ringers medium minimize peptide adsorption to the tubings. During processing and handling, losses can be reduced by choosing the combination of materials (glass or plasticware) and solvents (hydrophobic media, high ionic strength, detergents) that yield the highest peptide recoveries. Since chromatographic or electrophoretic separations are frequently required in the identification and measurement of neuropeptides, the controls and precautions indicated above should be extended to these preparative steps. A discussion of these problems in relation to the opiate peptides, which illustrates the case, has been presented elsewhere (Bayón et al., 1983).

The greater chemical complexity of peptides, compared to classical neurotransmitters, and their lower levels in nervous tissue—and in perfusates—demand detection methods of high specificity and sensitivity. So far, only the stereochemical recognition of peptides by antibodies or receptors has allowed us to develop techniques, such as radioimmunoassay and receptor binding assay, that fulfill these requirements. From these, immunodetection has become the major tool for measuring neuropeptides, since it discriminates on the basis of chemical structure rather than biological activity. Immunodetection has been preferred also because it provides the possibility of correlating histological and biochemical aspects of peptide neurobiology, since antibodies are amenable to both localization and quantification techniques.

Problems inherent in the immunoassay of peptides are profusely spread throughout the literature; we have collected and discussed examples of these problems when analyzing perfusates containing opioid peptides that can be used as paradigms for those

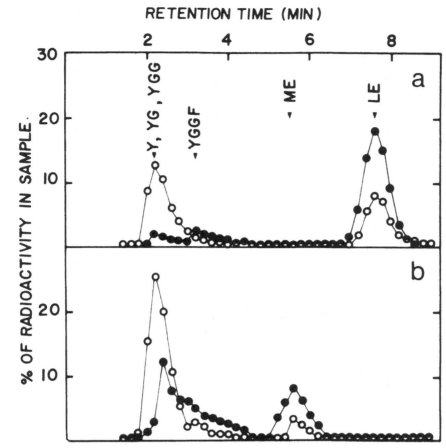

Fig. 9. High pressure liquid chromatography (HPLC) of a globus pallidus perfusate obtained during K^+-stimulated enkephalin release. The medium used in the perfusion contained trace amounts of pre-purified tritium-labeled enkephalins. (a) Degradation by tissue slices of (tyrosyl-3,5-^{3}H)-Leu-enkephalin added to the 50 mM K^+ perfusing medium ($\approx$4000 cpm in 2 mL) was 62% in the absence of bacitracin (○) and 23% with bacitracin (30 μg/mL) (●). (b) As in (a), but with addition of (tyrosyl-3,5-^{3}H)-Met-enkephalin ($\approx$6000 cpm in 2 mL). Degradation was 90% in the absence of bacitracin (○) and 63% with bacitracin (●). Tracer recoveries were almost complete (from Bayón et al., 1981b).

occurring when studying the release of other neuropeptides (Bayón et al., 1983). Strictly speaking, absolute chemical specificity is impossible to obtain with the unaided immunoassay: Neuropeptides frequently exist in families (opiates, tachykinins, groups of gut peptides) (Krieger et al., 1983; Valverde and Bayón, 1983) in

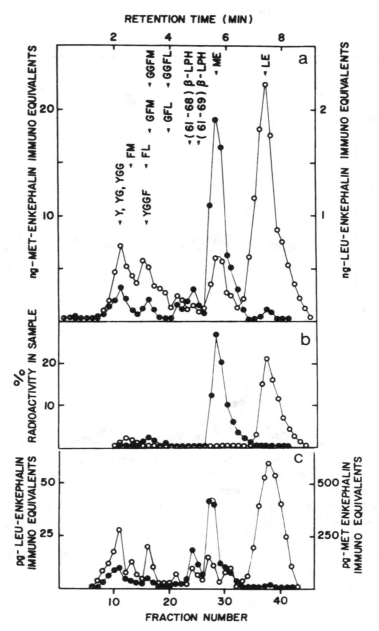

Fig. 10. HPLC fractionation of extracts from freshly dissected pallid-
al tissue (a and b) and high K⁺ perfusate obtained in vitro (c). The
µ-Bondapak-CN column was calibrated with unlabeled synthetic standards: ME, Met-enkephalin; LE, Leu-enkephalin; other peptides are rep-

which members have similar amino acid sequences that often cannot be completely distinguished by antibodies. Also, biosynthetic precursors or degradation fragments of neuropeptides may bind to antibodies raised against the bioactive molecule. Even if these related substances barely crossreact with the antibody, the interference may be significant when their concentration in perfusates is high. Figure 10 illustrates the case: Perfusates obtained in vitro from the pallidum, fractionated by high pressure liquid chromatography, and subsequently immunoassayed for both Met-enkephalin and Leu-enkephalin immunoreactivity, contain numerous components, also present in the perfused tissue, that are capable of displacing the labeled enkephalin from the antibodies. Also, this chromatogram shows that each radioimmunoassay reads, although to a different extent, both enkephalins. It is this limitation of immunoassays that leads researchers to use the expression "neuropeptide-like immunoreactivity," unless additional evidence of the identity of the peptide is available, such as chromatographic retention time, in this case.

Other interfering materials in radioimmunoassay may come from the perfusing media or the electrophoretic or chromatographic supports used in peptide fractionation. These interferences usually introduce an error in the measurements that shows up when dose–response curves—volume of perfusate vs measured peptide content—do not intercept the origin. Figure 11 exemplifies

---

←

resented using the single-letter amino acid nomenclature (Y, Tyr; G, Gly; F, Phe; M, Met; L, Leu). (a) Radioimmunoassays using both Met-enkephalin and Leu-enkephalin antisera were performed in each fraction of the chromatograms: Values are expressed as ng of Met-enkephalin (●) that would produce an equivalent immunodisplacement in the Met-enkephalin assay and, analogously, as ng of Leu-enkephalin (○) producing an equivalent displacement in the Leu-enkephalin assay. The content of Met-enkephalin and Leu-enkephalin in their respective peaks was estimated from the two radioimmunoassays and agreed within 5% variation. (b) Radioactivity profiles of samples, similar to those in (a), that were extracted in the presence of either tritiated Met-enkephalin (●) or tritiated Leu-enkephalin (○). The radioactivity is expressed as % of cpm injected into the column. (c) HPLC fractionation of the material released from globus pallidus slices during K^+ stimulation: Met-enkephalin (●) and Leu-enkephalin (○) radioimmunoassay profiles: (from Bayón et al., 1981b).

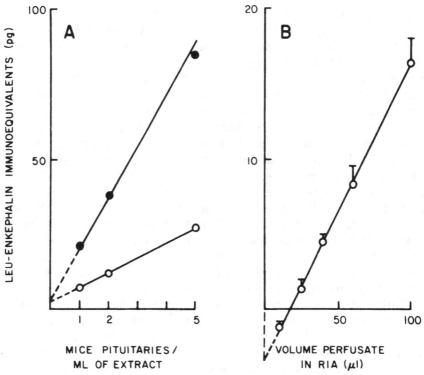

Fig. 11. Blank effects produced by samples in an enkephalin RIA.
(A) Positive blank effect caused by a decrease in maximum binding in the
RIA. Deproteinized extracts from one, two, and five mice pituitaries were
run in silica plates, recovered in 1 mL of assay buffer, and their enkephalin contents determined from 20 μL (○) and 100 μL (●) aliquots in dupli-
cates; the dose–response plots intercept the ordinate above the origin,
giving a positive blank value to be subtracted from the sample values. (B)
Negative blank effect caused by an increase in maximum binding in the
RIA. In this case different volumes of a Krebs-bicarbonate perfusate from
rat globus pallidus slices were assayed. The dose–response plot in-
tercepts the ordinate below the origin (see text) (from Bayón et al., 1983).

this type of interference, showing that the analysis of enkephalins
released into perfusates may give negative intercept values in
dose–response curves because of the presence of interfering materials (bicarbonates) in the perfusing medium. In contrast, the immunoassay of enkephalins extracted from silica plates produces
the opposite effect: dose–response curves intercept above the origin. These problems can be solved by processing the peptides used

in the standard curves of the radioimmunoassay in exactly the same way as the unknown samples.

Development of a sensitive assay is necessary for the accurate detection of the small amounts of peptides that are released. Since the sensitivity of an immunoassay is limited by the affinity of the antibody and the peptide, any effort to obtain a good antiserum is worthwhile. Also, the higher the specific activity of the tracer-ligand peptide, the better the sensitivity of the immunoassay. Several strategies used to prepare immunogens, various immunization schedules and the results of using different radiotracers have been presented in another paper (Bayón et al., 1983). The work of Cesselin et al. (1981) exemplifies a successful case in which sera and tracer allow the reliable analysis of perfusate samples to measure enkephalin release. A word of caution: When the sensitivity of an assay is not high enough, one is inclined to concentrate the perfusates in order to detect the peptide under study; but concentrating the peptide also concentrates salts, proteins, and anything else in the perfusate capable of interfering with the assay. High sensitivity has no substitute.

3. Strategies in Studying the Release of Neuropeptides

Neurochemistry has been largely the study of neurotransmission. Neurotransmitter release is a key aspect of this process, and the acceptance of a putative transmitter as such requires, among other proofs the demonstration of its release in the neuronal system under study (Werman, 1966). So the seemingly simple task of testing the liberation of a novel peptide generally focuses on the modifications observed under resting and stimulation conditions, physiologically or pharmacologically induced. Consideration of the mechanism of release receives formal attention only later. This segmentary view of the problem, although strategically necessary in the beginning, is untenable when optimizing the methodology for release studies: the fundamental chemical events underlying release processes do not just explain it, but are also the substrates at which electrochemical, physiological, and pharmacological stimuli control release. Failure to consider, explore, and manipulate these factors when designing perfusion experiments might keep one from recognizing the parameters that physiologically regulate this release. Worse, one may be unable to distinguish physiological release from the leakage produced by tissue damage, or even to detect release. Therefore, we will here address problems related to the observation of spontaneous and induced release and also to the experimental control of these events.

3.1. Spontaneous and Induced Release

The initial phase of a perfusion experiment—before any experimental change is attempted—is usually extended until stable release is attained. This basal release is taken as a control value to which induced changes can be compared. Little attention is therefore ordinarily paid to its nature or physiological significance; terms like "resting" or "spontaneous" release, although conventionally accepted, usually reflect a conviction or an assumption, not a description. Basal release, however, can be the subject of scientific inquiries and also of experimental artifacts. We have already indicated (*see* section 2) that a leakage of neuropeptides unavoidably occurs as a consequence of the tissue erosion caused by the circulation of perfusion fluids. When studying the physiological, spontaneous release of peptides, one must know the contribution of this leakage to the total basal release. The presence in the perfusing media of blockers of bioelectrical activity (e. g., tetrodotoxin) or inhibitors of the secretory processes (e. g., calcium channel antagonists) can help to reveal the contribution of the remaining leakage (*see* Glowinski, 1981). One must be cautious, however, since unknown physiological release mechanisms, insensitive to these blockers, may exist. Therefore, monitoring the temporal profile of leakage of the tissue contents from lesion (*see* section 2) is always necessary to distinguish physiological release from the consequences of cell damage. By taking into account these considerations and keeping control of the decrease in the releasable peptide pools during perfusion (*see* section 2), one is able to study the spontaneous, physiological release of peptides.

The spontaneous release of peptides is subject to physiological variations; endogenous changes, usually following biological rhythms, may modify their release. These periodic oscillations in neuropeptide release would be reported more frequently if more attention were given to spontaneous release. Ramírez (1985) has described spontaneous oscillations in LHRH release from brain within periods of minutes that could have been disregarded as experimental artifacts by less careful researchers. In our laboratory, a diurnal rhythm in enkephalin release from the pallidum was found when analyzing unexpected variations in its resting control values (Bayón et al., 1985). Thus spontaneous release should be carefully explored before attempting to modify it.

Changes in neuropeptide release can be induced through electrical, chemical, or physiological manipulations. Under experimental conditions, these stimuli usually have to be maximized (by either repetitive electrical stimulation with high currents or frequencies, high concentrations of chemical or pharmacological agents, or extreme physiological changes) in order to obtain correspondingly large release responses that can be detected with the analytical methods available (section 2). The physiological relevance and significance of these observations must be validated by complete dose–response curves and studies of the effect of specific antagonists and blockers of the stimulating agent (see Glowinski, 1981; Pittman et al., 1985). In addition to this problem, the interpretation of data obtained from induced-release experiments is constrained by limitations of the perfusion techniques in the anatomical and temporal resolution of the release events. In vitro, the magnitude of the stimulus and the extent of tissue affected is under control; during perfusion in vivo this is not so, since the perfusion medium containing the stimulating agents diffuses along its concentration gradient in an uncircumscribed volume of tissue around the area perfused. Therefore, different populations of nerve cells are exposed to varying concentrations of the stimulating agent(s). A study of the quantitative distribution of the drug in the perfused area is needed in order to have an idea of the extent of tissue affected (see Velasco et al., 1985). For the same reason, one has to verify the origin of the neuropeptides released into perfusates. Immunohistochemistry and in vitro release studies can help, but the interpretation of the data obtained in vivo requires a study relating variations in the amount of neuropeptide released to small changes in the implantation sites of the perfusion probe (Cesselin et al., 1981). Resolution of time-of-release events is also important in perfusion experiments. The time course of peptide release and its latency after stimulation and possible rebound effects give fundamental information on neurotransmission (Bourgoin et al., 1982). Resolution is severely limited, however, by dead spaces in the perfusion system, diffusion in the pipelines, and mixing of the perfusates in pumps and fraction collectors. These problems can be minimized by the use of small probes and perfusion chambers, small diameter pipelines, and collection of smaller perfusate fractions. Nevertheless, the interpretation of data must always take into consideration the fact that perfusion systems are slower than most release events.

3.2. Experimental Control of Peptide Release

The current "stimulus–secretion coupling" hypothesis provides a framework with which to explore and understand the basic roles that ion concentrations and ion permeabilities have in the release of neuropeptides from neural tissue (Terry and Martin, 1978; Dreifuss et al., 1980; Warberg, 1982). Knowledge of these factors is required to establish the conditions of peptide release studies for two types of reasons. First, because the ionic composition of perfusion media, usually taken from recipes, might not approach the actual composition of the extracellular environment of the tissue under study. Since this information is not readily accessible, adjusting the Ringer's composition to maintain the physiologically normal behavior of the neural elements seems the best strategy. For instance, slight changes in the Ca^{2+} ion concentration of a standard perfusing medium are able to shift the firing pattern of hypothalamic cells in a tissue slice closer to its in vivo behavior (Pittman et al., 1981). Second, the manipulation of ion concentrations and ion permeabilities allows for the stimulation or the inhibition of peptide release at will, through different mechanisms: by modification of the equilibrium potential of Na, K, or Ca *ions* and their relative contributions to the membrane potential, or even by acting directly on the Ca^{2+}-dependent exocytotic process. The work of Drouva et al. (1981) is an elegant example of the use of these manipulations to study peptide release (*see* Fig. 12). This kind of study also provides a strategy to analyze the mechanisms by which other neurotransmitters, pharmacological agents, neurotoxins, and so on modify peptide release—many of them directly or indirectly affecting ion permeabilities. For reviews on the neurotransmitter modulation of neuropeptide release, *see* Kordon et al. (1980), McKelvy et al. (1980), and Robbins and Reichlin (1982).

4. Accomplishments and Perspectives

Traditional perfusion techniques have allowed us to understand the basic principles and mechanisms of neuropeptide release, as well as many fundamental aspects of the modulation of

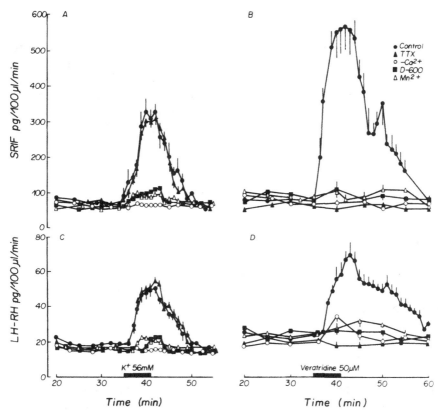

Fig. 12. Spontaneous and K$^+$- or veratridine-induced LHRH and somatotropin release inhibiting factor (SRIF) release from mediobasal hypothalamic slices superfused with either control Locke medium or Locke medium containing $5 \times 10^{-7}M$ tetrodotoxin, $10^{-4}M$ D-600, or 3 mM Mn^{2+}. Values represent the mean ± SEM from four experiments (from Drouva et al., 1981).

this release by drugs, chemical messengers, and neural circuitry. Furthermore, perfusion methods have allowed us to correlate neurochemical, electrophysiological, and behavioral events (*see* contributions in Bayón and Drucker-Colín, 1985a). Nevertheless, in the present decade new developments promise new applications. Recent improvements in chromatographic techniques (e. g., HPLC) and immunological instruments (monoclonal antibodies) have greatly increased sensitivity and specificity in the analysis of perfusates. New dialysis fibers will improve selectivity and peptide recoveries in perfusates while decreasing tissue damage. In vitro calibration of dialytrodes will allow truly quantitative measurements of peptide release.

Methodological alternatives to perfusion techniques are being explored: for instance, the *in situ* detection of released substances by electrochemical methods is rapidly progressing. Only oxidizable chemicals, however, such as catecholamines and indolamines, can be measured with it, and specificity is still unsatisfactory. Neuropeptides containing tyrosine, tryptophan, or cysteine have been detected, though not differentiated, with this method (Bennett et al., 1981). Thus we must await—or create—future developments in this and other methods that will diversify our possibilities in the study of peptide release.

Acknowledgments

I thank M. Sordo and L. M. Navarro for their help in preparing the manuscript. This work was supported in part by CONACyT, Mexico and Fundación R. J. Zevada.

References

Bayón A. and Drucker-Colín R., eds. (1985a) *In Vivo Perfusion and Release of Neuroactive Substances: Methods and Strategies.* Academic, Orlando.

Bayón A. and Drucker-Colín R. (1985b), Methodological Alternatives and Experimental Strategies for Studying the *In Vivo* Perfusion and Release of Neuroactive Substances in the Central Nervous System, in *In Vivo Perfusion and Release of Neuroactive Substances* (Bayón A. and Drucker-Colín R., eds.), Academic, Orlando.

Bayón A., Rossier J., Mauss A., Bloom F. E., Iversen L. L., and Guillemin R. (1978) In vitro release of methionine-enkephalin and leucine-enkephalin from the rat globus pallidus. *Proc. Natl. Acad. Sci. USA* **75,** 3503–3506.

Bayón A., Shoemaker W. J., Lugo L., Azad R., Ling N., Drucker-Colín R., and Bloom F. E. (1981a) In vivo release of enkephalin from the globus pallidus. *Neurosci. Lett.* **24,** 65–70.

Bayón A., Drucker-Colín R., Lugo L., Shoemaker W. J., Azad R., Ling N., and Bloom F. E. (1981b) Enkephalin Release From the Globus Pallidus: *In Vitro* and *In Vivo* Studies, in *Regulatory Mechanisms of Synaptic Transmission* (Tapia R. and Cotman C. W., eds.), Plenum, New York.

Bayón A., Shoemaker W. J., McGinty J. F., and Bloom F. E. (1983) Immunodetection of Endorphins and Enkephalins: A Search for Reliability, in *International Review of Neurobiology* vol. 24, Academic, New York.

Bayón A., Solano S., Anton B., Castaño I., and Diaz D. (1985) Push-Pull Perfusion Studies on the *In Vivo* Release of Proteins, Enzymes, and the Neuroactive Peptides—Enkephalins—From the Rat Brain, in *In Vivo Perfusion and Release of Neuroactive Substances* (Bayón A. and Drucker-Colín R., eds.), Academic, Orlando.

Bennett G. W., Brazell M. P., and Marsden C. A. (1981) Electrochemistry of neuropeptides: A possible method for assay and in vivo detection. *Life Sci.* **29,** 1001–1007.

Bourgoin S., Cesselin F., Artaud F., Glowinski J., and Hamon M. (1982) In vivo modulation by GABA-related drugs of Met-enkephalin release in basal ganglia of the cat brain. *Brain Res.* **248,** 321–330.

Cesselin F., Soubrie P., Bourgoin S., Artaud F., Reisine T. D., Michelot R., Glowinski J., and Hamon M. (1981) In vivo release of Met-enkephalin in the cat brain. *Neurosci.* **6,** 301–313.

Delgado J. M., Lerma J., Martin del Rio R., and Solis J. M. (1985) Use of Dialytrodes in Brain Neurochemistry, in *In Vivo Perfusion and Release of Neuroactive Substances* (Bayón A. and Drucker-Colín R., eds.), Academic, Orlando.

Dreifuss J. J., Mathison R. D., and Gähwiler B. H. (1980) Secretion of Neurohypophysial Peptides, in *The Role of Peptides in Neuronal Function* (Barker J. L. and Smith, Jr., T. G., eds.), Marcel Dekker, New York.

Drouva S. V., Epelbaum J., Héry M., Tapia-Arancibia L., Laplante E., and Kordon C. (1981) Ionic channels involved in the LHRH and SRIF release from rat mediobasal hypothalamus. *Neuroendocrinology* **32,** 155–162.

Franco-Bourland R. and Fernstrom J. D. (1981) *In vivo* biosynthesis of L-(^{35}S)Cys-arginine vasopressin, -oxytocin and -somatostatin: Rapid estimation using reversed phase high pressure liquid chromatography. *Endocrinology* **109,** 1097–1106.

Gaddum J. H. (1961) Push–pull cannulae. *J. Physiol.* **155,** 1–2P.

Glowinski J. (1981) *In vivo* release of transmitters in the cat basal ganglia. *Fed. Proc.* **40,** 135–141.

Gray E. G. and Whittaker V. P. (1962) The isolation of nerve endings from brain: An electron microscopic study of the cell fragments derived by homogenization and centrifugation. *J. Anat.* **96,** 79–88.

Greenfield S. A. and Shaw S. G. (1982) Release of acetylcholinesterase and aminopeptidase in vivo following infusion of amphetamine into the substantia nigra. *Neuroscience* **7,** 2883–2893.

Greenfield S., Cheramy A., and Glowinski J. (1983) Evoked release of proteins from central neurons in vivo. *J. Neurochem.* **40**, 1048–1057.

Hajos F. (1975) An improved method for the preparation of synaptosomal fractions in high purity. *Brain Res.* **136**, 387–392.

Hamberger A., Berthold C. H., Jacobson I., Karlsson B., Lehmann A., Nyström B., and Sandberg M. (1985) In Vivo Brain Dialysis of Extracellular Nontransmitter and Putative Transmitter Amino acids, in *In Vivo Perfusion and Release of Neuroactive Substances* (Bayón A. and Drucker-Colín R., eds.), Academic, Orlando.

Henderson G., Hughes J., and Kosterlitz H. W. (1978) In vitro release of Leu- and Met-enkephalin from the corpus striatum. *Nature* **271**, 677–679.

Hökfelt T., Johansson O., Ljungdahl A., Lundberg J and Schultzberg M. (1980) Peptidergic neurones. *Nature* **284**, 515–521.

Honchar M. P., Hartman B. K., and Sharpe L. G. (1979) Evaluation of in vivo brain site perfusion with the push–pull cannula. *Am. J. Physiol.* **236**, R45–56.

Hughes J., Smith T. W., Kosterlitz H. W., Fothergill L. H., Morgan B. A., and Morris H. (1975) Identification of two related pentapeptides from the brain with potent opiate agonist activity. *Nature* **255**, 577–579.

Iversen L. L., Iversen S. D., Bloom F. E., Vargo T., and Guillemin R. (1978a) Release of enkephalin from rat globus pallidus in vitro. *Nature* **271**, 679–681.

Iversen L. L., Iversen S. D., Bloom F. E., Douglas C., Brown M., and Vale W. (1978b) Calcium-dependent release of somatostatin and neurotensin from rat brain in vitro. *Nature* **273**, 161–163.

Jessell T. M. (1978) Substance P release from the rat substantia nigra. *Brain Res.* **151**, 469–478.

Jones, C. A. and Marchbanks R. M. (1983) The synthesis and release of [^{3}H-tyrosine^1]methionine^5-enkephalin from guinea pig brain slices. *J. Neurochem.* **40**, 357–363.

Klaff L. J., Hudson A. M., Paul M., and Millar R. P. (1982) A method for studying synaptosomal release of neurotransmitter candidates as evaluated by studies on cortical cholecystokinin octapeptide release. *Peptides* **1**, 155–161.

Kordon C., Enjalbert A., Hery M., Joseph-Bravo P., Rotsztejn W., and Ruberg M. (1980) Role of Neurotransmitters in the Control of Adenohypophyseal Secretion, in *Handbook of the Hypothalamus* vol. 2 (Morgane P. and Panksepp J., eds.), Marcel Dekker, New York.

Krieger D. T., Brownstein M. J., and Martin J. B. (1983) *Brain Peptides* John Wiley, New York.

Lenoir D., Battenberg E., Bloom F. E., and Milner R. J. (1985) The rat brain specific gene 1B236 is expressed in cultures of fetal brain. *Soc. Neurosci. Abst.* **11,** 1065.

Levi G., Banay-Schwartz M., and Raiteri M. (1978) Uptake, Exchange and Release of GABA in Isolated Nerve Endings, in *Amino Acids as Chemical Transmitters* (Fonnum F., ed.), Plenum, New York.

Marien M., Brien J., and Jhamandas K. (1983) Regional release of (^{3}H) dopamine from rat brain in vitro: Effects of opioids on release induced by potassium, nicotine, and L-glutamic acid. *Can. J. Physiol. Pharmacol.* **61,** 43–60.

McIlwain H. and Rodnight R. (1962) *Practical Neurochemistry* Churchill, London.

McKelvy J. F., Charli J.-L., Joseph-Bravo P., Sherman T., and Loudes C. (1980) Cellular Biochemistry of Brain Peptides: Biosynthesis, Degradation, Packaging, Transport and Release, in *The Endocrine Functions of the Brain* (Motta M., ed.) Raven, New York.

Myers R. D. (1972) Methods for perfusing different structures of the brain. *Meth. Psychobiol.* **2,** 169–211.

Nakanishi S., Inoue A., Kita T., Nakamura M., Chang A. C. Y., Cohen S. N., and Numa S. (1979) Nucleotide sequence of cloned cDNA for bovine corticotropin-β-lipotropin precursor. *Nature* **278,** 423–427.

Nakata Y., Kusaka Y., Yajima H., and Segawa T. (1981) Active uptake of substance P carboxy-terminal heptapeptide (5-11) into rat brain and rabbit spinal cord slices. *J. Neurochem.* **37,** 1529–1534.

Neouillón A., Cheramy A., and Glowinski J. (1977) An adaptation of the push–pull cannula method to study the in vivo release of (^{3}H)-dopamine synthesized from (^{3}H)-tyrosine in cat caudate nucleus: Effect of various physical and pharmacological treatments. *J. Neurochem.* **28,** 819–828.

Ohno N., Hashimoto K., Yunoki S., Takahara J., and Ofuji T. (1981) A perfusion method for examining arginine vasopressin release from hypothalamo-neurohypophyseal system. *Acta Med. Okayama* **35,** 27–35.

Philippu A., Przuntek H., and Roensberg W. (1973) Superfusion of the hypothalamus with gamma-aminobutyric acid: Effect on the release of noradrenaline and blood pressure. *Naunyn Schmiedebergs Arch. Pharmacol.* **276,** 103–118.

Pittman Q. J., Hatton J. D., and Bloom F. E. (1981) Spontaneous activity in perfused hypothalamic slices: Dependence on calcium content of perfusate. *Exp. Brain. Res.* **42,** 49–52.

Pittman Q. J., Disturnal J., Riphagen C., Veale W. L., and Bauce L. (1985) Perfusion Techniques for Neural Tissue, in *Neuromethods* vol. 1,

General Techniques (Boulton A. A. and Baker G. B., eds.), Humana Press, Clifton, New Jersey.

Raiteri M., Angelini F., and Levi G. (1974) A simple apparatus for studying the release of neurotransmitters from synaptosomes. *Eur. J. Pharmacol.* **25,** 411–414.

Ramírez U. D. (1985) The Push–Pull Perfusion Technique in Neuroendocrinology, in *In Vivo Perfusion and Release of Neuroactive Substances* (Bayón A. and Drucker-Colín R., eds.), Academic, Orlando.

Redburn D. A., Biela J., Shelton D., and Cotman C. W. (1975) Stimulus-secretion coupling *in vitro:* A rapid perfusion apparatus for monitoring efflux of transmitter substances from tissue samples. *Anal. Biochem.* **67,** 268–278.

Robbins R. and Reichlin S. (1982) In Vitro Systems for the Study of Secretion and Synthesis of Hypothalamic Peptides, in *Neuroendocrine Perspectives* vol. 1 (Miller E. and MacLeod R., eds.), Elsevier, Amsterdam.

Segawa T., Nakata Y., Yajima H., and Kitagawa K. (1977) Further observation of the lack of active uptake system for substance P in the central nervous system. *Japn. J. Pharmacol.* **27,** 573–580.

Sitges M., Possani L. D., and Bayón A. (1986) Noxiustoxin, a short chain toxin from the mexican scorpion *Centruroides noxius* induces transmitter release by blocking K^+ permeability. *J. Neurosci.* **6,** 1570–1574.

Terry L. C. and Martin J. B. (1978) Hypothalamic hormones: Subcellular distribution and mechanisms of release. *Ann. Rev. Pharmacol. Toxicol.* **18,** 111–123.

Valverde C. and Bayón A. (1983) Neuropeptides: Chemical Structure and Localization of the Producing Cells, in *Aminoácidosy Péptidos en la Integración de Funciones Nerviosas* (Pasantes H. and Arechiga H., eds.), UNAM, Mexico.

Velasco M., Velasco F., Pacheco M. T., and Estrada-Villanueva F. (1985) Efficacy of the Push–Pull Perfusion Techniques in the Study of Sleep, Epilepsy and Tremor, in *In Vivo Perfusion and Release of Neuroactive Substances* (Bayón A. and Drucker-Colín R., eds.), Academic, Orlando.

Warberg J. (1982) Studies on the release mechanism for hypothalamic hormones. *Acta Endocrin.* (suppl.) **250,** 1–48.

Werman R. (1966) Criteria for identification of a central nervous system transmitter. *Comp. Biochem. Physiol.* **18,** 745–766.

Yaksh T. L. and Yamamura H. I. (1974) Factors affecting performance of the push–pull cannula in brain. *J. Appl. Physiol.* **37,** 428–434.

Functional Identification of Peptide Receptors

Serge A. St-Pierre

1. Introduction

With the advent of modern chromatographic instrumentation and improved sequencing methods that allow for isolation and characterization of minute amounts of biological material, a host of biologically active peptides are now known. More recently, the dramatic development of genetic engineering has contributed to revealing the structure of complete genes coding for known, as well as novel, biologically active peptides, in addition to a large number of other peptide sequences of unknown significance. Even though precise physiological functions could be assigned to insulin, parathyroid hormone (PTH), calcitonin, growth hormone (GH), and growth hormone releasing factor (GHRH), corticotropin (ACTH) and corticotropin releasing factor (CRF), atrial natriuretic factor (ANF), oxytocin, and a few others, the role of more than one hundred peptides displaying various biological activities in vitro or in vivo is still awaiting definition. Moreover, recent advances in immunocytochemistry and receptor imaging have revealed the ubiquity of several among these "well-established" peptides and their corresponding receptors, for which the localization and action had been previously assigned to given target organs or tissues. Most remarkable are the peptides of the now so-called "brain-gut-heart" axis, for which coexistence among themselves and/or with classical neurotransmitters such as the catecholamines in nerve fibers has been recently demonstrated (Cuello, 1982).

2. Peptide–Receptor Interaction

The concept of the drug–receptor interaction, *"Corpora non agunt nisi fixata"* (roughly translated as "A substance cannot cause an action without binding first") as understood by Paul Ehrlich in 1913, was applied by Ariens (1964) 50 yr later to the receptor-occupation theory, and still has great significance today, since receptors can be studied as tangible molecular entities.

Even though the proteinaceous nature of receptors has been firmly established, our understanding of the mechanisms through which receptors bind ligands and undergo functional changes that trigger the sequence of biophysical and biochemical events leading to the effect is still largely incomplete. One of the main reasons receptor study is so difficult, compared to that of other biological macromolecules, such as enzymes or nucleic acids, is that, unlike the latter, receptors are fully functional only *in situ;* that is, bound in cell membranes to specific lipids that appear to be essential to their integrity. Pure preparations of enzymes or nucleic acids that are active in simple media, however, can be studied in vitro. In those instances in which receptor proteins have been extracted, it has been demonstrated in only a few cases, where the receptor lipidic environment could be reconstituted, that these proteins were identical to pharmacologically functional receptors. Therefore, functional aspects of receptors are better obtained otherwise, by means of indirect methods involving substances that are known to bind to receptors.

In the early decades of pharmacology, structure–activity studies of drugs merely served as classifying principles or guidelines to the synthesis of novel, therapeutically useful compounds. Presently, in analogy with the characterization of enzyme binding sites, receptor mapping is a novel application for structure–activity studies of biologically active molecules, since the cell response obtained for a series of analogs can give information about the molecular nature of the receptor surface. As a general rule, the more rigid the structure of the ligand, the more it can tell about the properties of its receptor, and the smaller the chance that it will fit different types of receptors.

The study of peptide receptors represents a formidable challenge. Contrary to the situation with most classical drugs or to other natural autacoids and hormones, peptides are large, flexible molecules, with an almost infinite number of conformations and functional group combinations given by their amino acid components. Even though these properties of peptides offer a unique chance to shape the natural molecules and change their biological properties at will, they are at the origin of numerous experimental difficulties.

As for other types of ligands, according to Ariens' occupation theory (1964) the peptide–receptor interaction itself is likely to involve two phases: (1) the binding of the peptide substrate to the receptor (occupation) and (2) the functional change of the receptor molecule (activation). However, occupation and activation are not necessarily mediated by the same group on the peptide molecule, and this has to be taken into account for structure–activity studies. The properties of peptides and their analogs to *occupy* and *activate* the receptors are evaluated in terms of *affinity* and *intrinsic activity*, respectively, and the various analogs can be classified as agonists, partial agonists, and antagonists. Even though the occupation theory is clearly an oversimplification of reality, it has been sufficient, in most cases, to interpret peptide–receptor interactions and related biological phenomena for the sake of receptor characterization.

3. Biological Response and Its Quantification

There are essentially two general categories of methods that can be used to obtain information regarding peptide–receptor interaction. The first category involves bioassays using either whole animals, isolated tissue preparations, or isolated cell systems. The second category is represented mainly by radioligand binding assay techniques.

Bioassays have been, and are still, widely used for the identification and classification of receptors and receptor subtypes. In vitro bioassays are most convenient for generating concentration–response curves for peptides. Several parameters related to the peptide-receptor interaction can be generated from such curves. The most commonly used is the EC_{50} (i.e., the concentration of peptide that produces half-maximal effect), which defines the potency of the peptide in a given system. Another parameter is the maximal effect elicited by the peptide, which is directly related to its intrinsic activity. In structure–activity studies, these parameters are compared for several related peptides interacting with the same receptor.

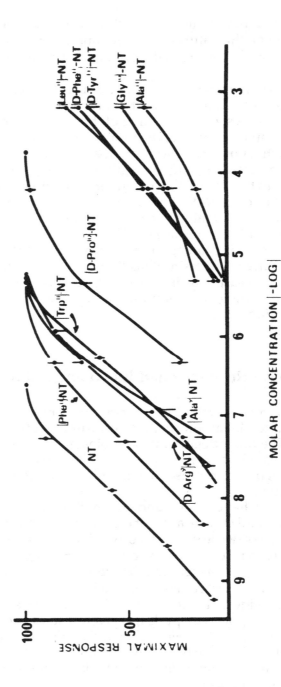

Fig. 1. Doses–response curves of neurotensin and analogs in the spontaneously beating guinea pig atria preparation.

Figure 1 illustrates the concentration–response curves for a series of neurotensin (NT) analogs, in the spontaneously beating guinea pig atria bioassay. It is seen that analogs [Phe^{11}]-, [D-Arg^9]-, [Ala^9]-, [Trp^{11}]-, and [D-Pro^{10}]-NT are full agonists of the parent compound, whereas other analogs with the curves on the right are partial agonists, being active only at high doses without provoking the maximal response in the tissue. The relative potency values obtained from the curves are found in Table 1. Such studies provide direct information about the structural features of the peptide molecule that are essential for activating the receptor. Structure–activity studies with in vitro bioassays also allow for identification of peptides that behave as full or partial agonists or antagonists. Finally, they make it possible to compare receptor specificity in different tissues and species.

Although they are "closer to reality," in vivo bioassays present several inconveniences as compared to in vitro bioassays. Complete dose–response curves are generally difficult to obtain, and they often require the use of a different animal for each peptide dose. The access of peptides to receptor sites in vivo is obstructed by factors like distribution and diffusion in extracellular space, degradation on their way to the receptor by ubiquitous proteases, and elimination from body fluids. The combination of these factors make it impossible to evaluate precisely the effective concentrations of peptide at the receptor level. For this reason, the parameter ED$_{50}$ (half-maximal effective dose) is used to compare drug potencies in vivo rather than EC$_{50}$. Furthermore, the factors mentioned above are likely to vary according to the administration route of the peptides, animal strains, and species, the nature of the anesthetic drug used, and so on. These factors may also affect some peptides differently than others, according to their structure, size, hydrophobicity, and enzyme resistance. Therefore, comparisons of peptide potencies among in vivo bioassays and between in vitro and in vivo bioassays may lead to discrepancies that should be interpreted with care.

The use of binding techniques for the identification of peptide receptors is reviewed in another chapter of this book. Briefly, radioligand binding assays are most useful in providing biochemical data concerning peptide–receptor interactions that precede the biological response. Such information cannot be obtained from bioassays in living tissues that have to be kept intact in physiological media. Such data are better derived from radioreceptor assay techniques. These techniques were first applied to the receptors for

Table 1
Primary Structure and Biological Activity of Neurotensin (NT) Analogs Substituted in Positions 9, 10, and 11 in the Spontaneously Beating Guinea Pig Atria Bioassay

	1	2	3	4	5	6	7	8	9	10	11	12	13	Relative potency
Neurotensin	pGlu	Leu	Tyr	Glu	Asn	Lys	Pro	Arg	Arg	Pro	Tyr	Ile	Leu	100
[D-Arg⁹]-NT	pGlu	Leu	Tyr	Glu	Asn	Lys	Pro	Arg	*Arg*	Pro	Tyr	Ile	Leu	4.1
[Ala⁹]-NT	pGlu	Leu	Tyr	Glu	Asn	Lys	Pro	Arg	*Ala*	Pro	Tyr	Ile	Leu	2.6
[D-Pro¹⁰]-NT	pGlu	Leu	Tyr	Glu	Asn	Lys	Pro	Arg	Arg	*pro*	Tyr	Ile	Leu	0.4
[Gly¹⁰]-NT	pGlu	Leu	Tyr	Glu	Asn	Lys	Pro	Arg	Arg	*Gly*	Tyr	Ile	Leu	0.01
[D-Tyr¹¹]-NT	pGlu	Leu	Tyr	Glu	Asn	Lys	Pro	Arg	Arg	Pro	*Tyr*	Ile	Leu	0.01
[Phe¹¹]-NT	pGlu	Leu	Tyr	Glu	Asn	Lys	Pro	Arg	Arg	Pro	*Phe*	Ile	Leu	15
[D-Phe¹¹]-NT	pGlu	Leu	Tyr	Glu	Asn	Lys	Pro	Arg	Arg	Pro	*Phe*	Ile	Leu	0.01
[Ala¹¹]-NT	pGlu	Leu	Tyr	Glu	Asn	Lys	Pro	Arg	Arg	Pro	*Ala*	Ile	Leu	0.01
[Leu¹¹]-NT	pGlu	Leu	Tyr	Glu	Asn	Lys	Pro	Arg	Arg	Pro	*Leu*	Ile	Leu	0.01
[Trp¹¹]-NT	pGlu	Leu	Tyr	Glu	Asn	Lys	Pro	Arg	Arg	Pro	*Trp*	Ile	Leu	3.2
[D-Trp¹¹]-NT	pGlu	Leu	Tyr	Glu	Asn	Lys	Pro	Arg	Arg	Pro	*Trp*	Ile	Leu	0.01
[Tyr(Me)¹¹]-NT	pGlu	Leu	Tyr	Glu	Asn	Lys	Pro	Arg	Arg	Pro	*Tyr*	*(Me)*Ile	Leu	0.03

insulin (Freychet et al., 1971; Cuatrecasas, 1971) and have been used since then to characterize the receptors for several peptides and other biologically relevant molecules. Moreover, variations in conventional radioligand binding techniques, such as autoradiography and covalent receptor labeling techniques, can be used to obtain information about the precise anatomical distribution of receptors in tissues and about their molecular structure, using biologically active peptides and analogs.

The main advantage of bioassays over binding assays is that bioassays can provide information about functional receptors by measuring the response that follows the peptide–receptor interaction, whereas binding assays can only measure the interaction itself. It has been shown that a number of cellular events generally take place between the moment of initial activation at the receptor level and the response measured by the bioassay. Because of the tissue heterogeneity in most bioassay preparations, and more remarkably in vivo, the peptide being evaluated frequently initiates its effect by interacting with receptors located on a group of cells distinct and/or distant from that producing the measured response. This so-called mediated effect occurs when the first group of cells releases, in response to the peptide, an endogenous substance capable of activating the second group. Using appropriate techniques and pharmacological tools, such as specific blockers and antagonists, it is often possible to identify the major events taking place between receptor activation and the biological response. The complexity of the bioassay system becomes in this case an important source of information about receptor localization and the mechanisms that couple the receptor activation to the biological response.

It is the purpose of this chapter to demonstrate how structure–activity studies of a given peptide can be used to construct the reciprocal image of its receptor. In most cases, biological data obtained from either bioassays or radioligand binding techniques will be reported merely as part of structure–activity studies, without further discussion on the mechanism of action. The reader is invited to refer to specialized works for more details on these techniques. The exhaustive review recently published by Kenakin (1984) should be consulted for information on the practical use of bioassays. Another chapter of this book is a review on binding studies.

4. Structure–Activity Studies of Peptides

Important information on the structural requirements of the receptor for a given peptide substrate can be obtained by designing chemical modifications of that peptide and evaluating the conformation and biological activity of the synthetic compounds. Prior to any synthetic work, the study of homologies within certain families of natural peptides can provide the basis for further structure–activity studies with analogs. This is the case, for instance, with the tachykinin family, which includes mammalian substance P, substance K, and batracian physalemin and eledoisin; the vasopressin and oxytocin family, with Arg-vasopressin, Lys-vasopressin, Arg-vasotocin, oxytocin, and several other natural members; the closely related CRF-sauvagine-urotensin family, formed of three peptides of mammalian, batracian, and fish origins; or the secretin family, in which peptides found in the digestive system of mammals, e.g., secretin, VIP, GIP, and glucagon, have profound sequence homologies.

Four main approaches are generally used for structure–activity studies of peptides. For clarity's sake, these approaches will be identified in the text as the "random" or "empirical" approach, the "predictive" approach of Chou and Fasman (1974), the "systematic" approach, inspired by Josef Rudinger's review article on the design of peptide hormone analogs (1971), and, more recently, the "amphiphilic secondary structure" approach, based on Emil Kaiser's theory of peptide–membrane interaction (Kaiser and Kezdy, 1984). Even though these methods share several similarities, they are based on different principles, and their application is described in the following pages.

4.1. Empirical Approach

The most generally conceived approach to evaluating the structure–activity relations of a new peptide can be fairly defined by the question, "Let's see what happens if . . .", which corresponds to a purely empirical method of exploring the functionality of a peptide. Although this approach has been, and is still, widely applied in several peptide chemistry laboratories, it will not be further discussed here, since it is not design, by definition. However, it must be recognized that chance observations have often served as starting points for lines of thought, and design, independent of the original concept.

4.2. Chou and Fasman's Predictive Method

In 1974, Chou and Fasman published a method that can be used to predict the secondary structure of a peptide or a protein from its known amino acid sequence, using statistical parameters related to the conformational preferences of amino acids in peptides and proteins of known conformation. The numerical values assigned to each amino acid side chain are directly related to its potential to be found in α-helical (P_α), β-sheet (P_β), or β-turn (f_i, f_{i+1}, f_{i+2}, f_{i+3}) regions of peptides or proteins. These values have been calculated by the authors using an exhaustive compilation of the frequencies of occurrence of amino acids in sequences corresponding to those conformations in several crystalline proteins analyzed by X-ray crystallography. Moreover, empirical rules have been established to help delineate the boundaries of the different structural zones.

The search for α-helical and β-sheet regions can be made by using the P_α and P_β values shown in Table 2, together with a set of empirical rules, where:

$$<P> = (P_1 + P_2 + \ldots + P_n)/n \qquad (1)$$

A summary of these rules for the determination of α-helix and β-sheet, respectively, is given below; the reader should consult Chou and Fasman's paper (1974) for further details.

1. Locate clusters of four helical residues out of six residues;
2. Extend the helical segment in both directions until terminated by tetrapeptides with $<P_\alpha> \leq 1.00$;
3. Any segment of six residues or longer in a native protein with $<P_\alpha> \geq 1.03$, as well as $<P_\alpha> > <P_\beta>$, is predicted as helical.

and

1. Locate clusters of three β residues out of five residues;
2. Extend the β segment in both directions until terminated by tetrapeptides with $<P_\beta> < 1.00$; any segment of five residues or longer in a native protein with $<P_\beta> \geq 1.05$, as well as $<P_\beta> > <P_\alpha>$, is predicted as β-sheet.

Another set of rules and P values have been established for the finding of β-turn regions in peptides and proteins. Since the forma-

Table 2

Assignment of the Helical and Beta-Sheet-Forming Potential of Amino Acids, According to Chou and Fasman (1974)

Helical residues	P_α	Beta-sheet residues	P_β
Glu	1.53	Met	1.67
Ala	1.45	Val	1.65
Leu	1.34	Ile	1.60
His	1.24	Cys	1.30
Met	1.20	Tyr	1.29
Gln	1.17	Phe	1.28
Trp	1.14	Gln	1.23
Val	1.14	Leu	1.22
Phe	1.12	Thr	1.20
Lys	1.07	Trp	1.19
Ile	1.00	Ala	0.97
Asp	0.98	Arg	0.90
Thr	0.82	Gly	0.81
Ser	0.79	Asp	0.80
Arg	0.79	Lys	0.74
Cys	0.77	Ser	0.72
Asn	0.73	His	0.71
Tyr	0.61	Asn	0.65
Pro	0.59	Pro	0.62
Gly	0.53	Glu	0.26

tion of β-turns involves four amino acids (Venkatachalam, 1968), the relative probability that a tetrapeptide will form a β-turn is the following:

$$<P_t> = f_i \cdot f_{i+1} \cdot f_{i+2} \cdot f_{i+3} \qquad (2)$$

where the f_I values are, respectively, the frequency of occurrence for a certain residue at the first, second, third, and fourth position of a β-turn. These values are given in Table 3. $<P_t>$ values equal to or greater than 0.5×10^{-4} have been found to correspond to reasonable cut-off values in predicting β-turns.

Even though these rules and the corresponding P values are purely empirical, they have been used extensively by peptide chemists to evaluate the secondary structure of several peptides

Table 3
Assignment of the β-Turn-Forming Potential of Amino Acids, According
to Their Position, as Calculated by Chou and Fasman (1974)

Amino acid	β-Turn forming potential			
	f_i	f_{i+1}	f_{i+2}	f_{i+3}
Ala	0.049	0.049	0.034	0.029
Arg	0.051	0.127	0.025	0.101
Asn	0.101	0.086	0.216	0.065
Asp	0.137	0.088	0.069	0.059
Cys	0.089	0.022	0.111	0.089
Gln	0.050	0.089	0.030	0.089
Glu	0.011	0.032	0.053	0.021
Gly	0.104	0.090	0.158	0.113
His	0.083	0.050	0.033	0.033
Ile	0.068	0.034	0.017	0.051
Leu	0.038	0.019	0.032	0.051
Lys	0.060	0.080	0.067	0.073
Met	0.070	0.070	0.036	0.070
Phe	0.031	0.047	0.063	0.063
Pro	0.074	0.272	0.012	0.062
Ser	0.100	0.095	0.095	0.104
Thr	0.062	0.093	0.056	0.068
Trp	0.045	0.000	0.045	0.205
Tyr	0.136	0.025	0.110	0.102
Val	0.023	0.029	0.011	0.029

and to modify natural peptides with intent to obtain better agonists
or antagonists. More recently, they have also been applied to
predict the antigenic segments of proteins, the zones of occurrence
of peptide–membrane interaction, the catalytic site of enzymes,
and other purposes. Even though several other authors have tried
to further refine Chou and Fasman's parameters and rules, the
original method remains the one most often used to predict the
conformation of peptides.

4.3. Systematic Approach of Rudinger

A set of systematic rules for the design of peptide hormone
analogs was provided by Josef Rudinger in 1971, based primarily

on his work in the oxytocin-vasopressin field. Since then, these rules have been intuitively and/or rigorously applied to the exploration of receptor sites for several peptides by various authors. Briefly, it is recommended that changes in peptide primary structure should be made initially one at a time, and that the native hormone or highly active analogs should be used as the basis for further structural change. An extension of this "rule of small changes" proposes that, in order to separate steric from chemically functional effects, structural changes should be made at first isosteric or isofunctional whenever possible. However, these rules do not imply that drastic structural changes should never be made. Especially in the initial stages of the investigation, such changes, made with minimal synthetic effort, can help to economically map out the lines or limits of more detailed structural exploration. Rudinger's approach for the design of peptide analogs is summarized and illustrated by several examples in the next pages.

4.3.1. Shortening at the Amino and Carboxyl Ends and Chain Extensions

For several peptides, series of C- and N-terminal fragments of decreasing length obtained through total synthesis or enzymatic degradation have been useful in pinpointing the portion of the chain recognized by the receptor. Early work (Bell et al., 1956) had demonstrated that the 24-amino-acid fragment obtained by pepsin digestion from natural porcine corticotropin (39 amino acids) was active. Also, fragment 1-34 of parathyroid hormone (Rasmussen, 1960), fragment 26-33 of cholecystokinin (Ondetti et al., 1970), and the C-terminal pentapeptide of gastrin (Anderson et al., 1964) were shown to display equal or even higher activity than the native peptide. More recently, synthetic work has demonstrated that C-terminal fragments 8-13 of neurotensin (St-Pierre et al., 1981) (Table 1), 18-27 of the gastrin-releasing peptide (Girard et al., 1984), or 27-33 of cholecystokinin (Crawley et al., 1984) contained the portion of the chain responsible for biological activity. On the other hand, removal of only one residue at the C-terminal end of these peptides significantly depresses their activity.

Similarly, N-terminal fragments of several peptides were found to be fully active. This is the case with opiate peptides Met- and Leu-enkephalin, which represent the active portion of beta-endorphin (or lipotropin$_{61-91}$) (Cox et al., 1976) and of dynorphins (Goldstein et al., 1979), respectively. Furthermore, C-terminal deletion analogs of beta-endorphin displayed inhibitory properties of the parent compound (Nicolas and Li, 1985). PTH_{1-34}, the *N*-terminal portion of parathyroid hormone, is fully active (Rosenblatt, 1984). Also, the *N*-terminal portion 1-29 of growth hormone releasing factor (Ling et al., 1985) bearing an amide function in *C*-terminal was shown to be nearly equipotent to GRF_{1-44} on a molar basis (Grossman et al., 1984). As noted in the above paragraph, none of these peptides can afford to lose one single residue in *N*-terminal without also losing most of its activity. The two situations above, in which biologically important chemical groups are located in definite portions of the chain, are illustrated in the top portion of Fig. 2.

Several small peptides such as bradykinin (Park et al., 1978; Regoli and Barabe, 1980) and angiotensin II (Regoli et al., 1974) seem to contain a critical number of residues in their sequence, since they cannot afford any further removal of amino acids, at either end of the chain, without suffering severe losses of biological activity. An illustration of this situation can be found in the lower part of Fig. 2, in which two distant parts of the chain are needed to activate the receptor. In the case of larger peptides such as glucagon (Hruby et al., 1981), VIP (Fournier et al., 1984), secretin (Voskamp et al., 1982), beta-endorphin (Ling and Guillemin, 1976), and PTH_{1-34} (Rosenblatt, 1984), the *N*-terminal residue appears to be critical for intrinsic activity, although more than one amino acid at the *C*-terminal end is important for affinity, as shown with $glucagon_{1-28}$ (Eugland et al., 1982) and PTH_{1-27}. Some of these observations might find an explanation in Kaiser's amphiphilic secondary structure theory that will be discussed in the next section.

4.3.2. Elision and Intercalation of Residues

Shortening of the peptide chain by omission (elision) of residues from within the sequence can have drastic effects on the steric relations of the side chains in the regions remaining on either side of the site of elision. It could be expected that in most cases such "frame shifts" lead to significant losses of activity. These changes can alter the interaction of the peptide with the receptor by perturbing hydrogen bonding, ion pairing, and global lipophilic or hydrophilic capacities. This observation is particularly true for linear peptides that cannot afford the removal of residues at either end of the chain, such as the examples mentioned above. For instance, among analogs of bradykinin with proline residues missing from positions 7, 2, or both 7 and 2, only the des-Pro^2 analog

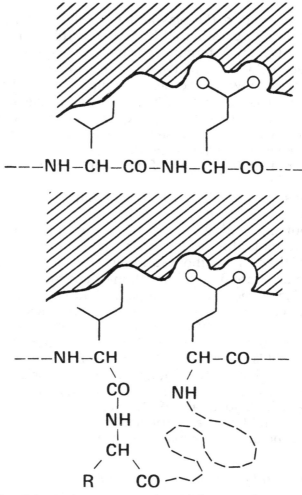

———NH–CH–CO–NH–CH–CO—·——

———NH–CH   CH–CO———
          CO   NH
          NH
          CH
      R   CO

Fig. 2. Schematic representation of the peptide–receptor interaction. (Top panel) Side chains involved in the interaction with the receptor belong to adjacent residues. (Lower panel) Side chains involved in the interaction belong to residues located in distant sections of the chain.

had some remaining bradykinin-like activity (Boissonnas et al., 1960). Obviously, in the case of other peptides that can tolerate removal of residues at one end or the other, elisions performed in the nonessential sequences will generally not affect biological activity. However, a similar decrease of potency as observed with the other group of peptides generally occurs when elisions are performed in the active region.

However, selected elisions of residues in the cyclic peptide somatostatin and linear beta-endorphin have given rise to interesting analogs in which biological activity was conserved or even increased. The successful design of cyclic somatostatin hexapeptides (Veber et al., 1984), which is the fruit of several years of structure–activity studies with the cyclic tetradecapeptide by several laboratories, is now the classical example of how "useless" portions of a peptide can be eliminated while preserving its biologically active conformation, provided that this conformation has been precisely determined. Since this modification of the peptide also involves substitution of side chains, it will be further discussed in the next section.

Elisions of single residues of Glu^{11}, Pro^{13}, Leu^{14}, Val^{15}, Asn^{20}, or Ile^{22} in beta-endorphin result in equipotent or slightly more potent analogs in vitro, [des-Ile^{22}]-endorphin being the most potent (122%) in the guinea pig ileum bioassay. The simultaneous removal of Glu^{11}, Leu^{14}, Asn^{20}, and Ile^{22} also gives rise to a better analog in vitro (138%). However, the potency of all the analogs was found to be lower than the 31-residue beta-endorphin in vivo (7–79%) in the analgesia test (Li et al., 1980).

The opposite operation to elision, that is the intercalation of amino acid residues within an active sequence, can also cause important "frame shifts" of the biologically important chemical groups and destabilize the interaction with the receptor. The few examples we have found showed no improvement, compared to the native peptide. For instance, although intercalation of a Gly residue in position 4 of bradykinin gives a fairly active analog (Nicolaides and Lipnik, 1966), doubling of the Tyr in angiotensin II (Riniker and Schwyzer, 1961) caused extensive or complete loss of activity.

4.3.3. Modification of Side Chains

Modification of side chains is by far the most popular approach to the modification of peptides for the purpose of structure–activity studies. In the general procedure recommended by Rudinger, replacements of target amino acids in a given sequence are made one at a time, after a preliminary study using C- and N-terminal fragments has pinpointed which portion of the chain is essential for biological activity. The reactive (chemically functional) side chain substituents (carboxyl, amine, hydroxyl, imidazole, indole), including the terminal amino and carboxyl groups, are naturally the first targets, since they are frequently directly responsible for interacting with the receptor. The role of such groups can be examined by omission or substitution.

Omission leaves the backbone intact, although the reactive group in the side chain is replaced with a neutral one, such as Ala or Gly. Replacement by Gly is equivalent to elimination of the whole side chain, whereas replacement by Ala eliminates everything beyond the beta-carbon atom. Because the steric properties of Gly are drastically different from those of all other amino acids, the use of a chiral residue such as Ala rather than Gly should give a better approximation of the original conformation of the peptide backbone. Selectively omitting functional side chains or end groups is usually more instructive than the omission of internal residues, discussed in the above section.

In general, substituting one side chain with another in a peptide can have either steric or functional effects. The former is related to the conformation or the conformational freedom of the peptide molecule and/or to the alteration of the topography of the molecular surface. Functionality is taken to mean not only reactivity, but also charge, hydrogen-bonding capacity, and lipophilic or hydrophilic properties. As mentioned above, Rudinger suggested that substitutions of side chains should be made, the first time, as "isosteric" or "isofunctional" as possible, in order to perform systematic mapping of the receptor. It is, of course, recognized that precisely isosteric relations are hardly ever achieved and that functionality in this context is a relative term. For example, the relevant functionality of a lysine side chain might, in different situations, be its charge, the hydrophilic character or hydrogen-bonding capacity of the amino group, or even the lipophilic properties of the four methylene groups. An examination of functionality will therefore often involve a series of graded substitutions. It is obvious that this type of approach will more often than not ultimately require the use of specially designed and prepared "nonnatural" amino acids, and therefore will be limited to the peptide synthesis laboratory with the required knowledge in organic chemistry, unless the derivatives become commercially available. In the following paragraphs, the relation between amino acid side chains will be considered from that point of view, and their respective contribution to the functional identification of the corresponding chemical functions on the receptor will be discussed.

4.3.3.1. BASIC AMINO ACIDS: LYS AND ARG. Replacement of Arg by other basic side chains, including Lys and ornithine, has been a popular structural modification in synthetic analogs. Lys and Arg are approximately isofunctional: Both side chains are protonated in the physiological pH range, though, of course, Arg is a much stronger base. They are also similar sterically in that both place the positive charge at about the same distance from the peptide chain, but the guanidino group is larger than the aminomethylene group replacing it in Lys. For this reason, as shown in Fig. 3, citrulline, isosteric with Arg but uncharged, and ornithine, with the same functional group as Lys, but a shorter side chain, should be included for more detailed comparisons. Other nonnatural amino acids and amino acid derivatives have been examined, including α, γ-diaminobutyric acid, homoarginine, homolysine, nitroarginine, and N-ϵ-trimethyllysine salts. Also, δ-guanidino-valeryl, δ-nitroguanidino-valeryl, ϵ-trimethylammoniocaproyl, and glycyl-glycyl can be used as substituents to either Lys or Arg positioned in the N-terminal.

The substitution of Arg[9] by Lys in bradykinin reduces the biological activity to 2% in vitro (rabbit jugular vein) and to less than 1% in vivo (rat blood pressure) (Regoli and Barabe, 1980). This substitution is less destructive in neurotensin at position 8, slightly altering the potency of [Lys[8]]-NT to 90% of NT in the guinea pig atria bioassay, whereas [Lys[9]]-NT retains 20% potency in the same test (St-Pierre et al., 1984). Replacing Arg[2] with Lys, ornithine, or citrulline in angiotensin II brings the potency down to 10, 20 (Regoli et al., 1974), and 2% (Schattenkerk and Havinga, 1965), respectively, in the rat blood pressure bioassay. In somatostatin, replacing Lys[4] with L-Phe generates a selective inhibitor of growth hormone release (Meyers et al., 1980).

4.3.3.2. ACIDIC AMINO ACIDS: ASP AND GLU.   The aminodicarboxylic acids (Glu and Asp) and their amides (Asn and Gln) form a convenient natural set in which the carboxamide and carboxyl groups are nearly isosteric and the length of the side chain varies by one methylene group, as shown in Fig. 4. The amides have been used to replace the acids in several cases, probably because they are more readily available and synthetically more convenient than the corresponding benzyl or *t*-butyl esters that are the usual starting materials in work with the acids. The side chain can be shortened by substitution with α-aminomalonic acid (removal of a methylene group) or alanine (carboxyl group). Approximately isosteric neutral side chains can be obtained by substitution with serine (hydroxyl) or β-cyanoalanine (cyano). A stronger negative group can be obtained with cysteic acid (sulfonic acid).

4.3.3.3. HISTIDINE.   Histidine is unique among protein-

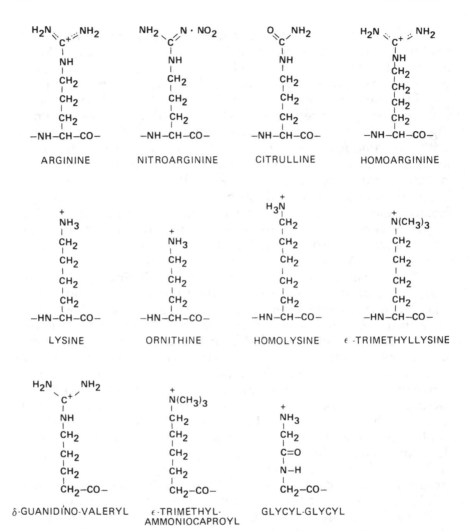

Fig. 3. Some of the possible substitutions for basic amino acids.

constituent amino acids because of the special properties of its imidazole group, e.g., its pK and its imide reactivity. His is implicated in the catalytic function of a number of enzymes and several examples exist in which it is essential for the biological activity of the peptides in which it is found. However, it cannot be replaced isosterically or isofunctionally by other natural amino acids. Derivatives such as β-(3-pyrazolyl)-alanine and β-(1-pyrazolyl) alanine have been designed to evaluate the importance

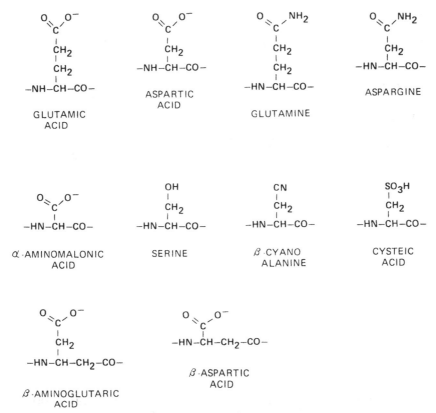

Fig. 4. Some of the possible substitutions for acidic amino acids.

of imide function in the ring (Fig. 5). Analogs of histidine resulting from the methylation of the π or τ nitrogens of the imidazole ring have also been prepared.

N-Terminal histidine appears to be of particular importance in polypeptides of the secretin-glucagon-VIP family. In secretin, replacement of His[1] by isosteric β-(3-pyrazolyl)-alanine, β-(2-thienyl)-alanine, (N-τ-methyl)-histidine, or alanine drastically reduces the stimulation of the exocrine pancreatic secretion in the rat to 6% for the former and to less than 1% for the others (Voskamp et al., 1982). On the other hand, changing His[1] to 3-methyl-His or to Phe in glucagon was less detrimental than in secretin, since the activation of rat liver adenylate cyclase decreased to only 30% as compared to the native peptide (Sueiras-Diaz et al., 1984). In the tripeptide TRH, chemical modification of the imidazole ring appears to be critical for TSH and prolactin release: [(N-τ-methyl)-

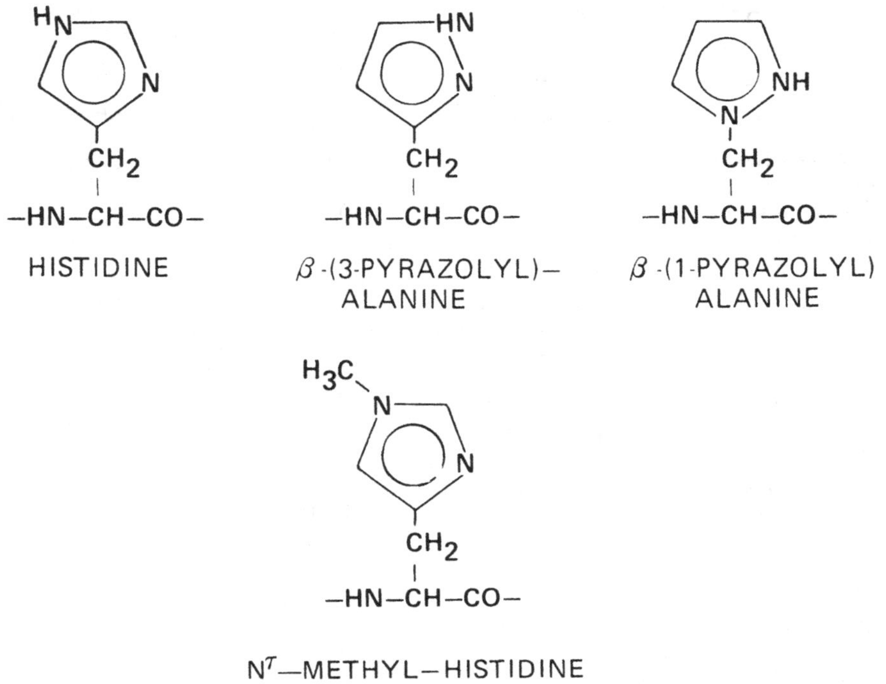

HISTIDINE

β-(3-PYRAZOLYL)–
ALANINE

β-(1-PYRAZOLYL)
ALANINE

N^7—METHYL—HISTIDINE

Fig. 5.   Some of the possible substitutions for histidine.

His^2]-TRH has 3–10 times the activity of the natural product, whereas [(N-π-methyl)-His^2]-TRH is only 0.1% active. Similarly, the histidine analog β-(3-pyrazolyl)-alanine increases the activity of TRH in rats and mice, and β-(1-pyrazolyl)-alanine decreases it to 5% (Schally et al., 1978).

4.3.3.4. AROMATIC AMINO ACIDS: TYR, PHE, AND TRP.   The substitution of Phe for Tyr represents omission of a reactive hydroxyl group. However, for a proper understanding of the consequences of Phe-Tyr replacements, it is necessary to separate steric and functional effects, and among the latter to distinguish between the properties of the hydroxyl group as an acidic, a hydrogen-bonding, or a reactive substituent, as well as its possible effect on the electron density of the aromatic ring. By a proper choice of suitably substituted derivatives, it may often be possible to do this.

A wide variety of ring substituents have been developed through the years, as shown in Fig. 6. Best known are polar groups such as halogens (F, Cl, Br, I), nitro, amino, methyl, or ethyl ethers located in the *para* position of the ring. The use of L-dopa (3,4-dihydroxyphenylalanine) as a substitute for Tyr can give good

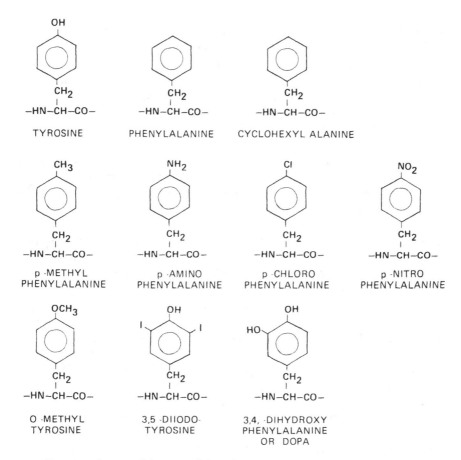

Fig. 6. Some of the possible substitutions for aromatic amino acids.

indication of the involvement of the *para* hydroxyl group in hydrogen bonding with the receptor. Neutral *p*-methyl-or *p*-ethyl-phenylalanine have also been used. The importance of the aromatic ring has been evaluated using cyclohexyl- or cyclopentyl-alanine, or natural aliphatic amino acids such as Leu, Ile, or Val. In most cases, aliphatic side chains cannot replace aromatic side chains without causing important losses of activity. However, antagonists have arisen from such a substitution in the cases of angiotensin II (Khosla et al., 1972; Regoli et al., 1974) and des-Arg[9]-bradykinin (Regoli and Barabe, 1980).

Tyrosine in position 11 of neurotensin is particularly sensitive to substitution, suggesting its strong involvement at the receptor level. Removing the hydroxyl group by substituting Phe results in surprising losses of potency in vitro, in the rat stomach strip (16%) and the guinea pig atria (15%) bioassays, whereas central in vivo properties (hypothermia, locomotor activity) of the analog are similar to NT following intracerebroventricular (icv) injection. However, the use of any aliphatic substituent such as Ala, Leu, Ser, cyclohexylalanine, or electronegative *para*-substituents of the aromatic ring abolishes biological activity, in vivo and in vitro (St-Pierre et al., 1984).

Growth hormone releasing factor was also shown to have an extremely sensitive N-terminal Tyr residue. Substitution of Tyr with Phe, Trp, O-methyl-Tyr, or Ala causes a drop in the GHRH stimulation of the production of GH by dispersed pituitary cells to 4% for the former and less than 1% for the others. Surprisingly, His is an acceptable substituent for Tyr in that position, being the natural N-terminal residue in murine GHRH. The resulting analog displays 35% of the human GHRH activity (Ling et al., 1985).

Phe is a critical residue in several peptides. In bradykinin, the substitution of either Phe[5] or Phe[8] alters the affinity of the peptide in one sense or the other, as shown in Table 4. In des-Arg[9]-bradykinin, its replacement by isosteric aliphatic residues such as Leu or cyclohexylalanine (Cha) gives rise to specific antagonists of the B_1-receptor of kinins (Regoli and Barabe, 1980). In PTH$_{1-34}$, replacing Phe[34] by Tyr creates a superagonist (140%) in the rat renal adenylate cyclase activity bioassay (Rosenblatt, 1984).

Tryptophan is also often found in the biologically active region of peptides, since the indole ring has quite unique aromatic properties. Its replacement by other aromatic amino acids, such as Tyr or Phe, or by aliphatic side chains, is rarely the cause of an increase of activity, more frequently destroying it partly or entirely. This is the case for accidental modifications of the ring resulting from side reactions in the course of synthesis, which have given rise mainly to alkylated or oxidized Trp. On the other hand, some substituents of the indole ring shown in Fig. 7 have been prepared for structure–activity studies. Methylation of the ring in positions 2, 4, 5, 6, and 7 has given rise to low-activity analogs in position 3 of LHRH (11, 6, 4, 2, and 7%, respectively) (Yabe et al., 1979). Naphthylalanine is generally an acceptable substituent for Trp, even though it is lacking the indole nitrogen.

4.3.3.5. ALIPHATIC AMINO ACIDS: LEU, ILE, VAL, ALA, MET.   Replacements of aliphatic amino acids have often been confined to the naturally occurring Ala, Val, Leu, and Ile. Comparison of Ile with Val and Ala can, indeed, give information about the im-

Table 4
Relative Potencies of Bradykinin Analogs Substituted in the Aromatic
Positions 5 and 8 In Vitro and In Vivo, as Compared to Bradykinin[a]

Compound	Bioassay	
	Rabbit jugular vein, relative affinity	Rat blood pressure, relative potency
Bradykinin (BK)	100	100
[D-Phe[5]]-BK	3.6	2.3
[Tyr[5]]-BK	0.58	0.64
[Trp[5]]-BK	0.74	0.38
[Cha[5]]-BK	49	43
[Tyr(Me)[5]]-BK	30	15
[Leu[5]]-BK	17	1.9
[D-Phe[8]]-BK	0.72	6.6
[Tyr[8]]-BK	7.80	5.0
[Trp[8]]-BK	24	125
[Cha[8]]-BK	49	100
[Tyr(Me)[8]]-BK	140	270
[Leu[8]]-BK	0.20	0.23
[Tyr(Me)[5,8]]-BK	42	56
[Cha[5,8]]-BK	0.47	17.5

[a]From Regoli and Barabe, 1980.

portance of side chain length, and comparison of Leu with Val and Ile can provide information on the role of beta-branching. For instance, substituting Leu[5] in dynorphin decreases its potency to 4% of that of the native peptide in vitro (guinea pig ileum), whereas substitution of Ile[8] by Ala (50% potency) is more acceptable (Turcotte et al., 1984). In substance P, the hydrophobic character of C-terminal Leu and Met aliphatic side chains appears to be important for the affinity of the peptide with the receptor, since [Ala[10]]- and [Ala[11]]-substance P have, respectively, 15 and 0.8% relative potency in the electrically stimulated rat vas deferens bioassay (Couture et al., 1979). Other nonnatural amino acids such as alpha-aminobutyric acid, norvaline, and norleucine, as shown in Fig. 8,

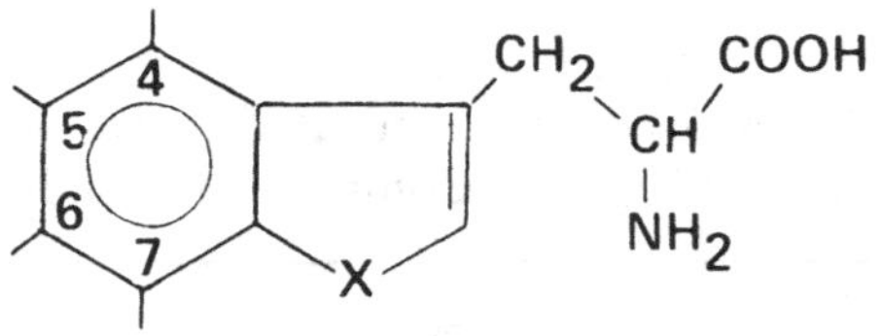

X = NH   TRYPTOPHAN
X = O   BENZOFURYLALANINE
X = S   BENZOTHIENYLALANINE
X = N–CH$_3$ N^{in} - METHYLTRYPTOPHAN
X = NH; 5-FLUORO   5 - FLUOROTRYPTOPHAN
X = NH; 4,5,6,7-FLUORO   4,5,6,7 - TETRAFLUOROTRYPTOPHAN

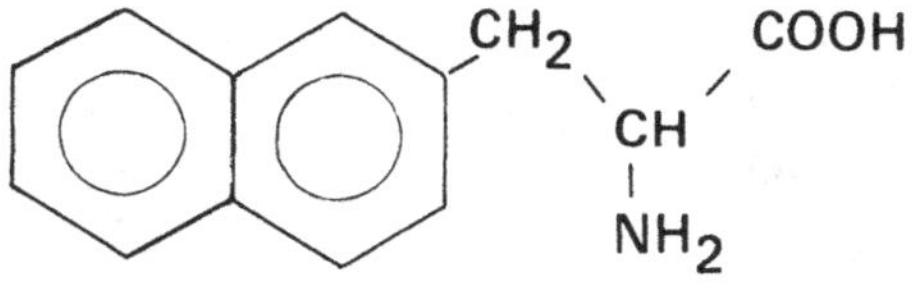

NAPHTYLALANINE

Fig. 7.   Some of the possible substitutions for tryptophan.

can give complementary information. Also, synthetic amino acids with a cyclic aliphatic side chain, such as cyclopentyl- or cyclohexylalanine, are often used as isosteres of Phe, as well as of Val or Leu.

Replacement of Met for norleucine (Nle) is usually inconsequential as far as biological activity is concerned. [Nle^{11}]-substance P maintains more than 50% potency of the native compound in most substance P bioassays in vitro and in vivo (Fournier et al., 1982). Double substitution of Met^8 and Met^{18} in the Tyr^{34} amidated analog of PTH$_{1-34}$ (300% potency of PTH$_{1-34}$ in the rat renal adenylate cyclase bioassay) increases the potency to 190% in the new analog (Rosenblatt, 1984). On the contrary, this substitution has been shown to improve the stability of several Met-containing peptides toward oxidizing agents and to facilitate their synthesis by decreasing the formation of side products. This is particularly true for the preparation of synthetic peptide pre-

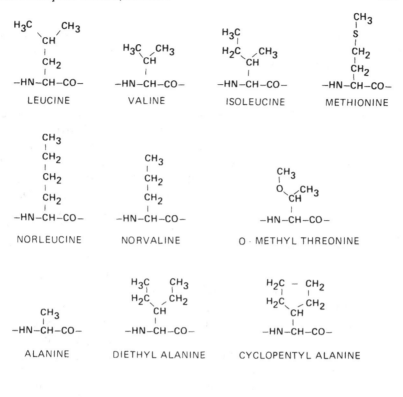

Fig. 8. Some of the possible modifications for aliphatic amino acids.

cursors for radioisotope labeling with ^{125}I, which can involve the use of oxidative reactions. However, aliphatic substituents with a branched side chain such as Leu, Val, or Ile are less acceptable to replace Met. On the other hand, the loss of the methylmercapto group can be evaluated by substituting alpha-aminobutyric acid for Met, and it often leads to a decrease of the biological activity; this suggests, in this case, the importance of the integral length of the hydrocarbon side chain for hydrophobic interaction with the receptor.

4.3.3.6. DISULFIDE BRIDGES: THE ROLE OF CYS.   Cysteine, and more particularly the disulfide bridges that it forms with congeners located elsewhere in the same chain, as in oxytocin and somatostatin, or in another chain of the molecule as in insulin, is usually involved in the stabilization of the tridimensional structure of cyclic peptides. Therefore, any substitution that gives rise to destruction of such bridges often abolishes biological activity partly or completely. Linear analogs can be obtained by substituting Cys with isosteric Ala or Ser, or by reducing and alkylating Cys-containing natural or synthetic peptides. The linear forms of monocyclic somatostatin, oxytocin, calcitonin, and atrial natriuretic factor, or polycyclic *Escherichia coli* heat-stable enterotoxin and *Conus geographus* toxin are all inactive, even at high doses. On the other hand, replacement of the disulfide bridge by isosteric carba or dicarba bonds (illustrated in Fig. 9) usually results in preservation of biological activity. Most of the results found in the literature suggest that Cys sulfur atoms themselves make little or no contribution to the mechanism of action of most cyclic peptides.

4.3.3.7. HYDROXYL SIDE CHAINS: SERINE AND THREONINE.   The importance of the hydroxyl group in Ser and Thr can be evaluated either by omission, using Ala, or by transforming the hydroxyl into an alkyl ether. The free SH group of Cys is useful for replacing the hydroxyl function, especially to gain information on the importance of the electron donor properties of the oxygen atom in the interaction with the receptor. No example exists, to our knowledge, in which isosteric substitution of Ser or Thr, aiming at eliminating the hydroxyl function, causes an important loss of activity.

4.3.3.8. CHAIN REVERSAL: THE ROLE OF PROLINE.   Although proline has no functional group, it is usually located in critically important regions of peptides because of the determinant conformational role of its pyrrolidone ring, which induces the folding of the chain on itself by means of various forms of turns. Pro is, indeed, found in most of the turns identified thus far in proteins by X-ray crystallography (Chou and Fasman, 1974). Its replacement by less rigid amino acids such as Gly or Ala in several biologically active peptides usually gives rise to important losses of biological activity, which have been interpreted as being the result of profound steric changes on both sides of the substitution. Synthetic analogs [Gly[10]]-neurotensin, [Ala[2]]-, [Ala[3]]-, or [Ala[7]]-bradykinin, or [Acpc[7]]-angiotensin II, which result from the replacement of a Pro residue, are all weak agonists of the reference peptide. Only a few cases have been reported, such as [Ala[2]]- and [Ala[4]]-substance

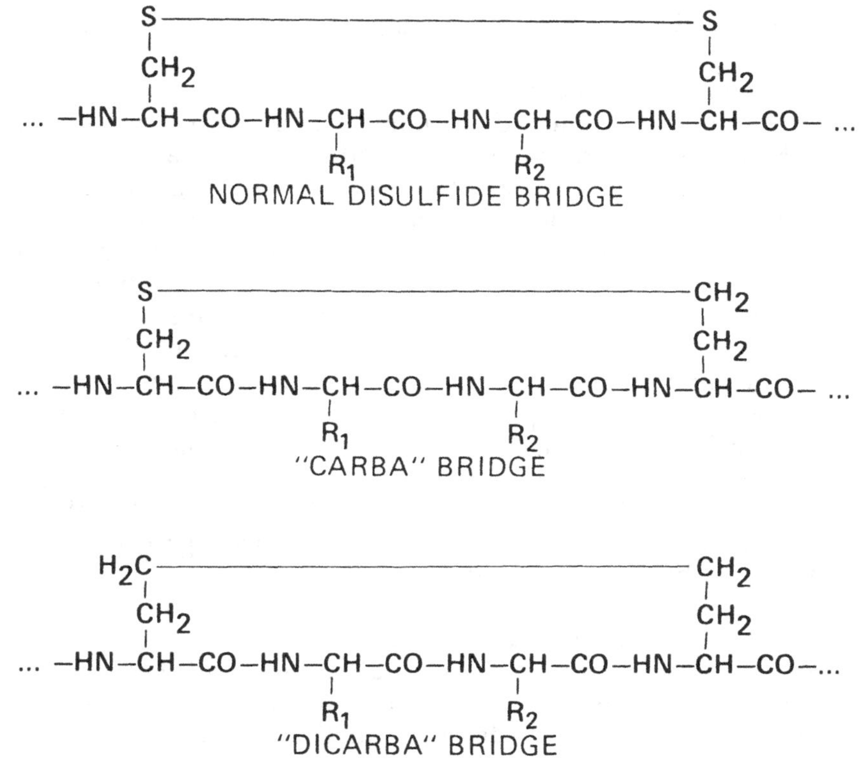

$$S\text{————————————}S$$
$$|\qquad\qquad\qquad\qquad\qquad\qquad\qquad|$$
$$CH_2\qquad\qquad\qquad\qquad\qquad CH_2$$
$$|\qquad\qquad\qquad\qquad\qquad\qquad\qquad|$$

... —HN—CH—CO—HN—CH—CO—HN—CH—CO—HN—CH—CO— ...

$$\qquad\qquad|\qquad\qquad\quad|$$
$$\qquad\qquad R_1\qquad\qquad R_2$$

NORMAL DISULFIDE BRIDGE

$$S\text{————————————}CH_2$$
$$|\qquad\qquad\qquad\qquad\qquad\qquad\qquad|$$
$$CH_2\qquad\qquad\qquad\qquad\qquad CH_2$$
$$|\qquad\qquad\qquad\qquad\qquad\qquad\qquad|$$

... —HN—CH—CO—HN—CH—CO—HN—CH—CO—HN—CH—CO— ...

$$\qquad\qquad|\qquad\qquad\quad|$$
$$\qquad\qquad R_1\qquad\qquad R_2$$

"CARBA" BRIDGE

$$H_2C\text{————————————}CH_2$$
$$|\qquad\qquad\qquad\qquad\qquad\qquad\qquad|$$
$$CH_2\qquad\qquad\qquad\qquad\qquad CH_2$$
$$|\qquad\qquad\qquad\qquad\qquad\qquad\qquad|$$

... —HN—CH—CO—HN—CH—CO—HN—CH—CO—HN—CH—CO—...

$$\qquad\qquad|\qquad\qquad\quad|$$
$$\qquad\qquad R_1\qquad\qquad R_2$$

"DICARBA" BRIDGE

Fig. 9. Some of the possible modifications for disulfide bridges.

P analogs, in which the substitution of Pro by a more flexible residue does not cause a significant loss of activity. The latter is not surprising, since the biologically significant chemical groups of substance P are located in the 5-11 portion of the sequence (Fournier et al., 1982). On the other hand, Pro-induced backbone rigidification can be mimicked, as we shall see in a further section, by introducing conformational restrictions in the chain, using *N*-substituted amino acids, lactams, cyclization, and other modifications.

4.3.3.9. CHAIN FLEXIBILITY: THE ROLE OF GLYCINE.   In contrast to the case with Pro, it is recognized that Gly plays an essential function in peptide chains by providing the flexibility required for proper orientation of critical chemical groups *vis-a-vis* the receptor. The important conformational role of Gly residues in peptides is well illustrated with the enkephalins, in which the substitution of Gly^2 and Gly^3 by the slightly more rigid residue L-Ala decreases the receptor affinity of the resulting analogs to 3 and 6%, respectively, in synaptic plasma membranes (Terenius et al., 1976). The use of Ser, alpha-aminobutyric acid, or Phe as substituents for any of these residues is even more detrimental to activity (Agarwal et al., 1977). However, substitutions of Gly with residues that can stabilize a given conformation are at the origin of the design of several superactive peptide analogs, as shown in the next section.

4.3.3.10. D-Amino Acids.   Substitution of Gly or L-amino acid residues will D-amino acids at appropriate sites of peptides has been at the origin, in several cases, of important modifications in their binding properties and biological activity. It is obvious that the side chains of enantiomeric amino acids are fully identical functionally, but in a peptide chain replacement of an L- by a D-amino acid usually leads to profound conformational changes. It is expected that such changes performed at either end of the peptide chain would least affect the overall topochemistry of the molecule and could be equivalent to merely displacing the terminal functional carboxyl or amino groups.

Several examples can be found in which the potency and the duration of action of peptides are significantly improved by substituting their N-terminal residue by a D-amino acid. For most of these analogs, it is still not clear whether the increase in biological activity resulted from a better fit with the receptor site, to an improved resistance to degradation by aminopeptidases, or to a combination of both. [D-Tyr[1]]-GHRH is a good example of the creation of a superagonist by replacing the N-terminal residue of a peptide with its "D" counterpart (Robberecht et al., 1985); on the other hand, [D-Tyr[1]]-enkephalin is an inactive analog of enkephalin in vivo or in vitro (Coy et al., 1976).

Substitutions of amino acids inside the chain with the D enantiomer have given rise to several interesting analogs, including antagonists. Following the description of the β-turn conformation in peptides by Venkatachalam in 1968, the synthesis of the superagonist of the hypothalamic peptide LHRH, [D-Ala[6]]-LHRH, was the first example of successful peptide design aimed at the stabilization of a hypothetical β-turn by a "D" residue (Monahan et al., 1973). Shortly thereafter, the synthesis of the superagonist of enkephalin, [D-Ala[2]]-enkephalinamide, was based on similar logic (Pert et al., 1976). In analogy, a long list of superagonists of several other peptides, which is beyond the scope of this chapter, has been obtained through various substitutions with "D" amino acids.

A large number of interesting analogs have been obtained by

using D-Trp as a mono- or a polysubstituent in various peptides. Most remarkable was the design of LHRH superagonists, among which are [D-Trp^6,Pro^9,N-ethylamide]-LHRH, which displays 140 times the potency of the parent peptide in vivo (Vale et al., 1977; Humphries et al., 1978), and more recently described reduced cyclic analogs of somatostatin (Veber et al., 1984). On the other hand, D-Trp-containing antagonists of some peptides have been reported. Neurotensin-induced glucagon release is specifically inhibited by a neurotensin analog [D-Trp^{11}]-NT (Ukai et al., 1982). A series of structurally related D-Trp-containing analogs of undecapeptide substance P have also been developed, [D-Pro^{2}D-Trp^{7,9}]-SP being the prototype (Engberg et al., 1981). However, the mechanism of action of these antagonists at the level of the target cell membrane is still ambiguous at this point, since extensively modified C-terminal fragments of substance P containing three D-Trp residues on a total of eight amino acids are potent substance P antagonists (Regoli et al., 1984).

Therefore, even though the information obtained by means of L- to D-amino acid substitution is often questionable as far the peptide-receptor interaction is concerned, it has proven very useful under well-defined circumstances to design enzyme-resistant derivatives and antagonists. However, further work must be done with most D-Trp superagonists and antagonists described so far in order to determine what, among increased lipophilicity, conformational stabilization, improved enzyme resistance, or a combination of all these factors is responsible for the new biological properties.

4.3.4. Modification of Terminal Reactive Groups

Chemical modifications of either C- or N-terminal groups have been used to slow down the degradation process of biologically active peptides by aminopeptidases and carboxypeptidases and/or to assess the functional role of terminal amine and carboxyl groups in the interaction of peptides with receptors. They can be performed through the omission or alteration of the chemically reactive function, by substitution, and/or by addition.

A classical example of successful end-group omission in peptide design is the elimination of the N-terminal amino group from oxytocin, which results in a dramatic increase of the uterotonic, avian depressor, and milk-ejecting activities of the parent peptide (Chan and du Vigneaud, 1962). Addition at the N-terminal can be performed by acylation with formyl, acetyl, or longer alkyls or aryls, using the corresponding activated acid. The addition of, or the substitution with, pyroglutamic acid, a natural cyclic rearrangement product of N-terminal L-glutamine found in several mammalian peptides, provides facile access to blocked terminal amines.

Natural peptides have either a free carboxyl or an amide group in the C-terminus. The latter generally results from the posttranslational cleavage of the peptide-Gly bond in the precursor by a specific enzyme. C-Terminal peptide amides can resist better than the corresponding free-COOH to proteolytic degradation by carboxypeptidases; when used as a synthetic modification of a naturally free-carboxyl terminal peptide, the amide is useful for evaluating the role of the negatively charged carboxylate in the interaction of the peptide with the receptor. For instance, free-carboxyl peptides such as neurotensin or bradykinin do not tolerate amidation (St-Pierre et al., 1984; Regoli and Barabe, 1980); however, enkephalinamide is a superagonist of enkephalin (Pert et al., 1976). Conversely, deamidated analogs of naturally amidated peptides such as gastrin-releasing peptide (Girard et al., 1984) and substance P (Fournier et al., 1982) are generally weak agonists of the parent compound. As mentioned above, part of this effect could be explained by a more rapid degradation of the analog by endogenous carboxypeptidases; the importance of the latter factor can be evaluated by adding specific inhibitors of these exopeptidases to the physiological buffer medium.

4.3.5. Changes in the Peptide Backbone

It is sometimes simplistically assumed that the backbone of peptide linkages merely serves as a sort of framework supporting the amino acid side chains, whose arrangement is then responsible for the topochemical features related to the binding and functional interaction of peptides with their receptor. However, evidence from X-ray crystallography data, mainly with enzymes, has established that the carbonyl and the amino functions of the peptide linkage indeed participate directly in the substrate-site interaction by means of hydrogen bonding, and peptide–receptor or peptide–antibody interactions should be submitted to similar mechanisms. Therefore, it is expected that structural changes in the backbone itself might give rise to most valuable information regarding the steric requirements of peptide–receptor interactions. Moreover, since the backbone is the main target for proteolytic enzymes, these modifications might be expected to confer on the peptide metabolic stability and to help in discerning between effects related to peptide destruction prior to peptide–receptor interaction and true conformational effects by modifying the nature of the interaction itself.

Modifications of the backbone have involved substitution on the imino nitrogen or alpha-carbon atom; intercalation of methylene or imino groups in the backbone; more or less isosteric replacements of atoms or groups of atoms within the backbone, including reversal of the direction of amide bonds; and, in a broader sense, attempts to stabilize particular conformations of the peptide chain through cyclization or other forms of rigidification of the peptide chain. Since an exhaustive list of such modifications in peptides is beyond the scope of this chapter, only some of them are shown in Figs. 10 and 11.

4.3.5.1. *N*-Alkylation.   This modification of amide bonds, which is sometimes motivated by the hope of obtaining metabolically stable analogs, profoundly alters both the stereochemistry and the chemistry (hydrogen-bonding capacity) of the backbone

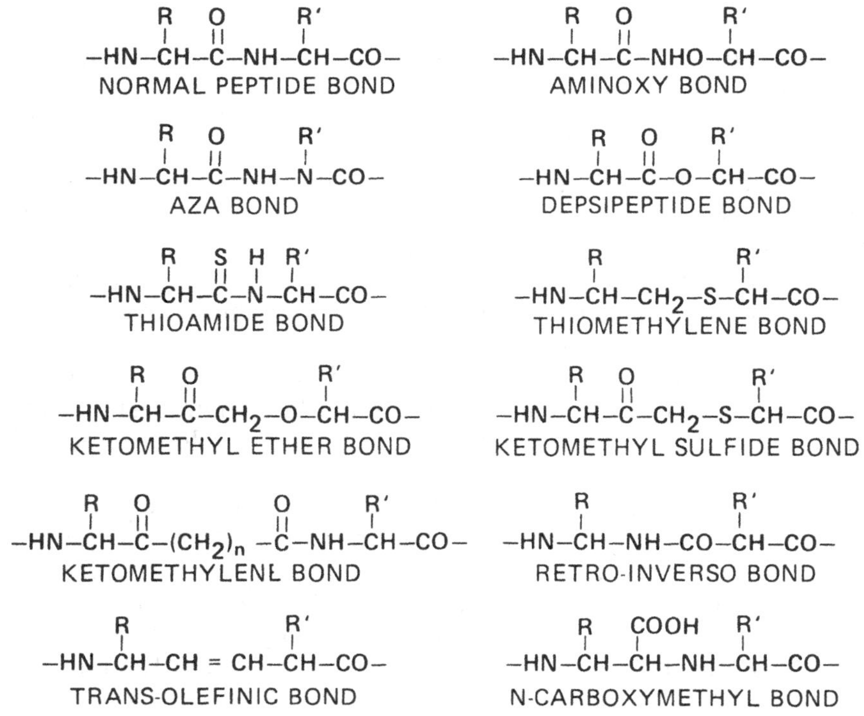

$$
\begin{array}{cc}
\underset{\text{NORMAL PEPTIDE BOND}}{-\text{HN}-\overset{R}{\overset{|}{\text{CH}}}-\overset{O}{\overset{||}{\text{C}}}-\text{NH}-\overset{R'}{\overset{|}{\text{CH}}}-\text{CO}-} &
\underset{\text{AMINOXY BOND}}{-\text{HN}-\overset{R}{\overset{|}{\text{CH}}}-\overset{O}{\overset{||}{\text{C}}}-\text{NHO}-\overset{R'}{\overset{|}{\text{CH}}}-\text{CO}-}
\end{array}
$$

$$
\begin{array}{cc}
\underset{\text{AZA BOND}}{-\text{HN}-\overset{R}{\overset{|}{\text{CH}}}-\overset{O}{\overset{||}{\text{C}}}-\text{NH}-\overset{}{\overset{}{\text{N}}}-\text{CO}-} &
\underset{\text{DEPSIPEPTIDE BOND}}{-\text{HN}-\overset{R}{\overset{|}{\text{CH}}}-\overset{O}{\overset{||}{\text{C}}}-\text{O}-\overset{R'}{\overset{|}{\text{CH}}}-\text{CO}-}
\end{array}
$$

$$
\begin{array}{cc}
\underset{\text{THIOAMIDE BOND}}{-\text{HN}-\overset{R}{\overset{|}{\text{CH}}}-\overset{S}{\overset{||}{\text{C}}}-\overset{H}{\overset{|}{\text{N}}}-\overset{R'}{\overset{|}{\text{CH}}}-\text{CO}-} &
\underset{\text{THIOMETHYLENE BOND}}{-\text{HN}-\overset{R}{\overset{|}{\text{CH}}}-\text{CH}_2-\text{S}-\overset{R'}{\overset{|}{\text{CH}}}-\text{CO}-}
\end{array}
$$

$$
\begin{array}{cc}
\underset{\text{KETOMETHYL ETHER BOND}}{-\text{HN}-\overset{R}{\overset{|}{\text{CH}}}-\overset{O}{\overset{||}{\text{C}}}-\text{CH}_2-\text{O}-\overset{R'}{\overset{|}{\text{CH}}}-\text{CO}-} &
\underset{\text{KETOMETHYL SULFIDE BOND}}{-\text{HN}-\overset{R}{\overset{|}{\text{CH}}}-\overset{O}{\overset{||}{\text{C}}}-\text{CH}_2-\text{S}-\overset{R'}{\overset{|}{\text{CH}}}-\text{CO}-}
\end{array}
$$

$$
\begin{array}{cc}
\underset{\text{KETOMETHYLENL BOND}}{-\text{HN}-\overset{R}{\overset{|}{\text{CH}}}-\overset{O}{\overset{||}{\text{C}}}-(\text{CH}_2)_n-\overset{O}{\overset{||}{\text{C}}}-\text{NH}-\overset{R'}{\overset{|}{\text{CH}}}-\text{CO}-} &
\underset{\text{RETRO-INVERSO BOND}}{-\text{HN}-\overset{R}{\overset{|}{\text{CH}}}-\text{NH}-\text{CO}-\overset{R'}{\overset{|}{\text{CH}}}-\text{CO}-}
\end{array}
$$

$$
\begin{array}{cc}
\underset{\text{TRANS-OLEFINIC BOND}}{-\text{HN}-\overset{R}{\overset{|}{\text{CH}}}-\text{CH}=\text{CH}-\overset{R'}{\overset{|}{\text{CH}}}-\text{CO}-} &
\underset{\text{N-CARBOXYMETHYL BOND}}{-\text{HN}-\overset{R}{\overset{|}{\text{CH}}}-\overset{\text{COOH}}{\overset{|}{\text{CH}}}-\text{NH}-\overset{R'}{\overset{|}{\text{CH}}}-\text{CO}-}
\end{array}
$$

Fig. 10.   Insertions within the peptide backbone.

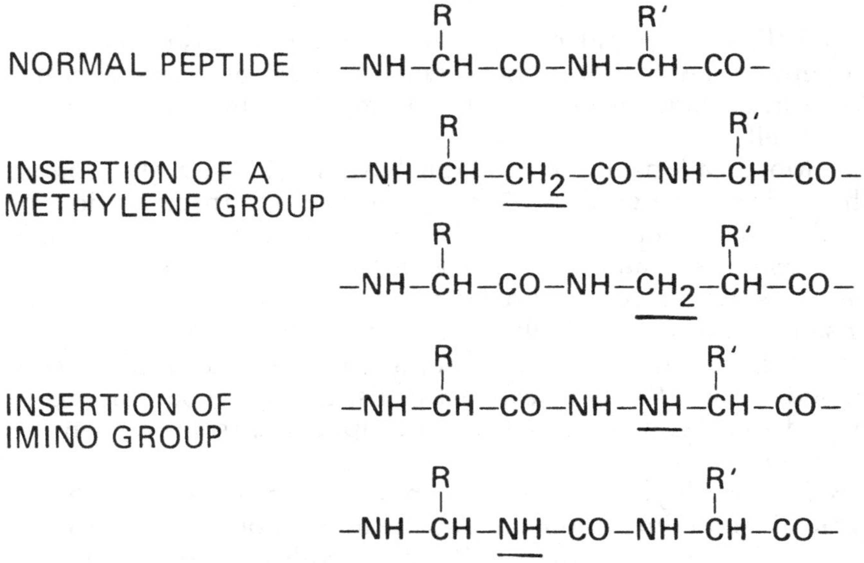

$$\text{NORMAL PEPTIDE} \quad -NH-\overset{\overset{\displaystyle R}{|}}{CH}-CO-NH-\overset{\overset{\displaystyle R'}{|}}{CH}-CO-$$

INSERTION OF A
METHYLENE GROUP

$$-NH-\overset{\overset{\displaystyle R}{|}}{CH}-\underline{CH_2}-CO-NH-\overset{\overset{\displaystyle R'}{|}}{CH}-CO-$$

$$-NH-\overset{\overset{\displaystyle R}{|}}{CH}-CO-NH-\underline{CH_2}-\overset{\overset{\displaystyle R'}{|}}{CH}-CO-$$

INSERTION OF
IMINO GROUP

$$-NH-\overset{\overset{\displaystyle R}{|}}{CH}-CO-NH-\underline{NH}-\overset{\overset{\displaystyle R'}{|}}{CH}-CO-$$

$$-NH-\overset{\overset{\displaystyle R}{|}}{CH}-\underline{NH}-CO-NH-\overset{\overset{\displaystyle R'}{|}}{CH}-CO-$$

Fig. 11. Some of the possible modifications of the amide bond in peptides.

amide group. It is therefore not surprising that the replacement of Gly with sarcosine, and Phe or Tyr with their N-methyl analogs, in several peptides strongly reduces, or even completely abolishes, biological activity. Such a result suggests a direct interaction of that portion of the peptide chain with the receptor. On the other hand, the opposite result, e.g., a gain in biological activity caused by N-alkylation of a given residue, could either indicate a stabilization of that bond against proteolytic enzyme cleavage or a beneficial effect of that local rigidification upon the interaction of the rest of the molecule with the receptor.

4.3.5.2. METHYLENE AND IMINO GROUP INSERTION. As in the case of amino acid intercalation in the peptide chain seen above, insertions of single methylene or imino groups would be expected to considerably alter the spacing of the side chains and the topochemistry of the molecule. The structural change is likely to have less effect if it changes the steric relations of, or within, a sequence that is not important for activity, or if it alters the spacing of only one side chain or an endgroup, as in substitutions for terminal amino acids of the peptide chain. Single methylene groups can be introduced in the peptide chain by using beta- rather than alpha-amino acids. Beta-alanine has been used most frequently for such backbone modifications (Rudinger, 1971). Also, beta-aspartic acid is often introduced accidently in Asp-containing peptides, following a sequence-dependent side reaction of intramolecular rearrangement that occurs to a variable extent during the course of synthesis in the presence of acid (Stewart and Young, 1984). An imino group can also be intercalated between the alpha-carbon and the carbonyl group to give a urea derivative (Rudinger, 1971). Such compounds have been obtained in several cases as the result of a side reaction during the azide coupling procedure of peptide synthesis rather than by design.

4.3.5.3. SUBSTITUTION OF THE PEPTIDE BOND. Subtle changes in the backbone can be induced when groups within the peptide chain are replaced approximately isosterically. Some of these changes are illustrated in Fig. 11. A well-known substitution of the peptide bond results from the replacement of amino acids by hydroxy acids, equivalent to replacing amide with ester linkages (Rudinger, 1971). Such depsipeptides are found naturally, valinomycin being a typical example.

The replacement of an amide bond by a thioamide bond, $CS—NH$, does not change the geometry of that particular bond, as shown previously by X-ray crystallography of model endothiopeptides. However, it is expected that the presence of a thioamide bond will influence the formation of secondary structure of the polypeptide chain because of the lowered tendency of sulfur as compared to oxygen to participate in conformation-stabilizing hydrogen bonding. The synthesis of enkephalin analogs containing such bonds has been recently described (Clausen et al., 1984; Benowitz and Spatola, 1985). The aza bond results from the replacement of an alpha-CH group with a nitrogen atom (Dutta et al., 1979). This bond is approximately isosteric with the amide bond, except that the arrangement of the side chain can correspond to that of either an L- or a D-amino acid. Apart from this, one would expect the topochemistry of the parent molecule to be largely unchanged, and active analogs should result. Aminoxy bonds were found to be poor substitutes for the amide bonds in several analogs of enkephalin, where aminoxyacetic acid was used to replace Gly^2 or Gly^3, and aminoxyphenylpropionic acid to replace Phe^4 (Salvadori et al., 1981). On the other hand, the N-carboxymethyl bond has been used successfully to prepare orally active inhibitors of the angiotensin-converting enzyme (Patchett et al., 1980).

The substitution of the peptide NH by CH_2 has also been achieved (Natarajan et al., 1984; McMurray and Dyckes, 1985).

This modification is equivalent to replacing part of the peptide backbone with a paraffin chain. Ketomethylene analogs of substance P with a potent effect as substance P-degrading enzyme inhibitors were prepared recently by Ewenson et al. (1986). The pseudopeptides were also full substance P agonists in the guinea pig ileum bioassay. The incorporation of a *trans*-olefinic pseudopeptide bond has also been achieved, using substance P as a model compound. It was found that the *trans* geometry of the bond was important for biological activity (Cox et al., 1980). Several other substituents for the amide bond can also be prepared, as shown in Fig. 11, but their insertion in peptides is difficult to achieve and has therefore been limited so far (Rudinger, 1971).

4.3.5.4. REVERSAL OF PEPTIDE BONDS.   A radical change in the backbone that might be expected to produce minor changes in side chain topochemistry is the reversal of the direction of peptide bonds. Retromodification of biologically active peptides has been widely applied in studying the relative importance of backbone vs side chains for biological activity. One of the objectives of such studies is to obtain topochemically complementary structures that interact efficiently with receptors or active sites and are less prone to biological degradation. Highly active analogs have resulted from the introduction of this modification within cyclic peptides, but attempts to introduce the same modification in linear peptides have met with limited success. In order to maintain the side chain topography in the case of acyclic peptides, the reversal of the peptide bonds is accompanied by a simultaneous inversion of the configuration at each chiral center, but also results in the reversal of end groups.

The reversal of end groups introduces the problem that the extended structures of the parent and its retro analog are not complementary. Attempts to surmount this problem by blocking the end groups or by eliminating them have generally been unsuccessful. Recently, Chorev and Goodman (1983) have approached this problem by introducing two modified residues into the peptide sequence. The modified segment starts with a gem-diaminoalkyl residue derived from the corresponding amino acid and terminates with a substituted malonyl residue, the side chain of which corresponds to that of the correspondent amino acid. The application of this so-called retro-inverso modification to enkephalinamide (Chorev et al., 1979), among others, has given rise to highly active analogs.

4.3.6. Conformational Stabilization

Most peptide hormones are small linear peptides that show considerable conformational flexibility. Even though the conformational space is more limited in cyclic disulfide-containing peptides, they can also present a manifold of possible low-energy conformations, as predicted by various methods of conformational calculations. Moreover, it is less than certain that the conformation adopted by the peptide in contact with its receptor will represent the lowest state of energy. One approach to limiting the uncertainties about the active conformation of a peptide is to introduce chemical constraints that greatly decrease the number of its possible conformers, while preserving or increasing its biological activity. The ideal situation is found when such constraints can enhance the desired biological response relative to undesired side effects, assuming that different conformational forms of the same peptide will interact at different receptors. An additional advantage of conformational constraints is the ability to simplify the structure by removal of those amino acids that serve solely to maintain the active conformation, but do not directly interact with the receptor. Some of the major approaches that have been used for conformational restrictions are listed in Table 5.

Since it has been demonstrated that β-turns or "hairpin" turns, which are a common form of secondary structure found in peptides and proteins, are responsible for changes in the direction of peptide chains, several attempts have been made by peptide chemists to stabilize this conformation, with the hope of obtaining more potent analogs of the parent compound. Several chemical modifications have been performed in order to impose conformational restrictions in peptide regions suspected to contain a β-turn. Such modifications are the substitution of selected amino acids by their corresponding D enantiomer, the N- or C-methylation of the backbone in regions of high β-turn probability, or the introduction of cyclic structures containing the bend, either with the use of lactam rings or cyclization between neighboring reactive side chains. In the case of naturally cyclic peptides such as somatostatin, further reduction of the ring size, from 12 to 6 residues, has resulted in more potent, long-acting analogs (Veber et al., 1984).

Remarkable success has been obtained with several peptides into which the aforementioned conformational restrictions have been introduced. However, in each case, it is difficult to assess

Table 5

Some Examples of Conformational Restrictions Introduced in Peptides by Side Chain Modifications and Their Biological Consequences

Conformational restriction	Peptide	Biological effect	Reference
D-Residue substitution for stabilization of β-turn	LHRH	Increased activity	Monahan et al. (1973)
Aib or Cyl substitution for stabilization of β-turn	Chemotactic Activity peptide	Activity	Iqbal et al. (1984)
D-Trp substitution	Neurotensin	Antagonist	Quirion et al. (1980)
	Neurotensin	Increased activity	Jolicoeur et al. (1981)
	Substance P	Antagonist	Fuxe et al. (1982)
	Substance P	Antagonist	Caranikas et al. (1982)
Penicillamine substitution for Cys	Oxytocin	Inhibitor, prolonged	Mosberg et al. (1981)
Dehydro amino acid substitutions	Enkephalin	Increased potency	English and Stammer (1978)
	Bradykinin	Increased potency	Fisher et al. (1978)
Introduction of lactam bridge	LHRH	Increased potency	Freidinger et al. (1980)
Isosteric cyclization of linear peptides	Enkephalin	Increased potency	DiMaio et al. (1982)
	Melanotropin	Increased potency and stability	Sawyer et al. (1982)
	β-Endorphin	High binding activity	Blake et al. (1981)
			Blake et al. (1985)
	Substance P	High activity	Ploux et al., (1985)
	Bradykinin	Increased potency and stability	Chipens et al. (1981)
Reduction of cycle size of naturally cyclic peptides	Somatostatin	Increased activity	Veber et al. (1984)

whether the observed increase in biological activity, as compared to that of the native peptide, is caused by the stabilization of the conformation recognized by the receptor, or by an improved resistance of the analog to degradation by proteases, which should allow more peptide molecules to reach the receptor site.

4.4. Amphiphilic Secondary Structure Approach

Cell membranes represent an anisotropic environment because of the amphiphilic nature of the separation existing between the biologic milieu and the lipid bilayer of the membrane. The essential principle behind Kaiser's hypothesis is that this amphiphilic environment can impose on the peptide a given secondary structure. Therefore, such a structure should be characterized by the amphiphilic distribution of the individual amino acid side chains. One "face" of the peptide structural arrangement should be occupied preferentially by hydrophobic amino acid side chains, such as Ile, Leu, Val, Phe, Trp, Tyr, and Met, whereas the hydrophilic side chains Asp, Glu, Arg, Lys, Thr, and Ser should be present on the other face. It is expected that such an arrangement, which should minimize the free energy contribution for each amino acid by as much as 1–3 kcal/mol, can cause secondary structures, with a low probability of occurrence in an exclusively aqueous environment, to become the particular conformation of the peptide in the membrane environment. That form of secondary structure could be any of those that are sterically allowed, such as the right- or left-handed α-helix, the 3_{10} and the π-helix, or the β-pleated sheet.

Searching for the possibility of those forms of secondary structures can be done merely on a sheet of paper by projecting the amino acids along an Edmunson wheel (Schiffer and Edmunson, 1967), with the pitch corresponding to the assumed helical conformation (α with 3.6 residues/turn, 3_{10} with 3 residues/turn, and π with 4.4 residues/turn), or alternately, on either side of a line, assuming the beta conformation, as shown in Fig. 12. A complementary verification of the prediction of amphiphilic structures can be made by using the Chou and Fasman's method (1974) discussed above. The predictive methods can also indicate which substitution should be made in the sequence in order to enhance the form of secondary structure that is the most likely to occur at membrane level, according to the amphiphilic theory. If synthetic peptide analogs comprising such substitutions display a higher

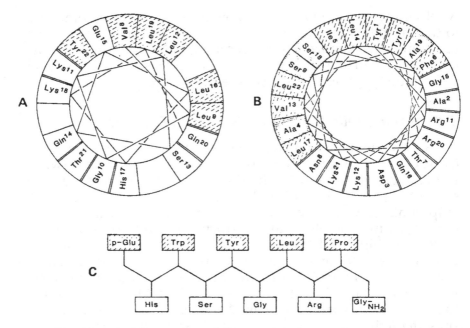

Fig. 12.   An illustration of the graphical procedure developed by
Schiffer and Edmunson (1967) and applied by Kaiser and Kezdy (1984) for
searching for amphiphilic secondary structures. (A) α-Helical conformation of salmon calcitonin fragment 8-22; (B) π-Helical conformation of
human GHRH fragment 1-22; (C) β-Pleated sheet conformation of
hypothalamic peptide LHRH.

affinity for the target cell membrane than the parent compound
and are more potent, it is likely that the conformation predicted by
the amphiphilic approach could represent the one that is adopted
by the peptide in contact with the membrane, if not with the
receptor.

    Testing of the hypothesis was done for some peptides, includ-
ing calcitonin, a natural peptide with potent hypocalcemic properties. It consists of an amino terminal loop formed by a disulfide
bridge between Cys^1 and Cys^7, followed by sequence 8-22 (-Val-
Leu-Gly-Lys-Leu-Ser-Gln-Glu-Leu-His-Lys-Leu-Gln-Thr-Tyr-)
with a high amphiphilic potential, as shown on Edmunson wheel
projection in Fig. 12a, and a highly hydrophilic C-terminal segment
23-32. To test the amphiphilic properties of sequence 8-22, a synthetic calcitonin analog was prepared, in which sequence 8-22 was
replaced by a totally different one with highly helical potential:

-Leu-Leu-Gln-Gln-Trp-Gln-Lys-Leu-Leu-Gln-Lys-Leu-Lys-Gln-Leu-. Even though most of the 15 residues of the sequence had been modified, the whole analog still retained high binding capacity to calcitonin receptors in rat brain and displayed potent hypocalcemic properties. Moreover, the CD spectrum of the peptide at an air-water interface showed a higher helical content than the native peptide, as demonstrated by the intensity of the band at 222 nm (Moe et al., 1983).

The amphiphilic secondary structure hypothesis was also successfully applied by Kaiser and collaborators to other peptides in the α-helix conformation, such as melittin (DeGrado et al., 1981), the π-helix conformation, such as beta-endorphin (Taylor et al., 1981), or the beta-strand conformation, such as apolipoprotein B (Osterman et al., 1984). A further extension of the theory could eventually be made to the design of enzymes and proteins with well-defined properties, as reviewed recently (Kaiser et al., 1985).

4.5. Visualization and Modeling of the Peptide-Receptor Interaction

Up to the late 1970s, peptide chemists have used various types of handheld molecular models to visualize the tridimensional structure of peptides, as based on data obtained by NMR and CD spectroscopy, X-ray crystallography, or theoretical conformational calculations. The consequences of chemical modifications transposed to a handheld model, such as the spacefilling CPK model, can be evaluated by assembling the proper spheres representing van der Waals radii and handling the new model in order to predict the possible changes in the secondary structure induced by the substitution. Even though handheld models have demonstrated their usefulness in early studies and still do, they cannot take into account all the parameters of the chain and therefore often provide misleading structural information. They also become cumbersome to handle in the case of larger peptides.

Recently, significant improvements in computer graphics hardware and software have allowed for display of structures and study of their tridimensional properties. Bond lengths, dihedral angles, and torsion angles can be inserted into programs, and the structures can be displayed and studied in two dimensions or as stereo pairs, and as stick or spacefilling tridimensional models, even with distinct colors for different atoms.

Once a structure has been created, several different manipulations and measurements can be carried out with the computerized modeling system. Intramolecular distances and angles can be determined precisely. The conformation can be modified as desired by rotating about any torsion angles, which can produce new conformations. Low-energy conformations can be found and their relative energies can be compared. Most useful for structure–activity and peptide–receptor interaction studies, two related structures can be superimposed to evaluate the conformational consequences of single chemical modifications of the parent molecule, and the corresponding tridimensional image of the receptor can be generated as a pseudoelectronic density map based on the volume occupied by the van der Waals spheres representing the peptide pharmacophore.

Following the pioneer work to develop the Merck Molecular Modeling System, which resulted notably in the design of reduced cyclic analogs of somatostatin (Veber et al., 1984), several other modeling systems and software packages have become commercially available. However, in order to handle large molecules such as peptides, those programs usually require elaborate and, therefore, costly computer facilities. Several review articles on molecular modeling have been published, and interested readers can consult them for further information (Gund et al., 1980; Humblet and Marshall, 1981; Freidinger and Veber, 1984; Marshall, 1984).

5. Peptide Synthesis

The principal reaction in the synthesis of peptide chains is acylation of the amino group of an amino acid by the carboxyl group of a second amino acid with formation of an amide bond. The presence of several functional groups in amino acid molecules and the necessity to maintain the integrity of the alpha-carbon chirality centers during coupling makes peptide synthesis enormously more complex than simple carboxamide formation.

Ever since Du Vigneaud and colleagues reported the first total synthesis of oxytocin in 1953 (du Vigneaud et al., 1953), peptide synthesis methodology has reached a high level of efficiency and sophistication through refinements that were continuously added to the basic strategy, mainly in terms of protecting groups and coupling methods. The solid-phase method, introduced a few years afterward, has resulted in the spread of the use of peptide synthesis outside the highly specialized organic chemistry laboratories where it had been confined before, and made peptides accessible to nonchemist investigators from various biomedical disciplines. One of the most important consequences of this remarkable breakthrough is that during the past 20 yr, more than 100 biologically active mammalian peptides have been discovered and chemically characterized. Most of them have been totally synthesized by the Merrifield method, providing investigators with an abundant source of synthetic, but authentic, material. In addition to natural peptides, several thousands of related analogs and various other peptides have also been prepared.

The purpose of this last section of the chapter is to provide the reader with a practical approach to the synthesis of peptide analogs. Since it is assumed that the majority of readers are not organic chemists, emphasis will be directed at the application of the solid-phase peptide synthesis, which requires less specialized equipment and personnel than the solution approach. The reader who requires more information on the synthetic strategies in solution is invited to consult specialized reviews written by Finn and Hofmann (1976) and Bodansky (1984). The next paragraphs will be directed primarily at describing the synthesis of biologically active peptides and analogs using the solid-phase approach.

5.1. Solid-Phase Peptide Synthesis

Solid-phase peptide synthesis (SPPS) was introduced by Bruce Merrifield in 1963 in an effort to overcome many of the problems of peptide synthesis in solution as it was practiced at that time. Its feasibility was first demonstrated by the synthesis of the simple tetrapeptide Leu-Ala-Gly-Val.

Briefly, the basic idea behind SPPS is to attach the amino acid corresponding to the carboxyl end of a peptide chain to an insoluble solid support and extend the chain toward the amino end by stepwise coupling of activated amino acid derivatives. Filtration and thorough washing of the solid phase removes soluble byproducts and excess reagents, but retains the peptide on the support. After completion of the chain, the peptide is removed from the solid support and purified.

The great advantage of using a polymer-supported peptide chain is that all the laborious purification at intermediate steps in the synthesis is eliminated, and simple washing and filtration of the peptide-resin is substituted. Completion of the reactions at each step can therefore be forced by using large excesses of re-

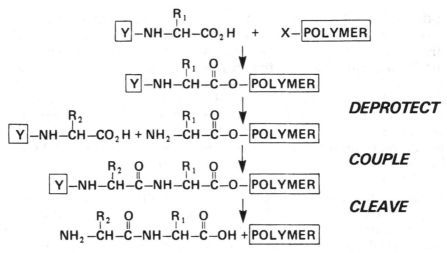

Fig. 13. Diagram of the principal operations in the solid phase synthesis of peptides. X represents a reactive group on the polystyrene matrix. Y corresponds to the labile protecting group of the α-amino function. R_1 and R_2 are peptide side chains.

agents. By means of a suitably designed reaction vessel, all the synthesis can be carried out in one container without any transfer of material from one vessel to another. The basic idea of SPPS is illustrated in Fig. 13.

5.1.1. The Support

The suitability of supports used in SPPS is critical to assemble rapidly homogeneous peptides in high yields. The qualities required for the support are mainly of a chemical and mechanical nature. Chemically, it has to be totally inert to various reagents, especially strong acids used in SPPS; it also has to be functionalized with chemical groups that are used to attach the peptide under facile conditions. Mechanically, it has to be made of homogeneous particles of a size and shape that permit easy handling and rapid filtration of solvents and reagents. Moreover, since it is used for heterogeneous synthesis, it has to be highly porous and allow for easy access of solvents and reagents to the peptide, in order to provide for reaction conditions that are as close as possible to solution. On the other hand, since it has to go through several cycles of mechanical agitation, it should remain intact even in a highly swollen state and therefore be supported by a proper degree of crosslinking.

The most common polymeric support used for SPPS is a low cross-linked polystyrene (copolymer of styrene with 1% divinylbenzene) matrix. Commercially available beads (100–400 mesh size) can swell extensively in solvents used for the synthesis (methylene chloride, dimethylformamide, dioxane). A wide variety of chemical functions can be linked chemically to the phenyl rings of polystyrene beads to attach the C-terminal carboxyl function of the peptide during the course of the synthesis. This functional group is symbolized by X in Fig. 13.

Ever since the first synthesis of a tetrapeptide by Merrifield, chloromethylated polystyrene-divinylbenzene, onto which the C-terminal amino acid can be condensed to form an ester bond, has been most widely used to assemble peptides of various sizes bearing a free C-terminal carboxyl function. However, it has been shown that this ester bond can be partly cleaved under acid conditions used at each deprotection cycle, resulting in losses of peptide. Even though short peptides (up to 15 residues, as an average) can be assembled in reasonable yields on chloromethylated polystyrene, low peptide recoveries observed after cleavage of longer ones has been at the origin of the development of more stable peptide-resin linkages. Among all the chemical functions proposed during the last 10 yr *p*-acetoxymethylphenylacetyl (Pam) resins developed originally by Mitchell et al. (1978) have now made their way as the most useful support with which to assemble longer peptides. They are also commercially available.

In recent years, an increasing number of biologically relevant peptides bearing an amidated C-terminal function have been characterized. Even though such peptides can be assembled on the Merrifield resin and obtained as amides through ammonolysis of the C-terminal ester bond (except for peptides containing Asp or Glu), they can be more conveniently prepared on polystyrene supports bearing the benzhydrylamine (Pietta et al., 1974) function. The amide function is obtained directly following acidolysis of the peptide-resin bond at the end of the synthesis. Several other types of polymeric supports and peptide-support bonds have been developed and used during the last 20 yr, and they are reviewed in Stewart and Young's book on SPPS (1984).

5.1.2. Protection of the Amine Function

Temporary protection of the α-amino function of amino acids is represented by Y in Fig. 13. This protecting group should be stable under regular coupling conditions, but has to be rapidly hydrolyzed under selective conditions that leave intact the amide bond, the peptide resin bond, and the protecting groups of the reactive side chains, prior to the addition of the next amino acid residue.

The most commonly used alpha-protecting group for the amino function in solid phase peptide synthesis during the last two decades has been the *t*-butyloxycarbonyl (Boc) group. Its very satisfactory lability/stability characteristics, its solubility in SPPS solvents (CH_2Cl_2, DMF), in addition to the ready commercial availability of the corresponding amino acid derivatives, have contributed to its popularity. The Boc group can be rapidly hydrolyzed by strong acids, such as trifluoroacetic acid (TFA) solutions (30–50%, vol/vol) in SPPS solvent (CH_2Cl_2), to give two gases (isobutylene and carbonic anhydride), leaving the acid salt of the amine. This salt can be easily displaced with a dilute solution of a tertiary amine (*N,N*-triethylamine or *N,N*-diisopropylethylamine), in order to liberate the *N*-terminal end of the peptide before introducing the next amino acid derivative.

Several other alpha-amino protecting groups labile under various conditions are known. A more complete list of these derivatives is given in the book by Stewart and Young (1984).

5.1.3. Formation of the Peptide Bond

Obtaining quantitative coupling reactions yields at each addition of an amino acid residue is the most important individual goal in SPPS. In order to achieve completion, coupling reactions are usually carried out by using an excess of the activated amino acid derivative. This excess, unlike the situation with solution methods, can easily be eliminated by mere washing operations at the end of the coupling reaction.

Amino acid derivatives, with a few exceptions, are usually activated just prior the coupling step using a proper coupling agent. The most important characteristics for a coupling agent are general applicability, ease of use (particularly with automatic peptide synthesizers), rapidity and efficiency of coupling reactions, and freedom from racemization and other side reactions. The coupling agent originally used by Merrifield in his model synthesis (1963) was *N,N*-dicyclohexylcarbodiimide (DCC), and it remains the most widely used. However, inadequacies of DCC-mediated coupling reactions have led to investigations to develop better coupling procedures.

Coupling reactions mediated by DCC are generally rapid because of the high reactivity of the *O*-acyl urea species and therefore make attractive the stepwise approach in SPPS. The basic routine procedure for DCC couplings recommends the use of two- to four-fold excesses of Boc amino acid and DCC over peptide amino groups at room temperature. Couplings with this procedure, which can be used with most amino acid derivitives, are generally complete within 10 to 15 min. However, for some critical peptide sequences, the relative rate of the coupling reaction, such as with sterically hindered amino acid side chains, forces one to prolong coupling times for much longer periods, and even to repeat them. Other sequence-dependent factors involving the interaction of the amino terminus with itself or with the polymer matrix can also slow down the coupling rate. Various additives such as *N*-hydroxysuccinimide and the more recently developed *N,N*-hydroxybenzotriazole can be used with DCC to promote coupling and decrease racemization.

Preformed Boc-amino acid symmetrical anhydrides (SAs) have been found to give the best results of all methods tried for many difficult sequences. They are generally generated just prior to coupling using one equivalent of the Boc amino acid derivative and 0.5 equivalent of DCC, in comparable molar excess over peptide amino groups as above. Following the ~ 10 min required to form the anhydride, the resulting dicyclohexylurea can be filtered out of the solution, and the filtrate used for coupling.

Coupling *N*-protected Asn and Gln has to be done in such a way as to prevent the dehydration side reaction of the carboxamide side chain into a nitrile, which occurs to a large extent when DCC is used as a coupling agent. Even though they give rise to slow coupling rates, *para*-nitrophenyl active esters have been mainly used in large molar excesses as activated species of Asn and Gln to avoid the formation of nitrile. The coupling of *N*-protected Asn and Gln can also be performed by using the DCC/HOBT procedure.

Since no purification of the peptide is possible to achieve before the synthesis has been completed, every single coupling step has to be brought up to 100% yield, in order to avoid the formation of truncated peptide sequences that are difficult, if not impossible, to remove from the desired peptide during subsequent purification steps. A choice of destructive and nondestructive methods exists to monitor the extent of coupling after each step. Among these, quantitative amino acid analysis, which involves the interruption of the synthesis for at least a day, picric acid monitoring, which can be used to evaluate spectrophotometrically the proportion of uncoupled amino groups, and the ninhydrin test, which can determine qualitatively, in 5 min, the extent of coupling by giving a yes or no answer, are the best known methods. The latter, also called Kaiser's test, is routinely used in most SPPS laboratories.

5.1.4. Cleavage of Peptides From the Resin

Different acidic, basic, or neutral reagents can be used to cleave peptides from the resin support. The most popular and useful one is anhydrous liquid HF, which also liberates reactive side chains from most protecting groups used in SPPS. Even though some specific problems are related to the use of HF (toxicity, corrosion, use of special nonglass equipment), it provides rapid (30 min–1 h) cleavage and deprotection, generally with a minimum of harm to the peptide, as long as elementary precautions are taken and proper scavengers are added to the reaction. These precautions are based on more than 20 yr of experience in acidolytic cleavage of peptides from solid supports, which has provided a good knowledge of possible side reactions and has allowed for the development of new protecting groups that can minimize these side reactions.

HF cleavages can be carried out by using a commercially available ramp that allows for the transfer of HF *in vacuo* into reaction vessels and its disposal at the end of the reaction. After cleavage, the crude deprotected peptide is extracted from the resin in aqueous acidic solution and lyophilized.

5.2. Purification of Peptides

Methods for the purification of synthetic peptides are numerous, and their choice usually depends on (1) the nature of the peptide itself, (2) the efficiency of the synthetic strategy, which determines the proportion of expected peptide versus side reaction products and the extent of their resemblance with the peptide, and (3) the purity required of the final product, which is related to its application or use.

Since crude peptide mixtures obtained after HF cleavage normally contain low molecular weight contaminants, it is usual to elute them on a gel filtration column as the first purification step. The porosity of the gel is usually selected in order to let the peptide come out in the dead volume of the column (V_o), whereas the low-molecular-weight contaminants are retarded. One must be aware that abnormal results are often obtained with gel columns because of peptide–peptide and peptide–gel interactions that can cause peptides to elute generally later than expected and to remain contaminated with low-molecular-weight material.

The semipurified material obtained after this first step can be further refined using a purification procedure that is more specifically related to the peptide sequence, such as the net charge or the partition coefficient in organic/aqueous phase systems, which results from its lipophilicity. At that stage, conformational considerations also have to be taken into account in order to optimize the resolution of the selected technique. Countercurrent distribution, partition chromatography, ion-exchange chromatography, and low resolution reverse-phase chromatography are most frequently used as a second purification step. Peptides with a satisfactory purity are often obtained after this step, provided that no closely related contaminants were present at the end of the synthesis.

The occurrence of such contaminants originating from deletions or from various side reactions, especially in long peptides, calls for further purification steps by means of high-performance liquid chromatography (HPLC) systems coupled to high-resolution preparative or semipreparative columns that contain reverse-phase or ion-exchange media. Generally, the more purification steps required, the lower the global yields because of losses of various origins, the most frequent one being the nonspecific attachment of peptide material to the walls of columns and lyophilizing jars.

5.3. Assessment of Peptide Purity

From structure–function studies of several biologically active peptides, it has become evident that even slight structural changes can provoke drastic modifications in the activity profile, which provides information as to the details of the peptide–receptor interaction. It follows that an unequivocal control of the purity of each synthetic peptide used in structure–function studies is essential. To certify a given peptide preparation as high purity and authenticity, a large spectrum of analytical tests of differentiated specificities has to be performed.

Traditional low-resolution methods for the characterization of peptides such as thin-layer chromatography (TLC) and electrophoresis are no longer acceptable, as they were in the past, as sole criteria of peptide purity. The advent of high-resolution HPLC has contributed to setting the criteria of purity for peptides to much higher standards. Also, the discovery of unexpected side reaction products that are formed during the various steps of the synthesis of some peptides has pushed peptide chemists to resort to more sophisticated characterization criteria than quantitative amino acid analysis of peptide hydrolysates; additional information can now be provided by fast-atom bombardment (FAB) mass spectrometry and sequencing techniques. The latter is particularly useful for detecting deletion side products or backbone modifications originating from internal rearrangements.

Further details as to solid phase synthesis and the purification and the characterization of peptides can be found in the book of Stewart and Young (1984).

5.4. Automatic Solid-Phase Peptide Synthesis

SPPS involves a large number of repetitive operations related to solvent handling (metering, transferring, draining) and agitation of the peptide-resin slurry. All these operations except coupling itself are performed on a routine basis from one coupling cycle to the other and they can be automated, using a peptide synthesizer. Essentially, an automatic peptide synthesizer is a sophisticated washing machine, consisting of an array of valves that control the flow of solvents and reagents in and out of the reaction vessel, connected to a program of sequential meter-transfer-mix-drain operations via an interface.

Since the development of the first automatic peptide synthesizer by Stewart in the 1960s, several different versions have become commercially available, the first and classic being the Beckman Model 990. Since the operation of this early model was based on Stewart's original pin-and-hole-type programmer, the unit has evolved to become programmable, and based on microcomputer technology. The pioneer in that field was the computer-operated synthesizer conceived by Vega. Although the plumbing of the instrument was somewhat different from the Beckman 990 model, it remained based on the same principle of nitrogen-pushed solvent and mechanical agitation of the resin. Although several other versions of peptide synthesizers based on more or less similar modes of operation became available from other commercial sources, the ultimate instrument, combining robotics with a new concept of solid phase support and chemistry, was commercialized by Applied Biosystems. This system has been successfully used during the past 2 yr in several laboratories for the synthesis of various large peptides with a minimum of side products and high yields. This machine represents the most advanced version of automatic peptide synthesizers. However, the cost of the machine itself and its operation are important limiting factors for several laboratories with restricted research budgets.

5.5. Multiple Peptide Synthesis

One of the main reasons structure–activity studies of peptides progress so slowly is probably that the synthesis of analogs, as done now, is extremely time-consuming, since every one of them is prepared on an individual basis. Generally, SPPS is carried out stepwise and one SPPS apparatus or automatic instrument has to be dedicated to the assembly of one peptide at a time. For instance, the time required to synthesize an analog of neurotensin (13 amino acids) is approximately 1 wk, using a normal manual operator schedule, or a few days, when operating in the automatic mode of a peptide synthesizer. For a peptide the size of CRF (41 amino acids) or $GHRH_{1-44}$, assembly can last for as much as a month. It means that, at present, an average peptide laboratory operating with one synthesis apparatus can assemble a maximum of 500–600 amino acids per year. This figure is higher for owners of Applied Biosystems instruments, which is one of the few machines known that can be operated reliably on a continuous basis. At best, a large number of analogs with a modification in *N*-terminus only can be assembled simultaneously using one single large batch of resin that is split at the point of variation. However, *C*-terminal analogs must still be assembled entirely one at a time, unless fragment condensation can be used.

Recently, a novel approach has been proposed by Houghten (1985) to facilitate the synthesis of a large number of peptides. The approach is based on the following principles. SPPS consists of a series of operations that are common to the synthesis of any peptide (deprotection, neutralization, washing), except for coupling, which is specific for a given peptide at every step of addition of a Boc amino acid to the chain. The common steps for one addition cycle of several peptides can be carried out simultaneously, provided the individual resin supports for each peptide can be kept separated from each other during these steps. Such conditions can be realized by placing samples of resin in individual packets that are made of a material that is selectively permeable to SPPS solvents and reagents, without letting the resin leak out during the operations. At the same time, this material must be inert, especially to strong acids, since the usual deprotection and cleavage steps require the use of TFA and HF.

Houghten (1985) has applied this principle to the assembly, on a small scale (10–20 mg of resin support), of more than 200 peptide analogs in individual polypropylene mesh packets. The packets were placed together in the reactor of a peptide synthesizer for the common operations of deprotection, neutralization, and washing. For the coupling step, each properly identified bag was put in the presence of the appropriate solution of activated Boc amino acid in individual reaction flasks. After the coupling period, the bags were returned to the reactor to go through the common operations of the next cycle. Once the peptides have been assembled, they can be individually cleaved, deprotected, and purified as discussed above.

It is predictable that this concept of multiple peptide synthesis will be applied by several laboratories in the near future to assemble the numerous peptide analogs needed for structure–activity studies. Also, it can be expected that new automatic instruments for multiple peptide synthesis will soon become commercially available.

6. Conclusion

The purpose of this chapter is to show how peptide structure–activity studies can be used to probe its receptor and to separate the variables that cause phenomena such as recognition and activation. However, such an indirect approach requires inferring a hypothetical picture of the receptor based on the measurable properties of a series of synthetic analogs designed in such a way as to pinpoint which chemical groups are responsible for the peptide's action.

We propose practical approaches to the design and the synthesis of peptide analogs while trying to keep in mind that the functional identification of peptide receptors using structure–activity relationships rests on an oversimplification of the receptor problem, which is the traditional concept of the pharmacophore, normally credited to Erlich at the turn of the century. It has, however, been possible through the years to come up with several refinements of this old concept, in order to take into account our new knowledge of phenomena that occur at the cellular level following the peptide-receptor, or more generally, the ligand–receptor interaction. Because the constraints on a molecule to induce the appropriate biological response are more stringent than simply high affinity for the active site, which could lead to inhibition only, one has a higher probability of success with the pharmacophore assumption with agonists than antagonists. However, the few specific peptide antagonists that are available have proven to be most useful for structure–activity studies with the parent peptide, mainly to help discern between direct and mediated biological effects.

Obviously, all problems cannot be solved with structure–activity studies of peptide analogs, and the image of the receptor that is obtained represents no more than a common binding mode that can be used to align the set of analogs in a common frame of reference. The union of the volume of the parts of each analog that are fixed during presentation of the pharmacophore defines a minimal volume available at the receptor. It is necessary to assume here that the receptor site is relatively static and that it meets with an also static peptide immobilized in one given conformation. No explanation can be given, using this model, for what happened before or will happen to the rest of the receptor molecule, to the cell membrane environment, or to the peptide ligand after the "key into the lock" event takes place.

However, for the further purpose of drug design based on peptide structure, the amount of information that can be collected by using structure–activity studies is usually adequate, and several examples can be found in the literature, as mentioned in the text above, in which this approach was rewarded with the design and the synthesis of potent analogs that have now reached the status of drugs.

References

Agarwal N. S., Hruby V. J., Katz R., Kles W., and Nirenberg M. (1977) Synthesis of leucine-enkephalin derivatives: Structure–function studies. *Biochem. Biophys. Res. Commun.* **76,** 129–135.

Anderson J. C., Barton M. A., Gregory R. A., Hardy P. M., Kenner G. W., MacLeod J. K., Preston J., Sheppard R. C., and Morley J. S. (1964) The antral hormone gastrin. Synthesis of gastrin. *Nature* **204,** 933–934.

Ariens E. J., ed. (1964) *Molecular Pharmacology* Academic, London.

Bell P. H., Howard K. S., Shepherd R. G., Finn B. M., and Meisenhelder J. H. (1956) Studies with corticotropin. II. Pepsin degradation of beta-corticotropin. *J. Am. Chem. Soc.* **78**, 5059–5066.

Benovitz D. E. and Spatola A. F. (1985) Enkephalin pseudopeptides: Resistance to in vitro proteolytic degradation afforded by amide bond replacements extends to remote sites. *Peptides* **6**, 257–261.

Blake J., Ferrara P., and Li C. H. (1981) Beta-endorphin: Synthesis and radioligand binding activity of analogs containing cysteine bridges. *Int. J. Pept. Prot. Res.* **17**, 239–242.

Blake J., Helmeste D. M., and Li C. H. (1985) Beta endorphin: Synthesis and biological activity of analogs with disulfide bridge. *Int. J. Pept. Prot. Res.* **25**, 575–579.

Bodansky M. (1984) *Principles of Peptide Synthesis*. Springer-Verlag, New York.

Boissonnas R. A., Guttman S., Jaquenoud P. A., Konzett H., and Sturmer E. (1960) Synthesis and biological activity of peptides related to bradykinin. *Experientia* **16**, 326.

Caranikas S., Mizrahi J., D'Orleans-Juste P., and Regoli D. (1982) Antagonists of substance P. *Eur. J. Pharmacol.* **77**, 205–206.

Chan W. Y. and du Vigneaud V. (1962) Comparison of the pharmacologic properties of oxytocin and its highly potent analogue Desamino-Oxytocin. *Endocrinology* **71**, 977–983.

Chipens G. I., Mutulis F. K., Batayev B. S., Klusha V. E., Misina I. P., and Myshlyiakova N. V. (1981) Cyclic analogs of bradykinin. *Int. J. Pept. Prot. Res.* **18**, 302–311.

Chorev M. and Goodman M. (1983) Partially modified retro-inverso peptides. *Int. J. Pept. Prot. Res.* **21**, 258–268.

Chorev M., Shavitz R., Goodman M., Minick S., and Guillemin R. (1979) Partially modified retro-inverso enkephalinamides: Topochemical long-acting analogs in vitro and in vivo. *Science* **204**, 1210–1212.

Chou P. Y. and Fasman G. D. (1974) Prediction of protein conformation. *Biochemistry* **13**, 222–245.

Clausen K., Thorsen M., Lawesson S. O., and Spatola A. F. (1984) Studies on amino acids and peptides. 6. Methods for introducing thioamide bonds into the peptide backbone: Synthesis of the four monothio analogues of leucine enkephalin. *J. Chem. Soc. Perkin Trans.* I, 785–798.

Couture R., Fournier A., Magnan J., St-Pierre S., and Regoli D. (1979) Structure–activity studies on substance P. *Can. J. Physiol. Pharmacol.* **57**, 1427–1436.

Cox B. M., Goldstein A., and Li C. H. (1976) Opioid activity of a peptide, beta lipotropin-(61-91), derived from beta-lipotropin. *Proc. Natl. Acad. Sci. USA* **73**, 1821–1823.

Cox M. T., Gormley J. J., Hayward C. F., and Petter N. N. (1980) Incorporation of trans-olefinic dipeptide isosters into enkephalin and substance P analogs. *J. Chem. Soc. Chem. Commun.* 800–802.

Coy D. H., Kastin A. J., Schally A. V., Morin O., Caron N. G., Labrie F., Walker J. M., Fertel R., Berntson G. G., and Sandman C. A. (1976) Synthesis and opioid activities of stereoisomers and other D-amino acid analogs of methionine enkephalin. *Biochem. Biophys. Res. Commun.* **73**, 632–638.

Crawley J. N., St-Pierre S., and Gaudreau P. (1984) Analysis of the behavioral activity of C- and N-terminal fragments of cholecystokinin octapeptide. *J. Pharmacol. Exp. Ther.* **230**, 438–444.

Cuatrecasas P. (1971) Properties of the insulin receptor of isolated fat cell membranes. *J. Biol. Chem.* **246**, 7265–7274.

Cuello A. C., ed. (1982) *Co-Transmission* Macmillan, Basingstoke.

DeGrado W. F., Kezdy F. J., and Kaisr E. T. (1981) Design, synthesis and characterization of a cytotoxic peptide with melittin activity. *J. Am. Chem. Soc.* **103**, 679–681.

DiMaio J., Nguyen T. M. D., Lemieux C., and Schiller P. W. (1982) Synthesis and pharmacological characterization in vitro of cyclic enkephalin analogs: Effect of conformational constraints on opiate receptor selectivity. *J. Med. Chem.* **25**, 1432–1438.

Dutta A., Furr B. J. A., and Giles M. B. (1979) Polypeptides. 15. Synthesis and biological activity of alpha-aza-analogs of luliberin modified in positions 6 and 10. *J. Chem. Soc. Perkin Trans.* **I**, 379–388.

du Vigneaud V., Ressler C., Swan J. M., Roberts C. W., Katsoyannis P. G., and Gordon S. (1953) The synthesis of oxytocin. *J. Am. Chem. Soc.* **75**, 4879–4885.

Engberg G., Svensson T. H., Rosell S., and Folkers K. (1981) A synthetic peptide as an antagonist of substance P. *Nature* **293**, 222–223.

England R. D., Jones B. N., Flanders K. C., Coolican S. A., Rothgeb T. M., and Gurd R. S. (1982) Glucagon carboxyl-terminal derivatives: Preparation, purification and characterization. *Biochemistry* **21**, 940–950.

English M. L. and Stammer C. H. (1978) D-Ala^2,dehydro-Phe^4-Met-enkephalinamide, a dehydropeptide hormone. *Biochem. Biophys. Res. Commun.* **85**, 780–782.

Ewenson A., Laufer R., Chorev M., Selinger Z., and Gilon C. (1986) Ketomethylene pseudopeptide analogues of substance P: Synthesis and biological activity. *J. Med. Chem.* **29**, 295–299.

Finn F. M. and Hofmann K. (1976) The Synthesis of Peptides by Solution

Methods with Emphasis on Peptide Hormones, in *The Proteins*, 3rd Ed., vol. II (Neurath H. and Hill R. L., eds.) Academic, New York.

Fisher G. H., Marlborough D. I., Ryan J. W., and Felix A. M. (1978) L-3,4-Dehydroproline analogs of bradykinin: Synthesis, biological activity and solution conformation. *Arch. Biochem. Biophys.* **189,** 81–85.

Fournier A., Couture R., Regoli D., Gendreau M., and St-Pierre S. (1982) Synthesis of peptides by the solid phase method. 7. Substance P and analogs. *J. Med. Chem.* **25,** 64–68.

Fournier A., Saunders J. K., and St-Pierre S. (1984) Conformational studies and biological activities of VIP and related fragments. *Peptides* **5,** 169–177.

Freidinger R. M. and Veber D. F. (1984) Design of Novel Cyclic Hexapeptide Somatostatin Analogs from a Model of the Bioactive Conformation, in *Conformationally Directed Drug Design* (Vida J. A. and Gordon M., eds.) American Chemical Society, Washington, DC.

Freidinger R. M., Veber D. F., Perlow D. S., Brooks J. R., and Saperstein R. (1980) Bioactive conformation of luteinizing hormone-releasing hormone: Evidence from a conformationally constrained analog. *Science* **210,** 656–658.

Freychet P., Roth J., and Neville D. M. (1971) Insulin receptors in the liver: Specific binding of ^{125}I insulin to the plasma membrane and its relation to insulin bioactivity. *Proc. Natl. Acad. Sci. USA* **68,** 1833–1837.

Fuxe K., Agnati L. F., Rosell S., Harfstrand A., Folkers K., Lundberg J. M., Anderson K., and Hokfelt T. (1982) Vasopressor effects of substance P and C-terminal sequences after intracisternal injection to alpha-chloralose-anesthetized rats: Blockade by a substance antagonist. *Eur. J. Pharmacol.* **77,** 171–176.

Girard F., Bachelard H., St-Pierre S., and Rioux F. (1984) The contractile effect of bombesin, gastrin releasing peptide and various fragments in the rat stomach strip. *Eur. J. Pharmacol.* **102,** 489–497.

Goldstein A., Tachibana S., Lowney L. I., Hunkapiller M., and Hood L. (1979) Dynorphin (1-13), an extraordinarily potent opioid peptide. *Proc. Natl. Acad. Sci. USA* **76,** 6666–6670.

Grossman A., Savage M. O., Lytras N., Preece M. A., Sueiras-Diaz J., Coy D. H., Rees L. H., and Besser G. M. (1984) Responses to analogues of growth hormone-releasing hormone in normal subjects, and in growth-hormone deficient children and young adults. *Clin. Endocrinol.* **21,** 321–329.

Gund P., Androse J. D., Rhodes J. B., and Smith G. M. (1980) Three-dimensional molecular modeling and drug design. *Science* **208,** 1425–1431.

Houghten R. A. (1985) General method for the rapid solid phase synthesis of large number of peptides: Specificity of antigen–antibody interaction at the level of individual amino acids. *Proc. Natl. Acad. Sci. USA* **82**, 5131–5135.

Hruby V. J., Agarwal N. S., Griffen A., Bregman M. D., Nugent C. A., and Brendel K. (1981) Glucagon structure–function relationships using isolated rat hepatocytes. *Biochim. Biophys. Acta* **674**, 383–390.

Humblet C. and Marshall G. R. (1981) Three-dimensional computer modeling as an aid to drug design. *Drug Dev. Res.* **1**, 409–434.

Humphries J., Wan Y. P., Folkers K., and Bowers C. (1978) Inhibitory analogs of the luteinizing hormone-releasing hormone having D-aromatic residues in positions 2 and 6 and variation in position 3. *J. Med. Chem.* **21**, 120–123.

Iqbal M., Balaram P., Showell H. J., Freer R. J., and Becker E. L. (1984) Conformationally constrained chemotactic peptide analogs of high biological activity. *FEBS Lett.* **165**, 171–174.

Jolicoeur F. B., Barbeau A., Rioux F., Quirion R., and St-Pierre S. A. (1981). Differential neurobehavioral effects of neurotensin and structural analogs. *Peptides* **2**, 171–176.

Kaiser E. T. and Kezdy F. J. (1984) Amphiphilic secondary structure: Design of peptide hormones. *Science* **223**, 249–255.

Kaiser E. T., Lawrence D. S., and Rokita S. E. (1985) The chemical modification of enzymatic specificity. *Ann. Rev. Biochem.* **54**, 565–595.

Kenakin T. P. (1984) The classification of drugs and drug receptors in isolated tissues. *Pharmacol. Rev.* **36**, 165–222.

Khosla M. C., Leese R. A., Maloy W. L., Ferreira A. T., Smeby R. R., and Bumpus F. M. (1972) Synthesis of some analogs of angiotensin II as specific antagonists of the parent hormone. *J. Med. Chem.* **15**, 792–795.

Li C. H., Yamashiro D., Tseng L. F., Chang W. C., and Ferrara P. (1980) Beta-endorphin omission analogs: Dissociation of immunoreactivity from other biological activities. *Proc. Natl. Acad. Sci. USA* **77**, 3211–3214.

Ling N. and Guillemin R. (1976) Morphinomimetic activity of synthetic fragments of beta-lipotropin and analogs. *Proc. Natl. Acad. Sci. USA* **73**, 3308–3310.

Ling N., Zeytin F., Bohlen P., Esch F., Brazeau P., Wehrenberg W. B., Baird A., and Guillemin R. (1985) Growth hormone releasing factors. *Ann. Rev. Biochem.* **54**, 403–423.

Marshall G. R. (1984) Structure–activity studies: A three-dimensional probe of receptor specificity. *Ann. NY Acad. Sci.* **439**, 162–169.

McMurray J. S. and Dyckes D. F. (1985) A simple and convenient method for the preparation of ketomethylene peptide analogs. *J. Org. Chem.* **50**, 1112–1115.

Merrifield R. B. (1963) Solid phase peptide synthesis. I. The synthesis of a tetrapeptide. *J. Am. Chem. Soc.* **85**, 2149–2154.

Meyers C. A., Coy, D. H., Murphy W. A., Redding T. W., Arimura A., and Schally A. V. (1980) (Phe[4])-Somatostatin: A potent, selective inhibitor of growth hormone release. *Proc. Natl. Acad. Sci. USA* **77**, 577–579.

Mitchell A. R., Kent S. B. H., Engelhard M., and Merrifield R. B. (1978) A new synthetic route to tert-butyloxycarbonylaminoacyl-4-(oxymethyl) phenylacetamidomethyl-resin, an improved support for solid phase peptide synthesis. *J. Org. Chem.* **43**, 2845–2852.

Moe G. R., Miller R. J., and Kaiser E. T. (1983) Design of a peptide hormone: Synthesis and characterization of a model peptide with calcitonin-like activity. *J. Am. Chem. Soc.* **105**, 4100–4102.

Monahan M. W., Amoss M. S., Anderson H. A., and Vale, W. (1973) Synthetic analogs of the hypothalamic luteinizing hormone releasing factor with increased agonist or antagonist properties. *Biochemistry* **12**, 4616–4620.

Mosberg H. I., Hruby V. J., and Meraldi J. P. (1981) Conformational study of the potent peptide hormone antagonist (1-penicillamine, 2-leucine)-oxytocin in aqueous solution. *Biochemistry* **20**, 2822–2828.

Natarajan S., Gordon E. M., Sabo E. F., Godfrey J. D., Weller H. N., Pluscec J., Rom M. B., and Cushman D. W. (1984) Ketomethyldipeptides. I. A new class of angiotensin converting enzyme inhibitors. *Biochem. Biophys. Res. Commun.* **124**, 141–147.

Nicolaides E. D. and Lipnik M. (1966) Synthetic bradykinin analogs. *J. Med. Chem.* **9**, 958–960.

Nicolas P. and Li C. H. (1985) Inhibition of analgesia by C-terminal deletion analogs of human beta-endorphin. *Biochem. Biophys. Res. Commun.* **127**, 649–655.

Ondetti M. A., Pluscec J., Sabo E. F., Sheehan J. T., and Williams N. (1970) Synthesis of cholecystokinin-pancreozymin. I. The C-terminal dodecapeptide. *J. Am. Chem. Soc.* **92**, 195–199.

Osterman D., Mora R., Kzdy F. J., Kaiser E. T., and Meredith S. C. (1984) A synthetic amphiphilic beta-strand tridecapeptide: A model for apolipoprotein B. *J. Am. Chem. Soc.* **106**, 6845–6847.

Park W. K., St-Pierre S., Barabe J., and Regoli, D. (1978) Synthesis of peptides by the solid phase method. III. Bradykinin, fragments and analogs. *Can. J. Biochem.* **56**, 92–100.

Patchett A. A., Harris E., Tristram E. W., Wyvratt M. J., Wu M. T., Taub D., Petersen E. R., Ikeler T. J., Broeke J., Payne L. G., Ondeyka D. L., Thorsett E. D., Greenlee W. J., Lohr N. S., Hoffsommer R. D., Joshua

H., Ruyle W. V., Rothrock J. W., Aster S. D., Maycock A. L., Robinson F. M., Hirschmann R., Sweet C. S., Ulm E. H., Gross D. M., Vassil T. C., and Stone C. A. (1980) A new class of angiotensin-converting enzyme inhibitors. *Nature* **288**, 280–283.

Pert C. B., Pert A., Chang J. K., and Fong B. T. W. (1976) (D-Ala^2)-Met-Enkephalinamide: A potent, long lasting synthetic pentapeptide analgesic. *Science* **194**, 330–332.

Pietta P. G., Cavallo P. F., Takahashi K., and Marshall G. R. (1974) Preparation and use of benzhydrylamine polymers in peptide synthesis. II. Syntheses of thyrotropin releasing hormone, thyrocalcitonin 26-32, and eledoisin. *J. Org. Chem.* **39**, 44–48.

Ploux O., Lavielle S., Chassaing G., Julien S., Marquet A., Beaujovan J. C., Bergstrom L., and Glowinski J. (1986) Conformationally restricted analogs of substance P. *Symposium on Substance P and Neurokinins*, Montreal, July 21–23.

Quirion R., Rioux F., Regoli D., and St-Pierre S. A. (1980) Selective blockade of neurotensin-induced coronary vessel constriction in perfused rat hearts by a neurotensin analog. *Eur. J. Pharmacol.* **61**, 309–312.

Rasmussen H. (1960) The purification of parathyroid polypeptides. *J. Biol. Chem.* **235**, 3442–3448.

Regoli D., Park W. K., and Rioux F. (1974) Pharmacology of angiotensin. *Pharmacol. Rev.* **26**, 69–123.

Regoli D. and Barabe J. (1980) Pharmacology of bradykinin and related kinins. *Pharmacol. Rev.* **32**, 1–46.

Regoli D., Escher E., Drapeau G., D'Orleans-Juste P., and Mizrahi J. (1984) Receptors for substance P. III. Classification by competitive antagonists. *Eur. J. Pharmacol.* **97**, 179–189.

Riniker B. and Schwyzer R. (1964) Synthetische analoge des hypertensin. V Alpha-L, Beta-L, alpha-D und beta-D-Asp^1-Val^5-hypertensin II. *Helv. Chim. Acta* **47**, 2357–2374.

Robberecht P., Coy D. H., Waelbroeck M., Heiman M. L., De Neef P., Camus J. C., and Christophe J. (1985) Structural requirements for the activation of rat anterior pituitary adenylate cyclase by GRF: Discovery of (N-Ac-Tyr^1,D-Arg^2)-GRF (1-29)-NH$_2$ as a GRF antagonist on membranes. *Endocrinology* **117**, 1759–1764.

Rosenblatt M. (1984) Parathyroid Hormone: Intracellular Transport, Secretion, and Receptor Interaction, in *Peptide and Protein Reviews*, vol. 2, (M.T.W. Hearn, ed.) Marcel Dekker, New York.

Rudinger J. (1971) The Design of Peptide Analogs, in *Drug Design*, vol. II (Ariens E. J., ed.) Academic, New York.

Salvadori S., Menegatti E., Sarto G., and Tomatis R. (1981) Aminoxy analogs of leu-enkephalin. *Int. J. Pept. Prot. Res.* **18**, 393–401.

Sawyer T. K., Hruby V. J., Darman P. S., and Hadley M. E. (1982) (Half-Cys[4], half-Cys[10])-alpha-melanocyte-stimulating hormone: A cyclic alpha-melanotropin exhibiting superagonist biological activity. *Proc. Natl. Acad. Sci. USA* **79**, 1751–1755.

Schally A. V., Coy D. H., and Meyers C. M. (1978) Hypothalamic regulatory hormones. *Ann. Rev. Biochem.* **47**, 89–128.

Schattenkerk C. and Havinga E. (1965) Studies on polypeptides. IV. Syntheses and activities of some angiotensin-like peptides. *Rec. Trav. Chim. Pays-Bas* **84**, 653–658.

Schiffer M. and Edmunson A. B. (1967) Use of helical wheels to represent the structures of proteins and to identify segments with helical potential. *Biophys. J.* **7**, 121–135.

Stewart J. M. and Young J. D. (1984) *Solid Phase Peptide Synthesis*, 2nd Ed., Pierce Chemical Company, Rockford, Illinois.

St-Pierre S. A., Kerouac R., Quirion R., Jolicoeur F. B., and Rioux F. (1984) Neurotensin, in *Peptide and Protein Review*, vol. 2 (M.T.W. Hearn, ed.) Marcel Dekker, New York.

St-Pierre S. A., Lalonde J. M., Gendreau M., Quirion R., Regoli D., and Rioux F. (1981) Synthesis of peptides by the solid phase method. 6. Neurotensin, fragments and analogs. *J. Med. Chem.* **24**, 370–376.

Sueiras-Diaz J., Lance V. A., Murphy W. A., and Coy D. H. (1984) Structure–activity studies on the *N*-terminal region of glucagon. *J. Med. Chem.* **27**, 310–315.

Taylor J. W., Osterman D. G., Miller R. J., and Kaiser E. T. (1981) Design and synthesis of a model peptide with beta-endorphin-like properties. *J. Am. Chem. Soc.* **103**, 6965–6966.

Terenius L., Wahlstrom A., Lindeberg G., Karlsson S., and Ragnarsson U. (1976) Opiate receptor affinity of peptides related to Leu-enkephalin. *Biochem. Biophys. Res. Commun.* **71**, 175–179.

Turcotte A., Lalonde J. M., St-Pierre S. A., and Lemaire S. (1984) Dynorphin-(1-13). 1. Structure-function relationships of Ala-containing analogs. *Int. J. Pept. Prot. Res.* **23**, 361–367.

Ukai M., Itatsu T., Shibata A., Rioux F., and St-Pierre S. (1982) Inhibition of neurotensin-induced glucagon release by a neurotensin analogue *Experientia* **38**, 1222–1224.

Vale W., Rivier C., and Brown M. (1977) Regulatory peptides of the hypothalamus. *Ann. Rev. Physiol.* **39**, 473–527.

Veber D. F., Saperstein R., Nutt R. F., Freidinger R. M., Brady S. F., Curley P., Perlow D. S., Palaveda W. J., Colton C. D., Zacchei A. G., Tocco D. J., Hoff D. R., Vandlen R. L., Gerich J. E., Hall L., Mandarino L., Cordes E. H., Anderson P. S., and Hirschmann R. (1984) A superactive cyclic hexapeptide analog of somatostatin. *Life Sci.* **34**, 1371–1378.

Venkatachalam C. M. (1968) Stereochemical criteria for polypeptides and proteins. V. Conformation of a system of three linked peptide units. *Biopolymers* **6**, 1425–1436.

Voskamp D., Kranenburg P., and Beyerman H. C. (1982) The role of histidine in the gastrointestinal hormone secretin. *Rec. Trav. Chim. Pays-Bas* **101**, 393–396.

Yabe Y., Miura C., Horikoshi H., Miyagawa H., and Baba Y. (1979) Synthesis and biological activity of LH-RH analogs substituted by alkyltryptophans at position 3. *Chem. Pharm. Bull.* **27**, 1907–1911.

Identification of Neuropeptide Receptors

Daniel M. Dorsa and Denis G. Baskin

1. Introduction

Recently there has been an explosion of information on peptides that may act as neurotransmitters or neuromodulator substances in the brain and periphery. Many investigators interested in the effects of neuropeptides have made use of radioligand binding techniques to identify potential receptors in target tissues. In this chapter we discuss general approaches that have been used to study peptide receptors, with special emphasis on neuroanatomical aspects of these methodologies.

2. Characteristics of Receptors for Neuropeptides

2.1. Comparison of Peptide and Nonpeptide Transmitter Receptors

Receptors for neuropeptides are similar to those of other amino acid transmitters in basic characteristics such as saturability, reversibility, and specificity. Generally speaking, as a group they tend to have higher binding affinity (K_d values of approximately $0.1–10 \times 10^{-9}M$) in comparison to amino acid transmitters (K_d values of approximately $10^{-6}M$) when measured with radiolabeled versions of their endogenous ligands. The reason for this difference is not yet clear. It has been suggested that the relatively low affinities of dopamine and norepinephrine (noradrenaline) for their receptors may be related to the high synaptic concentration of these transmitters. Why, then, do peptides exhibit higher affinities? Perhaps the synaptic concentration of peptides is lower than is the case for classical transmitters. The difference in affinity may also be an indication of possible differences in functions of peptidergic synapses. It may be that the "background leak" rate of peptidergic neurons is low and that peptide receptors display high affinity because neuronal activation leads to rapid changes in synaptic peptide concentration, but at levels considerably below those of other nonpeptide transmitters. This notion is further supported by the fact that synaptic reuptake is a prominent feature of amino acid transmitter neurons, but apparently not of most peptidergic ones. A neuron that can recapture its released transmitter can "afford" to leak transmitter at a higher rate.

2.2. Peptide Receptor Subtypes

It is clear from studies of receptors for amino acid and amine transmitters that isotypes or subtypes of receptors exist both centrally and peripherally. In catecholaminergic systems, the existence of $\alpha_{1,2}$ and $\beta_{1,2}$ receptors and subtypes of receptors for dopamine (D_1, D_2, D_3) has been well documented. Subtypes of acetylcholine receptors (nicotinic and muscarinic) also exist. These subtypes can be distinguished on the basis of potency of various agonists and antagonists in displacing radiolabeled ligand from binding. They can also be distinguished by their functional characteristics, in that certain subtypes appear to function as autoreceptors (α_2, D_2) (e.g., receptors regulating the rate of secretion of presynaptic transmitter into the synaptic cleft). Anatomical distinctions can also be made, such as pre- or postsynaptic locations, but these differences seem to be less useful because some autoreceptors (such as α_2 adrenergic receptors) may exist both pre- and postsynaptically (Schmitt and Schmitt-Jubeau, 1984). Finally, receptor subtypes can be distinguished on the basis of postreceptor events mediating the binding signal to intracellular components, as is exemplified by the adrenergic α and β receptor systems. A clear distinction exists between β_1 receptors, which use cyclic-AMP-dependent protein kinase as second messenger, and α_1 receptors, which are not linked to adenylate cyclase and instead affect phosphatidylinositol turnover and alter Ca^{2+} flux in target cells (Creba et al., 1983).

In the case of peptides, the existence of receptor subtypes has been well documented for opiate peptides. Table 1 summarizes the characteristics of receptor subtypes for opiates. These receptors are

Table 1
Opiate Receptor Subtypes

Type	Prototypic agonist	Endogenous ligand	Bioassay
Mu (μ)	Morphine	β-Endorphin, met-enkephalin	Smooth muscle, guinea pig ileum
Kappa (κ)	Ethylketo cyclazocine	Dynorphin, α-neoendorphin	Smooth muscle, guinea pig ileum
Sigma (σ)	N-Allylnor metazocine, phencyclidine	Endopsycosin	Spinal dog preparation
Delta (δ)	Leu-enkephalin	Leu-enkephalin, others	Smooth muscle mouse, vas deferens
Epsilon (ε)	β-Endorphin	β-Endorphin	Rat colon

defined on the basis of tissue bioassays (usually smooth muscle), displacement potency of various peptide ligands, as well as exogenous opiates, and the potency of naloxone (a general antagonist) in reversing both the binding and in vitro and in vivo effects of opiate agonists (*see* Martin, 1984, for review). The primary reason so much is known about this class of neuropeptides is that exogenous agonists and antagonists of opiate receptors have been available for many years. The interaction of peptide ligands with these receptors is just now being clarified.

Vasopressin (VP) is another neuropeptide for which there is evidence of receptor subtypes. Michell et al. (1979) postulated that at least two subtypes of vasopressin receptors exist. The first is the pressoric, or V_1, receptor, which is found in vascular smooth muscle and hepatic tissue. The second is the antidiuretic, or V_2, receptor, which has so far been found only in the renal medulla. The V_2 receptor is linked to adenylate cyclase activity and is thought to influence water transport by initially increasing cAMP levels within renal tubular epithelial cells. In contrast, the V_1 receptor is not cyclase linked and alteration of phosphatidylinositol turnover and Ca^{2+} flux serves as a second messenger system in this case (Thomas et al., 1984). In addition, thanks to the elegant work done over a number of years by Manning and Sawyer (1984), numerous analogs of vasopressin with agonist and antagonist activity have become available, allowing for further characterization of the two arginine[8]-vasopressin (AVP) receptor subtypes. Agonists of the V_2 receptor, such as deamino D-arginine[8]-vasopressin (dDAVP), can clearly distinguish between the two receptor sites. Antagonists of the V_1 receptor [such as 1-(β-mercapto-β,β-cyclopentamethylene proprionic acid), 2-(O-methyl)tyrosine-Arg[8]-vasopressin] also are several times more potent in labeling vasopressin receptors in the liver than in the kidney.

Vasopressin receptors have been reported to be present in several other tissues, including the anterior pituitary (Antoni, 1984), testis (Meidan and Hsueh, 1985), adrenal cortex (Balla et al., 1985), and brain (Dorsa et al., 1983; Cornett and Dorsa, 1985; Baskin et al., 1983b; Biegon et al., 1984). The brain receptor will be discussed in the last section of this chapter. The other peripheral receptors have been similar in most respects to the V_1 subtype. The VP receptor in the pituitary may be somewhat different than the V_1 receptor, however, because certain V_1 antagonists display reduced potency in displacing ^{3}H-AVP in this tissue (Antoni, 1984; Lutz-Bucher and Koch, 1983).

Recent evidence suggests that receptors for cholecystokinin (CCK)-related peptides in pancreatic and brain tissue also demonstrate heterogeneity. The pancreatic CCK receptor seems to require the entire C-terminal octapeptide (CCK_8), containing a sulfated tyrosine for binding. Brain receptors for this class of peptides, however, seem to be less specific. Desulfated CCK_8, CCK_4, and gastrin are also relatively potent competitors for binding (Hays et al., 1980; Innis and Snyder, 1980; Praissman et al., 1983; Saito et al., 1981; 1980; Steigerwalt and Williams, 1981).

Insulin receptors have been reported to exist in mammalian brain tissue (Havrankova et al., 1978; Baskin et al., 1987). Heidenreich et al. (1983) have examined the physiochemical properties of insulin receptors in brain and found that the molecular weight, antigenicity, and carbohydrate composition of the brain receptor was different than that observed for peripheral-type insulin receptors, such as in adipose tissue. Interestingly, the specificity of insulin receptors in the brain is quite similar to those in peripheral receptors (Baskin et al., 1987; *see* section 5.2).

In the somatostatin system, an interesting story is emerging relating to receptors for various endogenous analogs of this peptide. It has long been known that both the tetradecapeptide so-matostatin (SS14) and the larger form, somatostatin 28 (SS28), are biologically active. Recent evidence from Tran et al. (1985) suggests that each form has separate receptors that show specificity for one or the other peptide. By labeling receptors using iodinated SS14 and SS28, they found that receptors in pituitary and pancreas showed monophasic displacement curves using SS analogs, but that brain membranes showed biphasic displacement characteristics, indicating the presence of two different receptor populations that are differentially distributed in various brain regions.

In summary, neuropeptides, like other neurotransmitters, may have complex receptor systems mediating their actions in the central nervous system (CNS). Our discussion of peptides is not intended to be exhaustive, but rather represents the trend in current information that is accumulating at an impressive rate. The dichotomy between central and peripheral receptor subtypes for neuropeptides is not a new concept, however. De Wied and coworkers, using behavioral assays, have been able to show with several peptides (including vasopressin, ACTH, and endorphins) that modification of the amino acid structure of the parent molecules can completely dissociate the central and peripheral activities of the molecules (De Wied et al., 1972). The existence of specific receptors for these fragments in the CNS has not yet been convincingly demonstrated, however.

2.3. Peptide Radioligands

Labeling neuropeptide receptors with antagonist would be a preferable approach for binding studies, since antagonists afford significant advantages in metabolic stability, including lack of binding to nonreceptor sites, such as enzymes and transporters, and insensitivity to receptor regulatory transitions. Unfortunately only a few antagonists of neuropeptide receptors have been synthesized. Potent antagonists for leutinizing hormone-releasing hormone (LHRH) (such as [D-Phe^2, Pro^3, D-Phe^6]-LHRH) have been synthesized and used for binding studies (Humphries et al., 1978). Antagonists of substance P receptors ([D-Pro^2, D-Trp^{7,9}]-substance P) and for angiotensin receptors ([Sar^1, Val^5, Ala^9]-angiotensin II) (Engberg et al., 1981) have also been reported. Several potential antagonists of the V$_1$-type vasopressin receptor have been available for several years, and are mentioned in section 2.2 of this article.

Several radionuclides are currently being used to study peptide receptors using radioligand binding techniques. Tyrosine-containing peptides can, of course, be iodinated using chloramine-T or lactoperoxidase methodologies. This usually provides a high-specific-activity label, and has been successfully used for peptides such as insulin, ACTH, VIP, and others. Peptides containing more than one tyrosine must frequently be chromatographically purified to monoiodoforms, such as monoiodoinsulin (18). Standard iodination procedures can destroy activity of some peptides that contain methionine or cysteine resulting from oxidation (Heyward et al., 1979). Iodination may not be a usable approach for some peptides containing tyrosine if the tyrosine is important for receptor recognition or if iodination dramatically alters the tertiary structure of the peptide. Iodination of arginine^8-vasopressin, for example, which contains tyrosine in its ring structure, destroys the biological activity of the molecule. Tritiation and ^{35}S labeling of cysteine residues have provided other useful peptide receptor probes, many of which are now commercially available.

Selection of a label may also be influenced by the method in which it is to be used. For example, the low energy β emission of tritium, although useful for autoradiography with tritium-sensitive film, may require up to 10 wk of exposure to allow for full image formation for analysis. In contrast, the higher specific activity possible with ^{125}I makes exposures of only a few days possible. This will be discussed later.

3. Membrane Preparations and Standard Radioligand Binding Techniques

Many investigators have used crude or semipurified membrane preparations from brain tissue to identify and characterize CNS peptide receptors. These methods have been well developed and several excellent reviews on the techniques are available (Levitzki, 1984; Laduron, 1984). In most of these studies the brain is either homogenized whole or dissected into gross anatomical regions such as hypothalamus, midbrain, hindbrain, cortex, and so on. The advantages of this approach are that data on pharmacologic characteristics of binding sites for peptides (for example, specificity and affinity data) in the CNS can be rapidly obtained. By dissecting the brain, gross anatomical information is also afforded the investigator.

Radioligand binding studies of neuropeptide receptors pose several problems peculiar to this class of transmitters. An excellent review of this topic has been published (Hanley, 1985). Peptides have a propensity to adsorb nonspecifically to many materials commonly used in binding assays, including glass and plastic polymers. Adsorption to glass fiber filters, which are commonly used to separate bound and free ligand, can be a significant problem. Several methods of circumventing this problem have been used, including presoaking filters in bovine serum albumin or other large inexpensive proteins. We have used filters soaked in unlabeled ligand in some of our assays, but this can be an expensive method. Polyethylene imine has also been used with some success to reduce nonspecific binding (Lee et al., 1983). We have also found that addition of ethanol in low concentrations to binding assays reduces nonspecific binding of vasopressin (Dorsa et al., 1983). Other precautions that are advisable include use of siliconized test tubes and pipets to minimize adsorption to glass surfaces.

Tracer degradation can be a major problem in peptide assays, especially in brain membrane preparations in which protease activity can be quite high. Use of specific inhibitors of suspect peptidases or low-temperature incubations (e.g., 4°C) are frequently required to minimize this problem.

Many neuropeptide assays seem to be extremely ion-dependent, and this must be thoroughly examined by the investigator during the process of setting up an assay. For example, binding of VIP (Taylor and Pert, 1979) and vasopressin (Jard, 1983; Dorsa et al., 1983) requires Mg^{2+}, and that of substance P is facilitated by Mn^{2+} (Lee et al., 1983).

There are several disadvantages to membrane binding techniques, however, especially as applied to the brain. The brain is the most anatomically complex organ of the body. The first level of complexity involves the cell types. It is usually assumed, although not proven, that most of the binding present in crude brain membrane preparations is bound to neuronal membranes. It is likely however, that glial and stromal components also contribute to binding. Binding to nonneuronal elements can be avoided in part, however, by use of more purified preparations, such as synaptosomes, which are thought to represent principally vesicular pre- and postsynaptic membrane components.

The second disadvantage is specific to certain peptides. The density of receptors in grossly dissected brain regions may be extremely low. This makes it difficult to increase the signal-to-noise ratio to a level high enough for binding studies. In our own experience with vasopressin binding to brain membranes, we found average densities of binding sites in grossly dissected brain regions to be from 15 to 50 fmol/mg membrane protein. This is significantly lower than the densities for other classical transmitters, which may be hundreds of fmol or pmol/mg protein. The consequence of this situation is that, in the rat, grossly dissected brain tissue parts must be pooled together to obtain enough membrane protein to carry out a standard saturation analysis or specifity study of peptide binding. This is usually not a problem for the initial characterization of binding, but when the investigator asks questions about regulation of affinity and number of binding sites for peptides, pooling of animals increases greatly the difficulty in obtaining meaningful data. In addition, when expensive animal models such as aged rats (which cost approximately $90 per animal if obtained from the National Institute on Aging) are being used, the cost can be prohibitive.

One possible reason for relatively low peptide receptor density in the brain is the highly specific anatomical distribution of many brain receptors. For example, using autoradiographic approaches (to be described below) to localize receptors for vasopressin, we have found that within a complex, but relatively small, structure such as the amygdala, vasopressin receptors are almost exclusively localized in the central nucleus of that region. Therefore use of tissue that includes the entire structure merely includes neural tissue that is not relevant to the binding, usually resulting in increased nonspecific binding. This might be circumvented by microdissection of the central nucleus itself using the Palkovits (1975) punch technique, but harvesting enough punches to yield enough membrane protein is not a practical approach in most circumstances.

Finally, since it has already become clear that some peptide receptors exist in isotypes (e.g., opiate receptors), it is important to bear in mind that these subtypes may exist in discrete anatomical locations, perhaps in nuclei not too distant from one another. If so, both would be included in membranes from grossly dissected brain pieces. Heterogeneity of labeling might be avoided by carefully selecting ligands that bind specifically to one class of binding site, but when using the endogenous peptide ligands themselves, this is difficult. For example, ^{3}H-endorphin binds avidly to mu, delta, and epsilon opiate receptors, all of which are present throughout the brain.

It is for the above reason that some investigators have adopted techniques of in vitro binding of radioligands to tissue sections in conjunction with quantitative autoradiography as an approach for identifying peptide receptors in the CNS.

4. Methods for Anatomical Localization of CNS Peptide Receptors

This section discusses several approaches to the anatomical localization of peptide receptors. The aim here is to review the general principles of these techniques and point out aspects that deserve particular attention for the investigator who is interested in quantitative measurements. The emphasis will be on autoradiographic techniques for receptor localization because this approach is relatively easy to use and can yield accurate quantitative data for pharmacological characterization of peptide binding sites. The outstanding value of autoradiographic techniques for localizing receptors is the ability to visualize binding sites in an anatomically complex organ such as the brain. Autoradiography also permits pharmacological characterization of a receptor in the anatomical environment in which it normally functions. Basic autoradiographic techniques can reveal the anatomical localization of a binding site, but identifying a receptor requires a rigorous quantitative approach for characterizing the pharmacology of binding. Immunocytochemical methods for anatomical localization of receptors with antireceptor antibodies cannot characterize the binding properties of a receptor, and hence will not be discussed here.

4.1. In Vivo Autoradiography

4.1.1. Basic Principles

Localization of peptide receptors by in vivo autoradiography involves injection of a labeled ligand into the bloodstream and sacrifice of the animal after a short period of time. The basis for this approach is the assumption that a labeled ligand will bind to receptors in tissues under relatively physiological conditions. Generally the animal is placed under anesthesia and a bolus of the tracer is injected into the carotid artery (in the case of a brain perfusion) or left cardiac ventricle. After about 2–5 min, the animal is perfused with a fixative solution such as Bouin's or glutaraldehyde, which presumably crosslinks the ligand to its receptor and surrounding proteins. The organ is then embedded in paraffin and processed for autoradiography by conventional liquid emulsion methods. Localization of ligand binding is done by observing the autoradiographic grains at high magnification with a microscope.

4.2. Advantages of In Vivo Autoradiography

One merit of the *in vivo* approach is the relative simplicity of the technique. A labeled and fixed organ such as the brain can be embedded in paraffin and serial sections can be processed for autoradiography dipping in liquid emulsion. After exposure and development, the tissue slice can be stained to produce preparations with excellent morphological detail. Radioactivity lost during tissue processing can be determined by counting tissue samples taken at different stages of the processing procedure. The location of ligand binding can be visualized with high morphological resolution to specific cells by light microscopy. Adapting in vivo autoradiography for electron microscopy permits subcellular localization of binding sites, with the added advantage of being able to follow the internalization and intracellular processing following surface binding.

In vivo autoradiography has been used for light microscopic detection of brain binding sites for peptides in the blood (van Houten and Posner, 1981) and cerebrospinal fluid (Baskin et al., 1983a). Examples of electron microscopic localization of peptide binding sites, both of cells in situ and in culture, are shown by use of ferretin-labeled insulin (Jarrett and Smith, 1975; Nelson et al., 1978) and [^{125}I]-insulin (Bergeron et al., 1977; 1980a,b; Gorden et al., 1980; Patel et al., 1982). Similar approaches have been used to localize binding sites in vivo for other peptides.

4.3. Disadvantages of In Vivo Autoradiography

The in vivo autoradiographic approach has several drawbacks that limit its usefulness for characterizing peptide receptors, particularly those in the central nervous system. The principal limitation is the difficulty of carrying out ligand binding under equilibrium conditions that are required for establishing the pharmacological binding parameters. The short in vivo exposure of ligand to the tissue precludes saturating the receptors. In addition, it is practically impossible to accurately measure the time and temperature dependency of binding, reversibility, and the relative affinity of series of agonists or antagonists for the receptor, and to separate rapidly the bound from free ligand. It is feasible to coinject varying amounts of unlabeled ligand with the labeled ligand and assess relative degree of competition by grain counting. But this approach is cumbersome and the results are equivocal because of the lack of equilibrium conditions and the possible loss of unknown amounts of labeled ligand from specific binding sites during tissue-processing procedures. Unfortunately, lengthening the in vivo exposure time to achieve equilibrium may result in artifacts from degradation of the tracer ligand and ligand internalization. Further, a major drawback of in vivo autoradiography with peptides for brain receptors is that blood-borne tracers may not have access to receptors on the other side of the blood–brain barrier. In summary, the in vivo autoradiographic approach is useful insofar as it can show probable sites of uptake of peptides from the blood or CSF and, when coupled with coinjection of unlabeled ligand, reveal putative specific binding sites. It is not, however, the best approach for determining if a binding site is a receptor.

4.4. In Vitro Autoradiography

Visualization of peptide receptors by in vitro autoradiography is done by incubating tissue slices containing the receptor in a solution containing the labeled ligand and then placing the tissue slice in contact with a photographic emulsion (Kuhar, 1982). In practice this is generally accomplished with cryostat sections of frozen tissue. The usual procedure is to perfuse the organ *in situ* with ice-cold saline and then remove it surgically. Perfusing with a weak (0.1%) paraformaldehyde solution has been found to improve morphological preservation without greatly changing the specific binding ratios of some peptides (Young and Kuhar, 1979). A mild fixation may be useful when cellular resolution is a goal, such as is possible with emulsion-coated coverslips, or when the principal aim is anatomical localization of binding sites. Ideally, fixation should probably be avoided when the goal is to characterize binding affinity or site number. In our laboratory, in vitro autoradiography is routinely carried out using unfixed tissue.

After removal, the organ is chilled further in saline at 0°C for several minutes and is then rapidly frozen. This can be done with crushed dry ice, but we prefer to immerse the organ in isopentane cooled to –35°C with dry ice or in liquid Freon 22 (–40°C). Care must be exercised to prevent the organ from cracking, which will cause difficulty in preparing cryostat sections. Usually about 10–15 s in Freon 22 is sufficient for a rat brain. This is followed by burying the brain in crushed dry ice for 10 min, then storing in sealed plastic bags at –70°C.

4.4.1. Preparation of Slide-Mounted Sections

Sections (10–20 μm thick) are cut in a cryostat and thaw-mounted onto gelatin "subbed" slides. The optimum temperature for sectioning various organs differs. We find that –12°C is optimum for rat brain. We use a Bright cryostat, which permits the brain to be warmed to –12°C while keeping the chamber and knife at a lower temperature and has an orientable specimen holder. For brain sections it is advantageous to have an orientable specimen holder so that the sections can be oriented to match brain atlas figures. Sections are placed on precooled subbed slides and allowed to warm by touching the opposite side of the slide with a fingertip. The section is then dried on a warming plate at about 37°C for a few minutes and then returned to the cryostat chamber. Slides are stored in plastic slide boxes (the size that holds 25 slides each) with desiccant and kept at –70°C until used.

4.4.2. Assay Procedures

The radioreceptor assay is carried out on the slide-mounted tissue slice. The optimum binding conditions (time, temperature, ligand concentrations, and so on) can be determined from experiments in which labeled sections are wiped from the slide and counted for radioactivity or from membrane binding assays. These conditions will vary depending upon the ligand and the aim of the experiment.

Incubations can be carried out by placing the slides in a coplin-type staining jar for an appropriate period of time. An alternative procedure that we use frequently is to lay the slides in a tray lined with wet paper towels and cover the sections with the ligand solution in a volume of about 100 μL/slice. The tray is covered to form a humid chamber that retards evaporation during the incubation period. The tray-processing procedure is recommended when only small volumes of trace solutions are available. At the end of the incubation period, slides are rinsed in ice-cold buffer to remove unbound ligand and briefly dipped in distilled water to remove buffer salts. Slides are dried either by placing them on a hot plate at 60–70°C for a few seconds, or air drying with a cool stream of air.

We have not observed marked differences in the autoradiographic results between these drying procedures, and the former is much quicker. Dried labeled slides are then used for autoradiography. Sections that are counted for radioactivity are wiped with filter paper after the buffer rinse, and the filter paper is counted in a scintillation or gamma counter.

4.4.3. Suggested Sequence of Studies

An investigator planning to initiate an autoradiographic study of a peptide receptor can save time by carrying out preliminary studies in which the tissue section is wiped and counted for radioactivity. Initial protocols should be set up using binding conditions that have been successful with membrane preparations, if this information is available. We have found it useful to first do an assay in which pairs of slide-mounted sections are exposed to labeled ligand ("hot") or to labeled ligand mixed with 100–1000-fold excess unlabeled competitor ("hot/cold"). The labeled ligand concentration should be no higher than the estimated K_d of the ligand. These sections are then wiped and counted for radioactivity and the percent specific binding calculated. Specific binding is the binding in the presence of the unlabeled ligand (total binding) minus binding in the presence of excess unlabeled competing ligand (nonspecific binding). This result will indicate whether the binding conditions are likely to reveal binding sites by autoradiography. This protocol can then be repeated with autoradiography in order to determine if the binding is localized. The autoradiography should be done even if the percent specific binding is very low, even 10–20% total. This result alone could discourage further studies. A high level of binding in a small anatomical region may, however, contribute a small fraction of total counts bound to a large tissue slice. As we found with localization of vasopressin binding sites in the brain (Baskin et al., 1983b; Dorsa et al., 1983, 1984), autoradiography may reveal that most of the specific binding is present in a few small and discrete anatomical sites. Once it has been established that binding sites are present and can be visualized, it is then necessary to proceed with protocols that demonstrate time/temperature dependency, reversibility, saturability, and affinity characteristics. For a receptor concentrated in a small anatomical region, these characteristics may have to be determined by quantitative autoradiography. This requires measuring the localized binding by densitometry.

4.4.4. Autoradiography Procedure

Two autoradiographic approaches have been used successfully to visualize the location of radioactivity bound to tissue slices by in vitro incubations. The most common technique currently in use involves LKB Ultrofilm because of its simplicity and ease of quantitative analysis by densitometry (Ehn and Larsson, 1979; Wamsley and Palacios, 1983; Baskin and Dorsa, 1986). The alternative method, which uses emulsion-coated coverslips (Young and Kuhar, 1979), is also widely used for visualizing the location of receptor sites. Herkenham and Pert (1982) have elegantly demonstrated the high resolution and superior morphology that can be obtained with the use of the liquid emulsion method when combined with dry formaldehyde vapor fixation. This technique is more technically demanding and less convenient for quantitation by computer densitometry (which is more suitable for grain counting). It generally requires a darkfield stereomicroscope for visualizing the binding sites, but has the advantage that it includes the tissue slice and can thus provide accurate anatomical correlation of the binding sites. The LKB Ultrofilm method, in contrast, is better suited for pharmacological analysis by computerized densitometry and has been widely used for this purpose (Geary and Wooten, 1983; Penney et al., 1981; Quirion et al., 1981; Rainbow et al., 1982; 1984; Unnerstall et al., 1982; Palacios et al., 1981; Greenmeyer et al., 1984; Corp et al., 1986; Baskin and Dorsa, 1986; Baskin et al., 1986a,b; Davidson et al., 1985; Bohannon et al., 1986; Biegon et al., 1984). These approaches are complementary and should be considered as separate tools of the anatomical pharmacologist, providing different kinds of information about the receptors *in situ*.

4.4.5. LKB Ultrofilm Method

For autoradiography with LKB Ultrofilm, labeled and dried slide-mounted sections are placed in a cassette for X-ray film and covered with a sheet of LKB Ultrofilm. The film must be handled in the dark or under appropriate safelight conditions. We conserve film by scoring the slides with a diamond point and breaking off the end of the slide that contains the tissue slice, which is used for autoradiography. In this way more tissue slices can be processed with a sheet of film. The film can be trimmed to small pieces if just a few slides are used, but it must be handled with extreme caution. LKB Ultrofilm does not have the protective coating over the emulsion that is typical of X-ray film. The emulsion can thus be easily scratched and abraded (because of this, LKB Ultrofilm cannot be processed in conventional X-ray film processors). Radioactivity standard slides are included with the labeled tissue sections. The X-ray cassette is closed and wrapped in black plastic sheets to prevent fogging from light leaks (we have found that most X-ray cassettes are not light-tight), and kept at room temperature for processing. The manufacturer claims that there is no advantage to exposing LKB Ultrofilm at low temperatures or using intensifying screens. Exposure periods will depend upon the radioactive tracer used, its specific activity, and the concentration of labeled binding sites. We have found that exposures of 1–5 d are adequate for many [^{125}I]-labeled peptides, whereas tritium-labeled peptides require 1–2 mo exposure to obtain enough contrast to be useful for quantitative analysis by densitometry. Development of the exposed LKB Ultrofilm should be done in a tray with emulsion side up to prevent scratching the emulsion. The time and temperature of development should be rigidly controlled. The developer (we use D-19 undiluted) should be discarded after each use. Partially exhausted developer will result in underdeveloped images that are not useable for quantitation.

4.4.6. Emulsion-Coated Coverslips

The basic procedures for in vitro autoradiography with emulsion-coated coverslips have been described in detail previously (Young and Kuhar, 1979; Wamsley and Palacios, 1983). In our experience, this method is simplified by having coverslips of proper dimensions. The coverslips should be "0" thickness and about 0.5 mm longer than a conventional microscope slide. These have been available on special order from Corning (NY). One end of the coverslip is dipped in Kodak NTB2 emulsion and allowed to dry thoroughly in complete darkness. They are stored in black plastic slide boxes sealed with plastic electrical tape. We have found that these coated coverslips can be stored for at least 6 mo at 4°C without a noticeable increase in background. The coverslip is mounted flush with the end of the slide at the glued end. The extra length of the coverslip results in a slight extension protruding from the tissue end of the slide; this greatly facilitates mounting the coverslip and makes it easy to lift the coverslip for development under very dim safelight conditions. Fewer coverslips are broken with this arrangement. Once the coverslip is glued to the label end of the slide, a blank slide is placed over it and the assembly is clamped using spring clips for pressure. These sandwich assemblies are stored in light-tight boxes during exposure. After development the tissue can be defatted in xylene and stained. We have found that staining these sections may interfere with darkfield imaging, however, since the stained cells also reflect light, and we prefer to examine these preparations without staining. It is necessary to scrape off the emulsion on the outer surface of the coverslip prior to darkfield illumination with a stereomicroscope.

4.4.7. ^3H vs ^{125}I Ligands

Both tritiated and iodinated ligands can be used for autoradiography with LKB Ultrofilm. The emulsion will register tritium beta emissions with a resolution of about 0.04 mm (Gallistel and Nichols, 1983). With ^{125}I-labeled ligands, it is the Auger electrons and electrons of internal conversion, which are emitted at energies below 6 KeV, that supposedly interact with the emulsion to produce autoradiograms (Ullberg et al., 1982). However, the influence of the gamma emissions on autoradiographic image formation is not well understood. The densitometric response of the LKB Ultrofilm emulsion to ^3H and ^{125}I radiation sources appears not to differ significantly (Baskin et al., 1986a). Iodinated ligands may be unsuitable for receptor studies if the iodine interferes with the receptor binding site, as we found was the case with iodovasopressin. The availability of tritiated vasopressin allowed us to localize vasopressin binding sites in kidney and brain (Baskin et al., 1983b; Dorsa et al., 1983; 1984). The iodinated ligands have distinct advantages, however, for in vitro autoradiography with LKB Ultrofilm. The higher specific activities available with iodinated ligands mean exposure times of a few days, rather than the weeks or months often required with tritiated ligands. Further, the relatively greater optical densities that can be produced on the LKB film with ^{125}I for a receptor present in low concentration can facilitate measuring binding by densitometry. One of the difficulties faced when using iodinated ligands is establishing tissue radioactivity standards because of the 60-d half-life of ^{125}I. We have successfully used commercial plastic tritium standards that we calibrated for equivalent tissue ^{125}I radioactivity for this purpose (Baskin et al., 1986a). Also, the half-life decay in radioactivity of ^{125}I must be considered when making calculations based on optical densities produced by ^{125}I at different times after iodination of the ligand. Another point in favor of iodinated ligands is that tritium produces the artifact of unequal optical densities in grey vs white matter of the brain because of differential absorption of beta particle energy, whereas iodinated ligands do not have this drawback (Kuhar and Unnerstall, 1985).

4.4.8. Methods for Quantitative Analysis

There are two main approaches for quantitative evaluation of radioligand binding by autoradiography: grain counting and densitometry. Although valid data on binding can be obtained with either approach, densitometry has distinct advantages for receptor pharmacology. Grain counting is usually done by enumerating the number of individual autoradiographic grains within geometric sampling areas at relatively high microscopic magnifications. This is a tedious and cumbersome method, usually done manually, for obtaining raw data, and is not conducive to obtaining rapid results. Computer systems have been devised to count grain densities in microscopical fields, but these systems have not been widely adopted. Nevertheless, grain counting methods may be the only feasible method of quantifying binding in extremely small anatomical sites, such as individual cells. For measuring binding to regions such as fiber tracts, layers, and nuclei of the brain, densitometry with LKB Ultrofilm is the method of choice. In principle, densitometry is no different than grain counting, except that the grain density is integrated over a larger area. The basic approach in densitometry is to compare the optical density of a region in an LKB Ultrofilm autoradiographic image with the optical densities produced by a set of standards of known radioactivity. By reference to a standard curve or table, the concentration of radioactivity (and, hence, amount of ligand bound) can be obtained. Densitometry can be done with spot densitometers, photomultiplier tubes, or video digitizers. We have used computer-assisted video digitization as a means to measure optical density of anatomical regions of LKB Ultrofilm autoradiographic images, and to determine receptor characteristics from this densitometric data (Baskin and Dorsa, 1986; Baskin et al., 1984, 1986; Corp et al., 1986; Davidson et al., 1985; Petracca et al., 1985; Bohannon et al., 1986).

4.4.9. Computer Systems for Quantitative Autoradiography

Computer analysis of quantitative autoradiographic (QAR) images for obtaining quantitative information about the concentration of a radioactive tracer was pioneered by Goochee et al. (1980) for studies on 2-deoxyglucose uptake by the brain. Similar computer systems have been used to measure ligand concentrations in many in vitro autoradiographic studies. Recently more inexpensive microcomputers have been adapted for this purpose (Kuhar et al., 1984; Baskin et al., 1986a).

The basic components of a computer digitizing system for autoradiography include the following: a computer with an analog-to-digital converter (digitizer), video camera (either vidicon or solid state), CCTV monitor, lenses (or microscopes) for image formation, and printer. If whole brain slices are to be analyzed, then a light box with uniform illumination is required. Some commercial slide transparency copiers are suitable for this purpose. Radioactivity standards can be purchased from commercial sources and should be calibrated against tissue standards by the investigator. In addition, a set of calibrated neutral density filters is required for standardizing the digitizer operating conditions and film optical density. We have used a calibrated step tablet (Kodak) in which adjacent panels differ by 0.15 optical density units. For imaging whole rat brain slices, a macrolens or enlarger lens can be used, but small regions may require magnification by stereomicroscope optics in order to overcome limitations imposed by the resolution of the digitizer. Finally, software must be available to enable the system to produce usable data.

The basic needs of a computer system for densitometry can be met with inexpensive components assembled around an Apple IIe personal computer (Apple Computer, Cupertino, CA). We have used such a system, which includes a 6-bit video digitizer (DS-65 Digisector, MicroWorks, Del Mar, CA), a General Electric Model 2405 CID-type solid-state camera, and a program written in Pascal (Lewellen and Graham, 1983). The CID camera is mounted on a Zeiss stereomicroscope in order to obtain magnified images that are suitable for measuring optical densities of regions of interest without loss of contrast. The CID camera that we have used is well suited to LKB film densitometry because both its linear response to light and dynamic range of response to brightness closely match the useful optical density range of LKB Ultrofilm in autoradiographic applications (Baskin and Dorsa, 1986). Other solid-state cameras are also undoubtedly satisfactory for this purpose, but each should be carefully evaluated. It has been pointed out, however, that inexpensive vidicon cameras may be suitable for optical density measurements with 6-bit digitizers (Ramm et al., 1984).

It is important to realize that the *sine qua non* of a computer digital imaging system for quantitative autoradiography is an accurate measure of optical density. This can be accomplished at modest investment with inexpensive components, such as an Apple IIe. Highly sophisticated image analyzers are available commercially for $40–100,000 and more. These are impressive in operation and produce beautiful images, but the majority of the imaging capabilities of those systems are irrelevant for quantitative autoradiography. Useful features such as graphic overlays and pseudocolor images are now available with systems based on the IBM PC. One such system (DUMAS) has recently been assembled around the IBM PC and configured specifically for quantitative autoradiography, including radioligand binding applications (Baskin and Dorsa, 1986), by the Image Analysis Center of Drexel University (Gallistel and Tretiak, 1985). The advantage of this latter system is that the software has been developed specifically for autoradiography and is available for investigators elsewhere.

4.4.10. Resolution in Quantitative Autoradiography by Densitometry

Resolution becomes an important consideration when making optical density measurements on small regions of the LKB Ultrofilm autoradiographic image. This problem could be encountered when attempting to measure peptide binding in a layer of the cerebral cortex or a small nucleus. In microscopy resolution is generally conceived as the ability to distinguish fine detail (i.e., "point-to-point" image separation). For purposes of image digitization, resolution must be defined, however, in terms of accuracy of optical density measurements. The consequence of inadequate digitizer resolution is underestimated optical densities of small structures and, hence, inaccurate binding data. Unfortunately, few investigators have confronted this problem, and the validity of results of many quantitative autoradiographic studies are open to question. It is therefore important to ascertain whether a videocamera and digitizer actually measure the true optical density of small objects, and to determine the linear dimensions below which inaccurate optical densities will be recorded. Since this error becomes severe with increasing spatial frequency, magnification (such as with microscope optics) may be essential to avoid errors. For practical purposes, resolution of spatial frequencies by a digitizer can be closely approximated by measuring its contrast transfer function (Castleman, 1979). This can be done with a target having increasingly fine spatial frequencies and software permitting linear scan grey-level profiles of the target to be generated. It is a simple procedure to measure digitizer gray-level output of a target image and determine the minimum object size that will be recorded at true gray level (Baskin and Dorsa, 1986). It is to be emphasized that the investigator must not assume that gray-level or optical density measurements obtained from a computer image analyzer are valid until the digitizer resolution has been ascertained.

4.4.11. Standardization of Quantitative Autoradiography

An important aspect of standardizing quantitative autoradiography for receptor binding is characterization of the response of the film to the amount of bound radioligand. This requires a basic understanding of photographic densitometry, and investigators unfamiliar with this area should consult elementary photography sources. The autoradiographic response of LKB Ultrofilm to ^3H and ^{125}I emission has been found by Kuhar and Unnerstall (1985), Baskin et al. (1986a) and Baskin and Dorsa (1986) to be nonlinear over the full optical density range of the film. This knowledge is essential to proper interpretation of the autoradiographic images and optical density measurements made on receptor autoradiographs. Preliminary studies should be carried out to determine optimum exposure time of the labeled tissue to LKB Ultrofilm so that maximum contrast is achieved without saturating the optical density capacity of the film. The relationship between radioactivity and optical density may be very close to linear over the range of optical densities actually encountered in receptor autoradiographs; this will vary with the ligand and organ being labeled. In vitro autoradiography can be used to detect peptide binding sites with LKB Ultrofilm, but characterizing these sites as true receptors requires quantitative measurements of the optical densities in the image. Results obtained by quantitative autoradiography with LKB Ultrofilm have questionable validity unless film response and exposure have been optimized.

In radioreceptor assays with membrane preparations it is common practice to normalize binding data to an amount of membrane protein present in the assay tube. In the autoradiographic approach to receptor binding measurements that we have described, there is yet no consensus of what should be the appropriate denominator for normalizing the data, nor is there an intuitively obvious "best choice" in the available options. One approach has been to standardize measurements using "tissue paste" standards. These are usually mashes of brain tissue containing known concentrations of tritium-labeled material. These brain paste standards are then cut on a cryostat to the same thickness as the brain slices being studied and placed along with them on the LKB Ultrofilm. A standard curve is then constructed in which optical density is related to concentration of radioactivity per weight of protein in the mash. Optical densities from the brain sections are then compared to this standard curve, and brain radioactivity is derived by interpolation. Another approach is to calibrate binding to a unit area of the tissue slice. This is done by labeling tissue slices with increasing amounts of radioactivity. After exposure of LKB Ultrofilm to these sections, they are scraped and counted for radioactivity; the radioactivity is expressed per unit area of the section (Baskin et al., 1986a; Mariash et al., 1982). Normalizing quantitative autoradiography data to unit area helps avoid errors caused by variation in protein content in different regions of an organ.

A variation of this method has recently become commercially available by use of plastic standards containing tritium of known concentrations. Some vendors have even "calibrated" these plastic standards to brain tissue equivalents. Again the use of these plastic standards may prove useful, especially for cross calibration of films, but the calibration to brain tissue needs to be addressed for each application by the investigator, using ^{3}H, since regional differences in quenching must be carefully determined. We have found ^{3}H plastic standards to be useful for calibrating tissue slices labeled with ^{125}I (Baskin et al., 1986a; Baskin and Dorsa, 1986). This allows one to avoid the problem of using brain paste standards with this relatively short-lived isotope.

5. Work With Vasopressin and Insulin Receptors

In order to demonstrate the methods and principles we have just discussed, we would like to make use of our experience in work with CNS vasopressin and insulin receptors.

5.1. Vasopressin

Although our primary interest has been in vasopressin receptors in the CNS, we used the kidney as a tissue source to develop our assay procedures for tissue slice and membrane binding. Since our other publications have adequately described the membrane

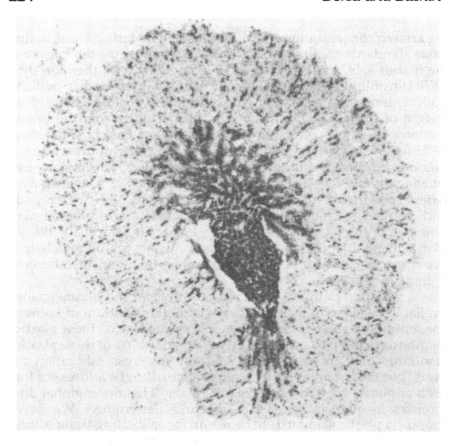

Fig. 1.   Autoradiographic image on LKB Ultrofilm showing binding of 2 nM ^{3}H-AVP to kidney slice. Highest binding is in the inner medulla. High binding is also present in outer medulla and glomeruli of cortex.

work (Dorsa et al., 1983; Dorsa et al., 1984; Cornett and Dorsa, 1985), we will restrict our discussion to tissue slice studies. Primarily through the efforts of Lisa Majumdar, MD, we were able to develop assay methods that gave reproducible results on 10 μm slices of kidney medulla. Direct counting methods were used to develop these techniques (Dorsa et al., 1983). This work was continued by Frances Petracca, who was able to perfect the autoradiographic methodology and generate LKB Ultrofilm images such as that shown in Fig. 1. This image was obtained by a 30-d exposure of kidney slices to LKB Ultrofilm.

   Figure 2 shows digital images obtained by digital image analysis (Apple IIe system). Both total binding and nonspecific binding

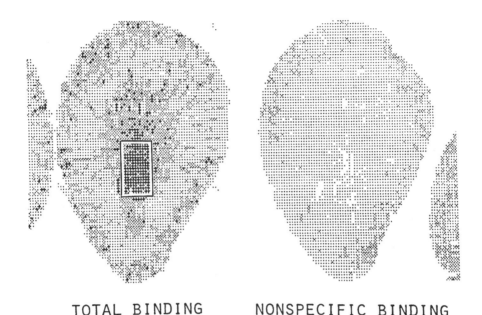

TOTAL BINDING   NONSPECIFIC BINDING

Fig. 2. Computer-generated digital images of autoradiographic film images similar to Fig. 1, showing total binding (left panel) of 2 nM ^{3}H-AVP alone and nonspecific binding (right panel) in the presence of 1 μM unlabeled AVP. The QAR digital imaging system measured the optical density of the medulla (region in rectangle) resulting from increasing the concentration of labeled AVP. These optical densities were converted to fmol bound/mm^2 by reference to a standard curve.

images are shown. The region sampled for quantitative measurements is demarcated by the rectangle. Optical densities are measured using this system and can be related to either dpm or fmol bound/mm^2 by reference to a standard curve. This curve was obtained by measuring the optical density of homogeneously labeled kidney or liver tissue of known radioactivity. Then by performing the assay using a range of ^{3}H-AVP concentrations (0.5–10 nM) sufficient to saturate renal receptors, QAR was used to develop the binding curve shown in Fig. 3. Figure 4 shows that Scatchard analysis of this data using an iterative curve-fitting procedure yielded a best fit for a single class of binding sites with equilibrium dissociation constant of 4.1 nM and B_{max} of 8×10^{-3} mol/mm^2.

We have also applied these techniques to localize and

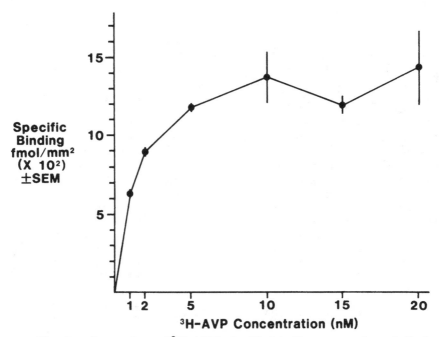

Fig. 3. Saturation of ^{3}H-AVP specific binding to renal medulla by QAR. Points represent total binding (^{3}H-AVP alone) minus nonspecific binding (^{3}H-AVP plus 1 μM unlabeled AVP) (n = 2–3 measurements per point).

characterize vasopressin receptors in the rat CNS (Petracca et al., 1983, 1986). We found that high densities of AVP binding sites could be localized to the lateral septum (Baskin et al., 1983b), nucleus of the solitary tract (Dorsa et al., 1983), central nucleus of the amygdala (Dorsa et al., 1984), and, as shown in Figs. 5 and 6, the olfactory tubercle and ventral tegmentum. These results are in good agreement with those of Biegon et al. (1984) and Van Leuwen and Walters (1983). We have also prepared membranes from the dorsal hindbrain (Cornett and Dorsa, 1985) and amygdala (Dorsa et al., 1984) and characterized ^{3}H-AVP binding in those areas. Saturation analysis of binding in both areas revealed a single class of binding sites with K_d = 0.7–1.2 nM and B_{max} in the range of 20–50 fmol/mg protein. Specificity of binding was similar in both areas and was consistent with a V_1-type receptor. This finding is in good agreement with that of Barberis (1983).

We have also used QAR to measure vasopressin binding in rat brain. Figure 7 shows digitized images of AVP binding to rat

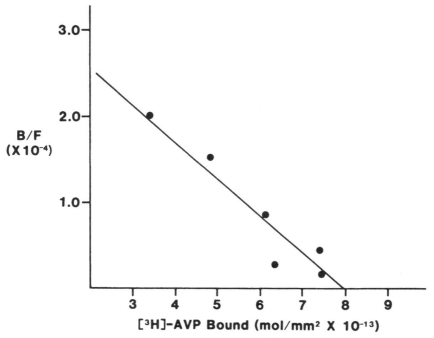

Fig. 4.   Scatchard analysis of binding of ^{3}H-AVP to renal medulla by QAR. The data fit a single site binding model with $K_d = 2.3 \times 10^{-9}$ M and $B_{max} = 8.0 \times 10^{-13}$ mol/mm^2 ($r = 0.95$).

forebrain sections. Total (left panels) and nonspecific (right panels) binding are represented. By measuring optical densities in various areas of brain sections incubated with either 2.5 nM ^{3}H-AVP (a saturating concentration) or ^{3}H-AVP plus 2.5 μm unlabeled AVP, we were able to measure specific binding of the peptide and obtain an estimate of the relative density of binding sites per unit area sampled. Figure 8 shows the results obtained after the optical densities were related to standards labeled with known amounts of ^{3}H-AVP. These results suggest that the olfactory tubercle and amygdala contain the highest density of binding sites of those areas studied, and the cerebral cortex the lowest. These values were not corrected for quenching of tritium, which may vary from region to region. However, Geary and Wooten (1983) have recently published their measurements of regional tritium quenching in quantitative autoradiography of the rat brain and found little difference in the degree of quenching in the areas we have examined.

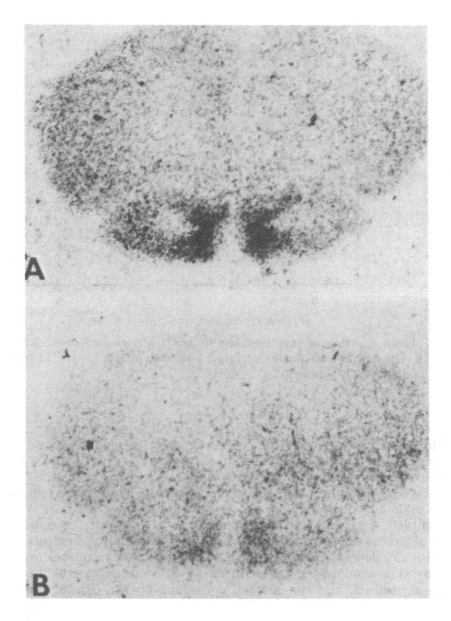

Fig. 5. Autoradiographic localization of ^{3}H-AVP binding sites in the olfactory tract of the rat brain using LKB film. (A) Total binding. (B) Nonspecific binding in the presence of 1000-fold excess unlabeled AVP.

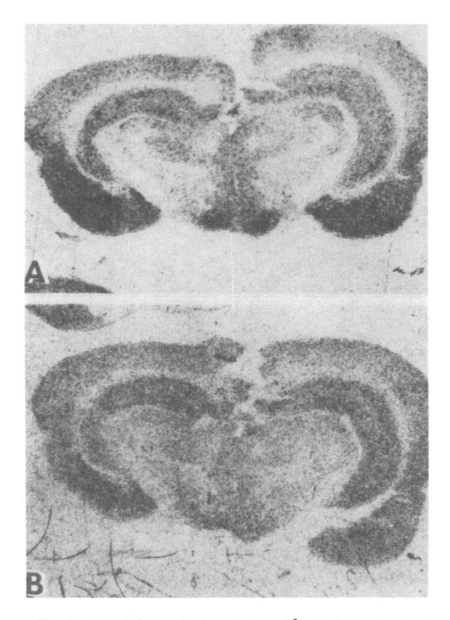

Fig. 6. Autoradiographic localization of ^{3}H-AVP binding sites in the ventral tegmentum of the rat brain using LKB Ultrofilm. (A) Total binding. (B) Nonspecific binding.

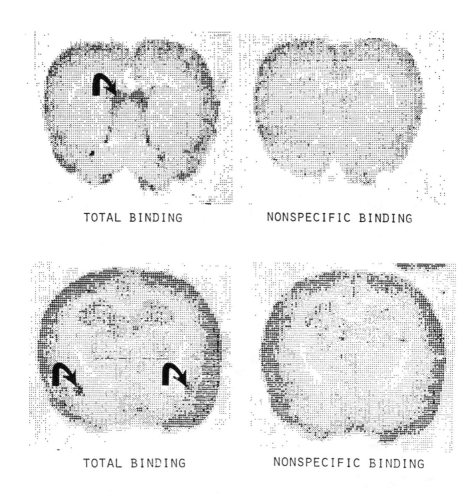

TOTAL BINDING   NONSPECIFIC BINDING

TOTAL BINDING   NONSPECIFIC BINDING

Fig. 7.   Digital images created by computer QAR system of auto-radiographic film images showing total (left panel) and nonspecific (right panel) binding of ^{3}H-AVP to lateral septum (top) and amygdala (bottom).

5.2. Insulin

As an example of quantitative autoradiography of an iodin-ated peptide, we show figures and data derived from our studies of insulin receptors in the unfixed rat brain. Much of this work was done by graduate students David Davidson, Eric Corp, and Bar-bara Brewitt in our laboratory. Figure 9 illustrates data on the time-course of 0.1 nM ^{125}I-insulin (human) binding to slide-

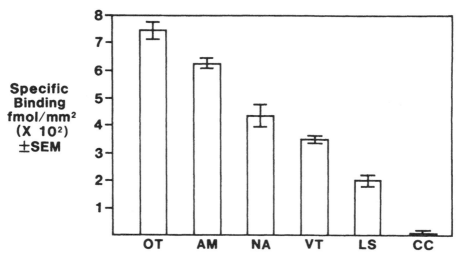

Fig. 8. Measurement of ^{3}H-AVP binding to several regions of rat brain by QAR. Abbreviations: OT, olfactory tubercle; AM, central amygdala; NA, nucleus accumbens; VT, ventral tegmentum; LS, lateral septum; CC, cerebral cortex (somatosensory).

mounted sections of olfactory bulb and hypothalamus and to the gelatin-coated slides at 22°C. The binding is specific binding (total minus nonspecific binding in the presence of 1 μM unlabeled porcine insulin) to two sections per slide. These were incubated for the periods shown and then wiped with filter paper, which was counted in a gamma counter. Bare glass slides without brain slices were also wiped and the filter paper similarly counted. These results indicate that the binding to brain slices reached equilibrium by 90–120 min. Accordingly, a 2-h assay was used for further binding studies. This type of protocol is necessary at the outset of an investigation to assure that binding is being done under conditions that produce results consistent with equilibrium binding. For example, the binding should not be significantly reduced with longer incubations. In Fig. 9, the binding remained constant up to 3 h. We have found, however, that similar binding conditions with liver slices result in a peak at 90–120 min, and thereafter there is a decline in binding. This suggests possible tracer or receptor degradation. Fortunately the brain reportedly has minimum insulin protease activity, so this is a less serious problem with brain slices. This procedure also revealed that the olfactory bulbs exhibited significantly higher binding of ^{125}I-insulin than the hypothalamus, even though the hypothalamus sections were six times greater in

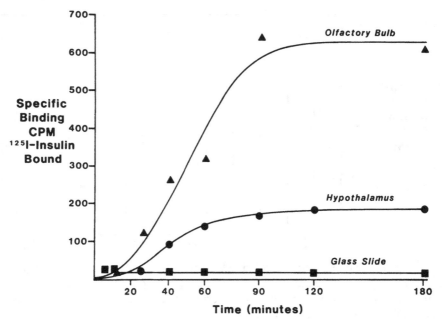

Fig. 9.   Time course of specific binding (total minus nonspecific) of
0.1 nM ^{125}I-insulin to slide-mounted cryostat sections of unfixed rat
olfactory bulb, hypothalamus, and gelatin-coated slides, at 22°C. Specific
binding (total of two sections per slide) was determined by wiping slides
with filter paper, which was counted for radioactivity. Each point is the
mean of results from three slides.

surface area (cortex and dorsal thalamus had been cut away).
Nonspecific binding to the glass did not change.

Once equilibrium conditions were approximated, binding for
purposes of autoradiography was done. Figure 10 shows the
iodoinsulin binding to slices of rat paired olfactory bulbs, as revealed with emulsion-coated coverslips and darkfield stereomicroscope illumination. As can be seen, binding in the external plexiform layer is visibly reduced in the presence of 1μM unlabeled
insulin. This figure shows an artifact, the brightness produced by
staining nuclei with hematoxylin. In Fig. 10B, the core of each
olfactory bulb (granular layers) is brighter than the EPL because the
stained nuclei reflect some light (the EPL is largely a synaptic
region). This effect is readily detected by microscopic observation,
but not apparent in a photographic print.

One limitation of the emulsion-coated coverslip technique is
the difficulty in quantifying binding by densitometric methods.

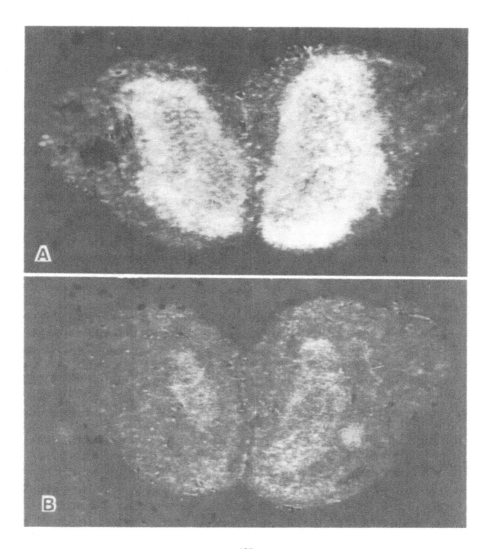

Fig. 10. Binding of 0.1 nM ^{125}I-insulin to adjacent slices (20-μm thick) of rat olfactory bulb revealed by autoradiography with emulsion-coated coverslip technique and darkfield illumination. High total binding present in the external plexiform layer (A) is blocked in the presence of 1 μM unlabeled insulin (B).

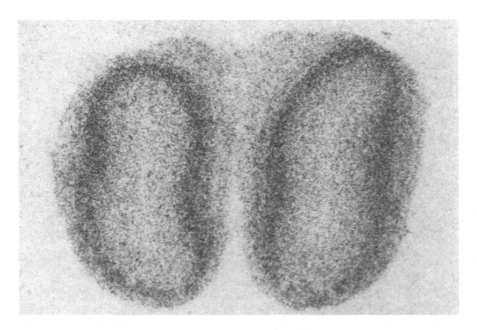

Fig. 11.   Binding of ^{125}I-insulin to external plexiform layer of olfactory bulb, revealed by autoradiography with LKB Ultrofilm. Total binding only is shown.

Autoradiographic images produced with LKB Ultrofilm are more suitable for this purpose. Figure 11 shows an LKB Ultrofilm image of a rat olfactory bulb labeled in 0.05 nM ^{125}I-insulin at 4°C for 18 h. Note that the optical density is highest in the EPL, which reflects the high binding of iodoinsulin in this region. Obviously, to measure insulin binding with membrane homogenates in this layer would be a heroic task indeed, and for practical purposes probably could not be done with existing methodology. In vitro autoradiography, however, allows binding in a single slice to be measured by densitometry. We have in fact accomplished this with computer densitometry using a microcomputer (Apple IIe) digital imaging system (Baskin et al., 1986a; Corp et al., 1986). Figure 12A

→

Fig. 12.   Binding of ^{125}I-insulin to EPL of olfactory bulb as measured by QAR. (A) Competition curve of labeled vs unlabeled insulins. (B) Scatchard analysis of ^{125}I-insulin binding to external plexiform layer of olfactory bulb as measured by QAR. Each point is the mean of duplicate measurements (adapted from Baskin et al., 1986a).

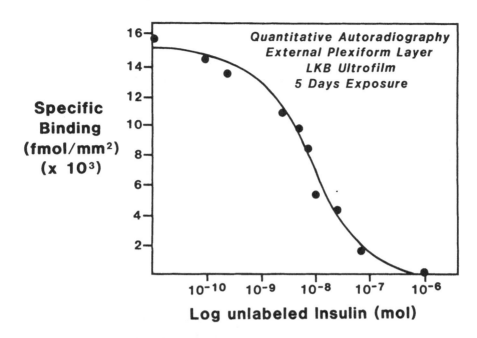

Specific Binding (fmol/mm²) (x 10³)

Quantitative Autoradiography
External Plexiform Layer
LKB Ultrofilm
5 Days Exposure

Log unlabeled Insulin (mol)

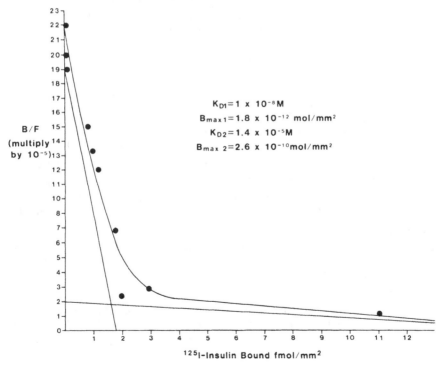

B/F (multiply by 10⁻⁵)

$K_{D1} = 1 \times 10^{-8}M$
$B_{max1} = 1.8 \times 10^{-12} \ mol/mm^2$
$K_{D2} = 1.4 \times 10^{-5}M$
$B_{max\ 2} = 2.6 \times 10^{-10} mol/mm^2$

^{125}I-Insulin Bound fmol/mm²

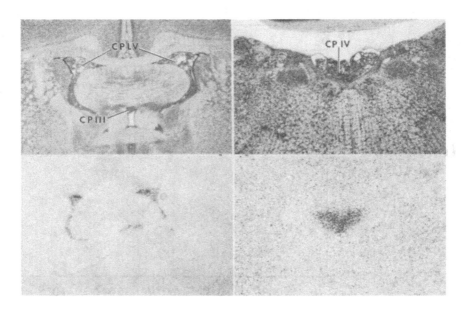

Fig. 13.   Autoradiographic images of ^{125}I-insulin binding to choroid plexus of rat brain. Top panels show stained paraffin sections of choroid plexus of third (CPIII), lateral (CPLV), and fourth (CPIV) ventricles. Lower panels show corresponding LKB Ultrofilm images of labeled choroid plexuses (left, x15; right, x30). Nonspecific binding was equal to the surrounding background levels seen here.

shows the dose-dependent reduction of iodoinsulin binding in the EPL in the presence of increasing concentrations of labeled insulin. Data were obtained by optical density measurements of olfactory bulb images on LKB Ultrofilm. A standard curve relating ^{125}I radioactivity vs film exposure relationship was characterized. The data in Fig. 12A is shown in a Scatchard plot format in Fig. 12B, with binding constants and site numbers derived from a two-site binding model. The interpretation of the curved Scatchard plot for insulin binding is a controversial subject that is beyond the scope of this article. The important point here is that pharmacological binding data were obtained by quantitative autoradiography for a region within the brain that would have been inaccessible for direct study by other methods.

    We have used a similar approach to investigate insulin binding to the choroid plexus of the rat brain (Davidson et al., 1985; Baskin et al., 1986b). Figure 13 shows LKB film images of labeled choroid

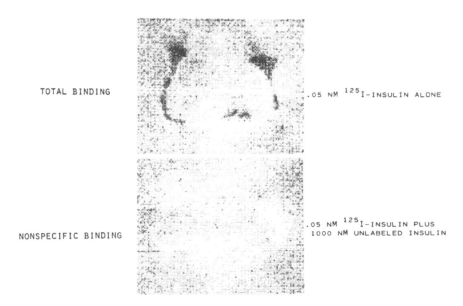

TOTAL BINDING

.05 NM ^{125}I-INSULIN ALONE

NONSPECIFIC BINDING

.05 NM ^{125}I-INSULIN PLUS
1000 NM UNLABELED INSULIN

Fig. 14. Digital images of choroid plexus labeled with 0.05 nM ^{125}I-insulin alone (top) and its reduction to background binding level in the presence of 1 µM unlabeled insulin (bottom).

plexus, along with stained slices for orientation. Figure 14 shows computer digital images of the lateral ventricle choroid plexuses; these images were processed by software to derive optical density and convert this to an amount of labeled insulin bound per unit area of tissue slice. Increasing the amount of unlabeled insulin produced a dose-dependent change in binding (measured as an optical density change by the computer) from that seen in the top panel to that shown in the bottom panel of Fig. 14. We have used this technique to characterize the choroid plexus insulin binding sites, as shown in Fig. 15. Competition of human iodoinsulin vs several related insulins and insulin-like molecules produced a series of curves, which reveals a rank order of IC$_{50}$ values that is similar to the relative potency of these insulins in bioassay tests. These results indicate that the choroid plexus binding site is a *bona fide* insulin receptor.

Herein lies an important conceptual point. It is normally not feasible to demonstrate with slide-mounted brain slices a corresponding dose-dependent and appropriate biological action of the ligand being studied. In the absence of this important pharmaco-

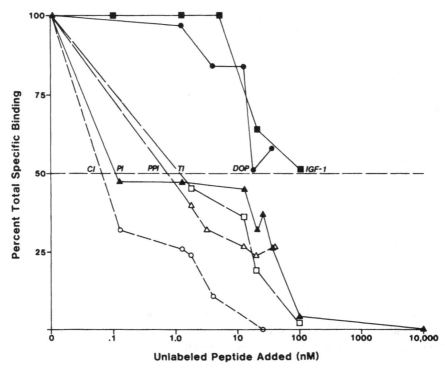

Fig. 15. Competition curves showing the displacement of 0.05 nM ^{125}I-insulin from choroid plexus by QAR measurements of digital images. The labeled insulin was displaced by unlabeled porcine insulin in a dose-dependent manner. The rank order of potencies of chicken (CI), porcine (PI), tuna (TI), porcine proinsulin (PPI), desoctapeptide insulin (DOP), and IGF-1 is similar to that of classical insulin receptors of nonneural tissues. Each point is the mean of two measurements.

logical criterion for identifying a receptor, we must depend upon affinity profiles of competing ligands, each with differing biological potency. With quantitative autoradiography, a competition analysis is thus *sine qua non* for calling a binding site a "receptor." In the absence of competition data, it is probably safer to refer to binding sites as "binding sites" rather than receptors. There has been a tendency for the term receptor to be used loosely in literature on in vitro autoradiography. Without quantitative autoradiographic data of the binding site *in situ* (e.g., by some densitometric method), binding sites probably cannot safely be called receptors.

Acknowledgments

Some of the work described in this chapter was supported by the Veterans Administration, NIH grant NS 20311, and NIH grant AM 17047 to the Diabetes Research Center of the University of Washington. The authors wish to thank Maxine Cormier for typing this manuscript.

References

Antoni F. A. (1984) Novel ligand specificity of pituitary vasopressin receptors in the rat. *Neuroendocrinology* **39**, 186–188.

Balla T., Enyedi P., Spat A., and Antoni F. A. (1985) Pressor type vasopressin receptors in the adrenal cortex; properties of binding, effects on phosphoinositide metabolism and aldosterone secretion. *Endocrinology* **117**, 421–423.

Barberis C. (1983) (^{3}H) Vasopressin binding to rat hippocampal synaptic plasma membrane kinetic and pharmacologic characterization. *FEBS Lett.* **162**, 400–405.

Baskin D. G. and Dorsa D. M. (1986) Quantitative Autoradiography and In Vitro Radioligand Binding, in *Functional Mapping in Biology and Medicine* (McEachron D., ed.), Karger, New York.

Baskin D. G., Woods S. C., West D. B., van Houten M., Posner B. I., Dorsa D. M., and Porte D. (1983a) Immunocytochemical detection of insulin in rat hypothalamus and its possible uptake from cerebrospinal fluid. *Endocrinology* **113**, 1818–1825.

Baskin D. G., Petracca F., and Dorsa D. M. (1983b) Autoradiographic localization of specific binding sites for ^{3}H-arginine^8-vasopressin with tritium sensitive film. *Eur. J. Pharmacol.* **90**, 155–158.

Baskin D., Petracca F., and Dorsa D. (1984) Quantitative receptor autoradiography and computer digital image measurement of ^{3}H-vasopressin binding in rat kidney and brain. *Soc. Neurosci. Abst.* **10**, Part 1, 557.

Baskin D. G., Davidson D., Corp E., Lewellen T., and Graham, M. (1986a) An inexpensive microcomputer digital imaging system for densitometry: Quantitative autoradiography of insulin receptors with ^{125}I and LKB Ultrofilm. *J. Neurosci. Meth.* **16**, 119–129.

Baskin D. G., Brewitt B., Davidson D., Corp E., Paquette T., Figlewicz D., Lewellen T., Graham M., Woods S., and Dorsa D. (1986b) Quantitative autoradiographic evidence for insulin receptors in the choroid plexus of the rat brain. *Diabetes*, **35**, 246–249.

Baskin D. G., Figlewicz D. P., Woods S. C., Porte D. Jr., and Dorsa D. (1987) Insulin in the brain. *Ann. Rev. Physiol.* **49**, 335–347.

Bergeron J., Levine G., Sikstrom R., O'Shaughessy D., Kopriwa B., Nadler N., and Posner B. (1977) Polypeptide hormone binding sites in vivo: Initial localization of ^{125}I-labeled insulin to hepaptocyte plasmalemma as visualized by electron microscopic radioautography. *Proc. Natl. Acad. Sci. USA* **74**, 5051–5055.

Bergeron J., Rachubinski N., Sikstrom R., Borts D., Bastian P., and Posner B. (1980a) Radioautographic visualization of in vivo insulin binding to the exocrine pancreas. *Endocrinology* **107**, 1069–1080.

Bergeron J., Rachubinski R., Searle N., Borts S., Sikstrom R., and Posner B. (1980b) Polypeptide hormone receptors in vivo: Demonstration of insulin binding to adrenal gland and gastrointestinal epithelium by quantitative autoradiography. *J. Histochem. Cytochem.* **28**, 824–835.

Biegon A., Terlous M., Boorhuis B., and DeKloet E. (1984) Arginine-vasopressin binding sites in rat brain: A quantitative autoradiographic study. *Neurosci. Lett.* **44**, 229–234.

Bohannon N. J., Figlewicz D. P., Corp E. S., Wilcox B. J., Porte D., Jr., and Baskin D. G. (1986) Identification of binding sites for an insulin-like growth factor (IGF-1) in the median eminence of the rat brain by quantitative autoradiography *Endocrinology* **119**, 943–945.

Castleman K. (1979) *Digital Image Processing*, Prentice Hall, Englewood Cliffs, New Jersey.

Cornett L. E. and Dorsa D. M. (1985) Vasopressin receptor subtypes in the dorsal hindbrain and renal medulla. *Peptides* **6**, 85–89.

Corp E. S., Woods S. C., Porte D., Jr., Figlewicz D. P., Dorsa D. M., and Baskin D. G. (1986) Localization of insulin binding sites in the rat hypothalamus by quantitative autoradiography. *Neurosci. Lett.* **70**, 17–22.

Creba J. A., Downes C. P., Hawkins P. T., Brewster G., Mitchell R. H., and Kirk C. J. (1983) Rapid breakdown of phosphatidyl inositol 4-phosphate and phosphatidyl inositol 4,5 biphosphate in rat hepatocytes stimulated by vasopressin and other Ca^{++} mobilizing hormones. *Biochem. J.* **212**, 733–747.

Davidson D., Corp E., Figlewicz D., Woods S., Porte D., Dorsa D., and Baskin D. (1985) Characterization of insulin receptors in the choroid plexus of the rat brain by quantitative autoradiography and computer densitometry. *Soc. Neurosci. Abstr.* **11**, 415.

de Wied D., Greven H. M., Lande S., and Witter A. (1972) Dissociation of the behavioral and endocrine effects of lysine vasopressin by tryptic digestion. *Br. J. Pharmacol.* **45**, 118–125.

Dorsa D. M., Petracca F. M., Baskin D. G., and Cornett L. E. (1984) Localization and characterization of vasopressin binding sites in the amygdala of the rat brain. *J. Neurosci.* **4**, 1764–1770.

Dorsa D. M., Majumdar L. A., Petracca F. M., Baskin D. G., and Cornett L. E. (1983) Characterization and localization of ^{3}H-arginine^8-vasopressin binding to rat kidney and brain tissue. *Peptides* **4**, 699–706.

Ehn E. and Larsson B. (1979) Properties of an antiscratch-layer-free x-ray film for the autoradiographic registration of tritium. *Science Tools (LKB)* **26**, 24–29.

Engberg G., Svensson T. H., Rosell S., and Folkers K. (1981) A synthetic peptide as an antagonist of substance P. *Nature* **293**, 222–223.

Gallistel C. and Nichols S. (1983) Resolution-limiting factors in 2-deoxyglucose autoradiography. I. Factors other than diffusion. *Brain Res.* **267**, 323–333.

Gallistel C. R. and Tretiak O. S. (1985) Microcomputer Systems for Analyzing 2-Deoxyglucose Autoradiographs, in *The Microcomputer in Cell and Neurobiology Research* (Mize R. R., ed.), Elsevier, New York.

Geary W. and Wooten G. (1983) Quantitative film autoradiography of opiate agonist and antagonist binding in rat brain. *J. Pharmacol. Exp. Therap.* **225**, 234–240.

Goochee C., Rasband W., and Sokoloff L. (1980) Computerized densitometry and color coding of [^{14}C] deoxyglucose autoradiographs. *Ann. Neurol.* **7**, 359–370.

Gorden P., Carpentier J., Freychet P., and Orci L. (1980) Morphologic probes of polypeptide hormone interactions. *J. Histochem. Cytochem.* **28**, 811–817.

Greenmeyer J., Young A. B., and Penney J. (1984) Quantitative autoradiographic distribution of L-[^{3}H] glutamate binding sites in rat central nervous system. *J. Neurosci.* **4**, 2133–2144.

Hanley H. (1985) Peptide Binding Assays, in *Neurotransmitter Receptor Binding*, 2nd ed. (Yamamura H., Enna S. J., and Kuhar M. J., eds.), Raven, New York.

Havrankova J., Roth J., and Brownstein M. (1978) Insulin receptors are widely distributed in the central nervous system. *Nature* **272**, 827–829.

Hays S. E., Bernfield M. D., Jensen R. F., Goodwin F. K., and Paul S. M. (1980) Demonstration of a putative receptor site for cholecystokinin in rat brain. *Neuropeptides* **1**, 53–62.

Heidenreich K., Zahniser N. R., Berhanu P., Brandenburg D., and Olefsky J. M. (1983) Structural differences between insulin receptors in the brain and peripheral target tissues. *J. Biol. Chem.* **258**, 8527–8529.

Herkenham M. and Pert C. (1982) Light microscopic localization of brain opiate receptors: A general autoradiographic method which preserves tissue quality. *J. Neurosci.* **2**, 1129–1149.

Heyward C. B., Yang Y. C. S., Ormberg J. F., Hadley M. E., and Hruby V.

J. (1979) Effects of chloramine T and iodination on the biological activity of metanotropin. *Hoppe Seylers Z. Physiol. Chem.* **360,** 1851–1859.

Humphries J., Wan Y. P., and Folkers K. (1978) Inhibitory analogs of the luteinizing hormone-releasing hormone having D-aromatic residues in position 2 and 6 and variation in position 3. *J. Med. Chem.* **21,** 120–123.

Innis R. B. and Snyder S. H. (1980) Distinct cholecystokinin receptors in brain and pancreas. *Proc. Natl. Acad. Sci. USA* **77,** 6917–6921.

Jard S. (1983) Vasopressin: Mechanism of receptor activation. *Progr. Brain Res.* **60,** 383–394.

Jarrett L. and Smith R. (1975) Ultrastructural localization of insulin receptors on adipocytes. *Proc. Natl. Acad. Sci. USA* **72,** 3526–3530.

Kuhar M. (1982) Localizing Drug and Neurotransmitter Receptors In Vivo With Tritium-Labeled Tracers, in *Receptor-Binding Radiotracers,* vol. I, (Eckelman W. and Colombetti L., eds.), CRC, Boca Raton, Florida.

Kuhar M. and Unnerstall J. (1985) Quantitative receptor mapping by autoradiography: Some current technical problems. *Trends Neurosci.* **8,** 49–53.

Kuhar M., Whitehouse P., Unnerstall J., and Loats H. (1984) Receptor autoradiography: Analysis using a PC-based imaging system. *Soc. Neurosci. Abstr.* **10,** 558.

Laduron P. M. (1984) Criteria for receptor sites in binding studies. *Biochem. Pharmacol.* **33,** 833–839.

Lee C. M., Javitch J. A., and Snider S. H. (1983) ^{3}H-substance P binding to salivary gland membranes. *Mol. Pharmacol.* **23,** 563–569.

Levitzki A. (1984) *Receptors: A Quantitative Approach.* (Levitzki A., ed.), pp. 1–40, Benjamin Cummings Publishing, New York.

Lewellen T. and Graham M. (1983) An inexpensive video digitizer. *Clin. Nucl. Med.* **8,** P39.

Lutz-Bucher B. and Koch B. (1983) Characterization of specific receptors for vasopressin in the pituitary gland. *Biochem. Biophys. Res. Commun.* **115,** 492–498.

Manning M. and Sawyer W. H. (1984) Design of selective agonists and antagonists of the neuropeptides oxytocin and vasopressin. *Trends Neurosci.* **7,** 6–9.

Mariash C., Seelig S., and Oppenheimer J. (1982) A rapid, inexpensive, quantitative technique for the analysis of two-dimensional electrophoretograms. *Anal. Biochem.* **121,** 388–394.

Martin W. R. (1984) Pharmacology of opioids. *Pharmacol. Rev.* **35,** 283–305.

Meidan R. and Hsueh A. J. W. (1985) Identification and characterization of vasopressin receptors in rat testis. *Endocrinology* **116,** 416–423.

Michell R. H., Kirk C. J., and Billah M. M. (1979) Hormonal stimulation of phosphatityl inositol breakdown, with particular reference to the hepatic effects of vasopressin. *Biochem. Soc. Trans.* **7,** 861–865.

Nelson D., Smith R., and Jarrett L. (1978) Nonuniform distribution and grouping of insulin receptors on the surface of human placental syncitial trophoblast. *Diabetes* **27,** 530–538.

Palacios J., Neihoff D., and Kuhar M. (1981) Receptor autoradiography with tritium sensitive film: Potential for computerized densitometry. *Neurosci. Lett.* **25,** 101–105.

Palkovits M. (1975) Isolated removal of hypothalamic brain nuclei of the rat. *Brain Res.* **59,** 449–450.

Patel Y., Amherdt M., and Orci L. (1982) Quantitative electron microscopic autoradiography of insulin, glucagon, and somatostatin binding sites on islets. *Science* **217,** 1155–1156.

Penney J., Pan H., Young A., Frey K., and Dauth G. (1981) Quantitative autoradiography of [³H] muscimol binding in rat brain. *Science* **214,** 1036–1038.

Petracca F. M., Baskin D., Diaz J., and Dorsa D. (1986) Ontogenic changes in Vasopressin binding site distribution in rat brain: An autoradiographic study. *Dev. Brain Res.* **28,** 63–68.

Petracca F. M., Baskin D. G., and Dorsa D. M. (1983) Autoradiographic localization of vasopressin binding in the rat brain. *Soc. Neurosci. Abstr.* **9,** Part 2, 1206.

Praissman M., Martinez P. A., Saladino C. F., Berkowitz J. M., Steggles A. W., and Finkelstein S. A. (1983) Characterization of cholecystokinin binding sites in rat cerebral cortex using a ^{125}I-CCK-8 probe resistant to degradation. *J. Neurochem.* **40,** 1406–1413.

Quirion R., Hammer R., Herkenham M., and Pert C. (1981) Phencylidine (angel dust) "opiate" receptor: Visualization by tritium-sensitive film. *Proc. Natl. Acad. Sci. USA* **78,** 5881–5885.

Rainbow T., Bleisch W., Biegon A., and McEwen B. (1982) Quantitative densitometry of neurotransmitter receptors. *J. Neurosci. Meth.* **5,** 127–138.

Rainbow T., Biegon A., and Berk D. (1984) Quantitative receptor autoradiography with tritium-labeled ligands: Comparison of biochemical and densitometric measurements. *J. Neurosci. Meth.* **11,** 231–241.

Ramm P., Kulick J., Stryker M., and Frost B. (1984) Video and scanning microdensitometer-based imaging systems in autoradiographic densitometry. *J. Neurosci. Meth.* **11,** 89–100.

Saito A., Sankaran H., Goldfine I. D., and Williams J. A. (1980) Cholecystokinin receptors in the brain: Characterization and distribution. *Science* **208,** 1155–1156.

Saito A., Goldfine I. D., and Williams J. A. (1981) Characterization of receptors for cholecystokinin and related peptides in mouse cerebral cortex. *J. Neurochem.* **37,** 483–490.

Schmitt H. and Schmitt-Jubeau H. (1984) Alpha-Adreno Receptors and Central Cardiovascular Control, in *Alpha and Beta-Adreno Receptors and the Cardiovascular System* (Kobinger W. and Alquist R., eds.), Exerpta Medica Pub., Amsterdam.

Steigerwalt R. W. and Williams J. A. (1981) Characterization of cholesystokinin receptors on rat pancreatic membranes. *Endocrinology* **109,** 1746–1753.

Taylor D. P. and Pert C. B. (1979) Vasoactive intestinal polypeptide: Specific binding to rat brain membrane. *Proc. Natl. Acad. Sci. USA* **76,** 660–664.

Thomas A. P., Alexander J., and Williamson, J. R. (1984) Relationship between inositol phosphate production and the increase in cytosolic free Ca^{++} induced by vasopressin in isolated hepatocytes. *J. Biol. Chem.* **259,** 5574–5586.

Tran V. T., Beal M. F., and Martin J. B. (1985) Two types of somatostatin receptors differentiated by cyclic somatostatin analogs. *Science* **228,** 492–495.

Ullberg S., Larsson B., and Tjälve H. (1982) Autoradiography, in *Biologic Applications of Radiotracers* (Glenn H, ed.), CRC, Boca Raton, Florida.

Unnerstall J., Niehoff D., Kuhar M., and Palacios J. (1982) Quantitative receptor autoradiography using ^3H Ultrofilm: Application to multiple benzodiazepine receptors. *J. Neurosci. Meth.* **6,** 59–73.

van Houten M. and Posner B. I. (1981) Cellular basis of direct insulin action in the central nervous system. *Diabetologia* **20,** (suppl.), 255–267.

Van Leuwen F. W. and Walters P. (1983) Light microscopic autoradiographic localization of (^3H) arginine vasopressin binding sites in the rat brain and kidney. *Neurosci. Lett.* **41,** 61–66.

Wamsley J. and Palacios J. (1983) Apposition Techniques of Autoradiography for Microscopic Receptor Localization, in *Current Methods in Cellular Neurobiology* vol. 1 (Barber J. and McKelvy J., eds.), Anatomical Techniques, New York.

Young W. S. and Kuhar M. J. (1979) A new method for receptor autoradiography: [^3H] Opioid receptors in rat brain. *Brain Res.* **179,** 255–270.

Biochemical Approaches to the Study of Peptide Actions

Pierre J. Magistretti

1. Introduction

1.1. General Considerations on the Role of Peptides in Central Neurotransmission

During the course of the last decade a considerable number of biologically active peptides has been identified in the nervous system; several of these recently discovered peptides fulfill the criteria for a neurotransmitter function (Hökfelt et al., 1980a; Snyder, 1980; Bloom, 1981; Krieger and Martin, 1981a, b; Krieger, 1983; Krieger et al., 1983). This remarkable expansion in the vocabulary of neurotransmission has generated questions that can be addressed, at least in part, with biochemical approaches.

The first biologically active peptides to be identified in the central nervous system (CNS) were vasopressin and oxytocin. They were isolated, sequenced, and synthesized by du Vigneaud and collaborators in the early fifties, and their role in endocrine regulation was subsequently clearly established (duVigneaud et al., 1953, 1954; duVigneaud, 1956). These two peptides are synthesized in cell bodies of neurons localized in the paraventricular and supraoptic nuclei of the hypothalamus, whose processes terminate in the posterior pituitary from where they are released into the bloodstream.

This first breakthrough in peptide biology was followed in the late sixties and early seventies by the discovery, by Guillemin and Schally, of a series of peptides, also of hypothalamic origin, controlling the release of peptide hormones from the anterior pituitary gland. Thus thyrotropin releasing hormone (TRH) (Boler et al.,

1969), somatostatin (SS) (Brazeau et al., 1973), and luteotrophic hormone releasing hormone (LHRH) (Amoss et al., 1971; Schally et al., 1971) were isolated, sequenced, and synthesized and their role in regulating the release of thyrotropin, growth hormone, and luteinizing hormone, respectively, was demonstrated (Guillemin, 1978a, b). Two recent additions to the list of releasing factors of hypothalamic origin are corticotropin releasing factor (CRF) and growth hormone releasing factor (GHRF) (Vale et al., 1981; Guillemin et al., 1982; Rivier et al., 1982, 1983; Spiess et al., 1983).

It is only in recent years, however, that the presence of biologically active peptides in extrahypothalamic areas of the CNS has been recognized. Opioid peptides, such as met- and leu-enkephalin were among the first to be identified outside the hypothalamus (Hughes et al., 1975). These were followed by a series of peptides whose presence had previously been demonstrated in the gastrointestinal tract, such as cholecystokinin (CCK), vasoactive intestinal peptide (VIP), secretin, bombesin, and pancreatic polypeptide (PP), or in the hypothalamus, such as TRH, SS, LHRH, and neurotensin (Snyder, 1980; Bloom, 1981; Grossman, 1981; Krieger and Martin, 1981a, b; Krieger, 1983; Krieger et al., 1983). Several extrahypothalamic areas of the CNS contain these biologically active peptides, as demonstrated by radioimmunoassay or immunohistochemistry. In several instances, marked regional variations in individual peptide distribution are observed. Thus, met- and leu-enkephalin are found in very high concentrations in the globus pallidus (Bloom and McGinty, 1981); cerebral cortex is enriched with VIP (Loren et al., 1979; McDonald et al., 1982; Morrison and Magistretti, 1983; Connor and Peters, 1984; Morrison et al., 1984; Magistretti and Morrison, 1985), SS (Finley et al., 1981; Morrison and Magistretti, 1983; Morrison et al., 1983; Johansson et al., 1984), and CCK immunoreactive material (Rehfeld, 1978; 1980; Larsson and Rehfeld, 1979; Dockray, 1980; Hendry et al., 1983; Morrison and Magistretti, 1983; Peters et al., 1983), and high amounts of neurotensin (Carraway and Leeman, 1976; Uhl and Snyder, 1976) are found in the basal ganglia.

Many of these peptides display specific homeostatic functions at the systemic level through cellular actions in those systems in which they had been originally identified. They are therefore generally viewed as hormones, or neurohormones, with specific roles in endocrine regulation in peripheral tissues, or, as in the case of the hypothalamic releasing factors, in the control of pituitary hormone release. Their recent identification in extrahypothalamic areas of the CNS, however, has raised questions and controversies as to their physiological role(s) in these areas. As to their actions at the cellular level, various hypotheses have been advanced (Fig. 1). One view is that they have the capacity to act as classical neurotransmitters, such as acetylcholine and certain amino acids. Electrophysiological studies have demonstrated, however, that with the possible exception of the enkephalins and substance P, the temporal and qualitative features of the effects of most peptides on neuronal excitability are not consistent with those expected for a classical neurotransmitter (Renaud et al., 1979; Zieglgansberger, 1980). It has thus been proposed that certain neuropeptides may act as neuromodulators, i.e., that they would have the capacity to facilitate or suppress the action of classical neurotransmitters on neuronal excitability without having a marked action by themselves (Siggins, 1981; Siggins and Bloom, 1981; Bloom, 1983; 1984). Convincing experimental evidence of such peptidergic effects on neurons of vertebrates is, however, still rare. It should be noted here that the coexistence of peptides and classical neurotransmitters in the same neuron has recently been observed. For example CCK and dopamine are colocalized in mesolimbic neurons (Hökfelt et al., 1980a, b, c), VIP and acetylcholine in bipolar neurons of the cerebral cortex (Eckenstein and Baughman, 1984), and serotonin (5-hydroxytryptamine, 5-HT) and substance P in neurons of the lower medulla oblongata (Hökfelt et al., 1978). These findings have been interpreted as a morphological substrate supporting a neuromodulatory function of neuropeptides (Hökfelt et al., 1980a). The role of neuropeptides in extrahypothalamic areas of the CNS, however, has been elusive thus far and somewhat resistant to experimental analysis.

It is generally agreed that a certain number of criteria for attributing a neurotransmitter role to a substance newly identified in the nervous system should be fulfilled. The substance should be localized to a specific population of neurons and should be releasable by chemical (other neurotransmitter or high extracellular K^+) or electrical stimulation. As for any secretory process, the release should be dependent upon the presence of physiological concentrations of extracellular Ca^{2+}. The metabolism, i. e., synthesis, degradation, and inactivation process (i. e., degradative enzymes or uptake mechanisms) should also be demonstrable. In addition to these criteria, which mainly address the presynaptic aspect of

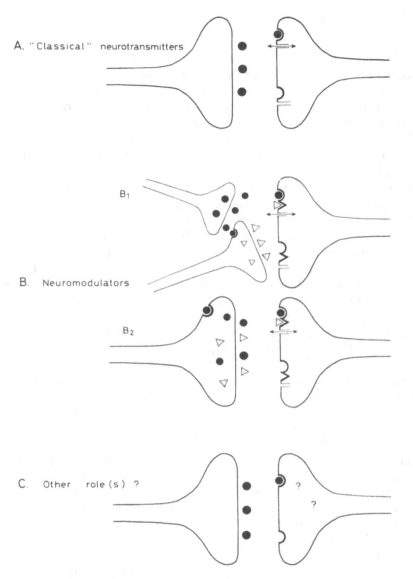

Fig. 1. Schematic representation of the possible role of neuropeptides in extra-hypothalamic areas of the CNS. Terminals (left) and postsynaptic elements (right). Symbols ●, neuropeptides; spackled triangle, "classical" neurotransmitter. A, The neuropeptide alters an ionic conductance (↔) in the postsynaptic element; B_1, The neuropeptide modulates the action of a "classical" neurotransmitter on a postsynaptic element, or its release from presynaptic terminals; B_2, Sequence of events similar to that represented under B_1, with the neuropeptide and the "classical" neurotransmitter being released from the same terminal (colocalization).

interneuronal communication, post-synaptic receptors and cellular actions activated by the putative neurotransmitter should also exist.

1.2. Biochemical Approaches to the Study of Intracellular Events Triggered by Peptide Receptor Occupation

In a sequence of events taking place after a neurotransmitter has been released into the synaptic cleft, the recognition of the neurotransmitter by specific postsynaptic receptors is the first event. Receptor occupation is subsequently transduced into specific intracellular signals whose characteristics will depend upon the degree and nature of the coupling between the receptor and the effector mechanisms. For example, receptors for certain amino acid neurotransmitters such as gamma-aminobutyric acid (GABA) and glutamate, as well as the nicotinic acetylcholine receptor, are coupled to ionic channels specific for a given ion (e. g., Na^+, Cl^-), which, when open, induce profound changes in the membrane potential of neurons (Cooper et al., 1982; Bloom, 1984; Iversen, 1984). Clearly, electrophysiological techniques, which are discussed elsewhere in this volume (chapters by Lukowiak and Murphy, and Pittman et al.), constitute the most appropriate methodology to assess the cellular actions of peptides whose receptors might be coupled to ionic channels.

Other intracellular signals, mediated by the occupation of receptors by peptides, can be conveniently examined by using sensitive biochemical approaches. First, variations in the levels of second messengers, such as cyclic nucleotides or free intracellular Ca^{2+}, can be evaluated (Bloom, 1975; Cheung, 1980; Kebabian and Nathanson, 1982; Drummond, 1983; Rasmussen and Barrett, 1984). Recent experimental evidence has emerged that points to a pivotal function for the metabolism of certain membrane phospholipids, such as phosphatidylinositol 4,5-bisphosphate and inositol trisphosphate (IP_3), as a molecular mechanism that mediates the increases in free intracellular Ca^{2+} concentration (Berridge and Irvine, 1984). Thus, modifications of membrane phospholipid turnover represent another index of a functionally relevant receptor occupation (Fig. 2).

Increases in the levels of second messengers will in turn trigger a series of complex intracellular regulatory mechanisms that are also amenable to biochemical analysis. For example, cyclic adenosine 3',5'-monophosphate (cAMP) binds to the regulatory subunit of a cAMP-dependent protein kinase whose catalytic subunit is

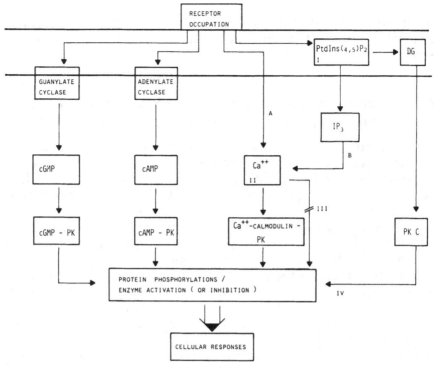

Fig. 2. Diagram representing various intracellular indices, triggered by receptor occupation, that can be assessed with biochemical techniques. (PK, protein kinase.) The arrows denote the relationships between individual steps in a sequence of events that can be assayed with biochemical approaches. No implication as to the nature of regulatory or feedback mechanisms between each step are made. I, [PdtIns(4,5)P₂]: phosphatidylinositol 4,5-bisphosphate is hydrolyzed into diacylglycerol (DG) and inositol trisphosphate (IP₃) by activated phospholipase C. II, Free cytosolic Ca^{2+} can increase as a consequence of (A) the entry of extracellular Ca^{2+} through receptor-operated Ca^{2+} channels or, (B) IP₃-mediated mobilization of Ca^{2+} from intracellular stores. III, Action of Ca^{2+} through intracellular effectors other than protein kinases. IV, Protein kinase C and Ca^{2+} can act synergistically to elicit a cellular response. Receptor-mediated activation (or inhibition) of adenylate cyclase is regulated by a GTP-binding protein.

thus released and can in turn activate or inhibit the activity of various enzymes (Figs. 2 and 3) via a cascade of phosphorylation steps. The phosphorylation state of specific proteins can be determined and changes in the activity of specific enzymes, as well in the levels of end-products of enzymatic reactions, can be moni-

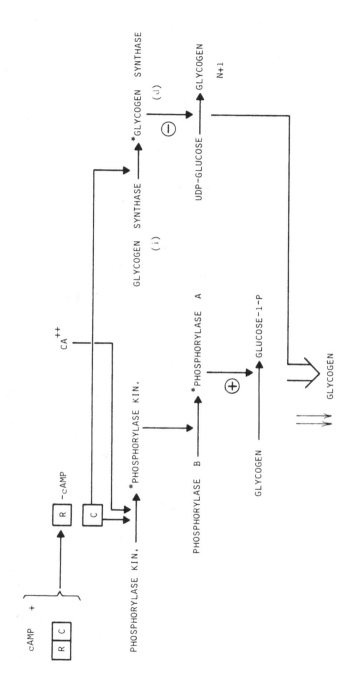

Fig. 3. cAMP-dependent phosphorylation cascade leading to a decrease in intracellular glycogen. R, regulatory subunit of protein kinase; C, catalytic subunit of protein kinase; *, phosphorylated protein. Note that a Ca^{2+}-dependent phosphorylation of phosphorylase kinase exists.

tored for those neurotransmitters whose receptors are coupled to adenylate cyclase (Cohen, 1982; Nestler and Greengard, 1983). Similar approaches can be followed for those intracellular processes regulated by changes in cyclic-GMP and in free intracellular Ca^{2+} concentrations (Fig. 2).

I have outlined some of the indicators of the cellular actions of neurotransmitters that can be assessed with biochemical methodologies. Activation of cAMP formation and modulation of enzyme activity has been examined and described in great detail for several CNS neurotransmitters, particularly the monoamines norepinephrine (noradrenaline, NA), dopamine (DA), 5-hydroxytryptamine (5-HT), and histamine (Bloom, 1975; Kebabian and Nathanson, 1982; Drummond, 1983). As for amino acid neurotransmitters, an example is represented by the elegant work of Baudry, Lynch, and collaborators, which has unraveled the relationships between the regulation of the number of L-glutamate binding sites, Ca^{2+}, and the activation of specific proteases in the hippocampus (Baudry and Lynch, 1980; Baudry et al., 1981; Siman et al., 1984). Furthermore, the same group has described the inhibitory effect of excitatory amino acids on amine-stimulated phosphatidylinositol metabolism (Baudry et al., 1986), and glutamate-stimulated inositol phosphate formation has also been reported recently (Sladeczek et al., 1985). Our state of knowledge of the cellular actions of the newly discovered neuropeptides can best be described by paraphrasing* the title of a play by the Italian playwriter Luigi Pirandello, "Six Characters in Search of an Author" into "A plethora of peptides in search of an action." This rather defeatist statement can, however, be tempered by those few instances in which the effect of a peptide on cellular regulatory processes has been observed in the nervous system.

In this chapter, I will try to describe some experimental strategies that can be used to examine the actions of peptides in the nervous system with biochemical techniques. For this purpose I will describe in detail two cellular actions of VIP in rodent cerebral cortex. VIP is a peptide that was isolated from the porcine duodenum by Said and Mutt in 1970 (Said and Mutt, 1970b), and subsequently identified in the central and peripheral nervous systems of several mammalian species, including man (Said, 1982). In the final sections of the chapter, I have included certain experimental

*A similar paraphrase—"various autacoids in search of a function"—has been used by W. W. Douglas for the physiological role of autacoids (Douglas, 1980).

approaches that have been successfully used to elucidate the actions of certain monoamines or amino acid neurotransmitters, even though no, or only few, examples of an action of neuropeptides on the cellular indices addressed by these strategies have been demonstrated. They are nevertheless briefly presented because of their potential usefulness. Rather than describing the methods in great detail (for which the reader is referred to the references section), I have placed the emphasis on the strengths and limitations of a given experimental approach. The examples presented are not exhaustive; they have been chosen to illustrate the type of results that can be obtained with a given strategy.

2. Tissue Preparations

The actions of peptides (and other neuroactive agents) amenable to biochemical analysis can be studied in vitro in two types of tissue preparations: homogenates and slices.

2.1. Homogenates

Homogenization of nervous tissue, followed by simple centrifugation or centrifugation on a gradient (e. g., sucrose, ficoll) yields various subcellular fractions on which to test the possible actions of a given neuroactive substance (deDuve, 1965; Rose and Sinha, 1969). The most commonly used preparations for neurotransmitter-specific neurochemistry are crude resuspended membranes and synaptosomes (deRobertis et al., 1961; Gray and Whittaker, 1962; Cotman, 1974). These tissue-preparation techniques are relatively simple and have been extensively described and validated (Siegel et al., 1981). One of the major drawbacks in the use of particulate fractions is the loss of the cellular integrity inherent in the preparative procedures. For reasons that are given below, slices of nervous tissue represent a valuable alternative for the assessment of the action of a neuropeptide and its potential interactions with other neuroactive substances.

2.2. Slices

The use of brain slices for neurochemical studies was pioneered by Henry McIlwain in the early fifties (Buchel and McIlwain, 1950; McIlwain, 1952). Slices can be prepared from virtually any area of the CNS. The optimal thickness, which allows for the penetration of nutrients across the slices, is generally between 200 and 500 μm (Dingledine, 1984). The advantages of such a preparation are multiple, since a remarkable degree of cellular integrity is maintained. Thus a number of perikarya and processes are intact, allowing for the persistence of certain synaptic contacts. In fact, McIlwain estimates the number of cells with which a single neuron of the cerebral cortex may synapse to be 10^3–10^5; this number roughly corresponds to the number of neurons that are present in the type of cerebral cortical slices most commonly used (1–10 mg of tissue) (McIlwain, 1984). In addition to morphological observations, metabolic studies have verified the presence of energy substrates (ATP, phosphocreatine, glycogen) in slices maintained in vitro in appropriate media, thus providing a biochemical reflection of the integrity of the tissue (McIlwain, 1951, 1953; McIlwain et al., 1952; LeBaron, 1955; McIlwain and Tresize, 1956; Li and McIlwain, 1957).

Another considerable advantage in using nervous tissue slices is the possibility, because of the relative cellular integrity achieved, to assess receptor-activated mechanisms that occur intracellularly at sites situated a distance from the cell membrane. Thus the mechanisms of (1) receptor activation, (2) transduction of the signal across the plasma membrane, and (3) the sequence of intracellular events that lead to a cellular response are all potentially preserved in the same preparation. Cellular integrity also provides the unique opportunity to evaluate possible interactions between neurotransmitters at any of the steps that occur "downstream" to receptor activation (Fig. 2). Some characteristics of nervous tissue slices that, depending on the question that the investigator intends to address, may represent serious limitations, should however be considered. First, the cellular elements comprising a slice are heterogeneous: different types of neurons are present, as well as astrocytes, oligodendrocytes, microglia, and cells of the vasculature. Thus, the resolution at the cellular level of a given action of a neuropeptide cannot be achieved. Second, the observed effect may be indirect and result from the release of another neurotransmitter elicited by the peptide under investigation. Pharmacological and lesion experiments may be of value in confirming or discarding this possibility (for an example, *see* section 3.3.2).

Another aspect to be considered when using slices is the penetration of the peptide across the tissue, and its possible degradation by peptidases. This consideration is of particular relevance when the potency of the peptide and the time course of its action are to be evaluated. The use of peptidase inhibitors (e.g., bacitra-cin, aprotinin) or alternative substrates for peptidase (e.g., bovine serum albumin) is therefore recommended. A detailed description of the procedure used to prepare slices of mouse cerebral cortex is presented below.

2.3. Preparation of Cerebral Cortical Slices

Mice are decapitated and the brain is rapidly removed and placed on dental wax on ice for dissection. The cerebral cortex is dissected out and placed in a modified Krebs-Ringer bicarbonate buffer (KRG) (at room temperature) containing NaCl, 120 mM; KCl, 5 mM; CaCl$_2$, 2.6 mM; MgSO$_4$, 0.67 mM; KH$_2$PO$_4$, 1.2 mM; glucose, 3 mM; NaHCO$_3$, 27.5 mM and previously gassed with O$_2$:CO$_2$ (95:5) to maintain a pH of 7.4. Usually 2–3 cortices per experiment are used (30–40 assays). The cortices are then placed, ventral surface facing down, on a McIlwain tissue chopper (Mickle Laboratory Engineering, Gromshall, Surrey, England) and slices of 250 μm are prepared. After a 90° rotation of the plate on which the cortices have been placed, slices of 250 μm are again prepared, and the tissue is resuspended in ice-cold KRG (6 mL per cortex) and pipeted 3–4 times with a thin-bore pipet in order to obtain a nearly complete dissociation of the slices. This procedure yields a homogeneous suspension of cortical parallelepipeds of 250 μm in cross-section that extend from pia to white matter (approximately 1200 μm in length). After replacing the buffer with the same amount of fresh KRG, the slices are incubated under vigorous shaking and continuous gassing with O$_2$:CO$_2$ (95:5) in a water-bath at 37°C for 15 min. The buffer is again replaced with fresh KRG and 270-μL aliquots (approximately 20 mg of tissue per mL) are pipeted into individual tubes. The substances under investigation can then be added, for varying periods of time (usually in 20–30 μL of buffer), and the index of their action (e.g., cAMP or glycogen levels) is then assayed as described in sections 3.2 and 3.3.

3. Cellular Actions of VIP in Rodent Cerebral Cortex

3.1. VIP

VIP is a 28-amino-acid peptide with structural homologies and certain biological actions in common with the following peptides: PHI (peptide with a histidine at the amino terminal and an isoleucine at the carboxy terminal), secretin, GHRF (growth hormone releasing factor), and glucagon. VIP shares 13 identical positions in amino acid residues with PHI, nine with secretin and GRF and five with glucagon (Fig. 4). Recently a peptide isolated from gila monster venom and denominated helodermin has been shown to possess 15 amino acid residues in common with VIP (Hoshino et al., 1984; Robberecht et al., 1984). Following the isolation from porcine duodenum by Said and Mutt (Said and Mutt, 1970b), VIP immunoreactivity has been localized to nerves throughout the gastrointestinal (Costa et al., 1980) and genitourinary tracts (Larsson et al., 1977) and in endocrine glands such as the pancreas (Larsson et al., 1978) and thyroid gland (Ahren et al., 1980). The presence of VIP has also been demonstrated in the CNS, where VIP is highly concentrated in the cerebral cortex (Loren et al., 1979; McDonald et al., 1982; Morrison and Magistretti, 1983; Connor and Peters, 1984; Morrison et al., 1984; Magistretti and Morrison, 1985). Within this CNS region, VIP is localized to a homogeneous population of intracortical, bipolar neurons oriented perpendicularly to the pial surface and arborizing only minimally (60–100 μm) in the horizontal plane, predominantly in layers I and IV–V (Fig. 5) (Morrison and Magistretti, 1983; Morrison et al., 1984; Magistretti and Morrison, 1985). VIP exerts various biological actions, which include vasodilation of peripheral, splanchnic, coronary, and pial vessels (Said and Mutt, 1970a; Thulin and Samnegard, 1978; Wei et al., 1980; Smitherman et al., 1982), smooth muscle relaxation throughout the gastrointestinal tract, and bronchodilation (Piper et al., 1970; Matsuzaki et al., 1980). In addition, VIP stimulates endocrine and exocrine pancreatic secretion (Schebalin et al., 1977; Peikin et al., 1978) and water and ion excretion in the intestine (Krejs et al., 1978; Davis et al., 1980), and promotes glycogenolysis in liver (Kerins and Said, 1973; Wood and Blum, 1982). At the cellular level, VIP stimulates the membrane-bound enzyme adenylate cyclase in several tissues, thus promoting an increase in the intracellular second messenger cAMP. In the CNS, VIP has also been shown to promote the formation of cAMP, through its interaction with specific membrane receptors (Deschodt-Lanckman et al., 1977; Quik et al., 1978; Taylor and Pert, 1979; Schorderet et al., 1981; Staun-Olsen et al., 1982; Magistretti and Schorderet, 1984; 1985). One of the intracellular actions of cAMP is to trigger a series of phosphorylations via an activated protein kinase (Fig. 3). One such series of cAMP-dependent phosphorylations regulates intracellular glycogen levels. This regulation is achieved by the phosphorylation of the enzyme phosphorylase, which leads to its conversion from a weakly active form (b) to an active form (a) (*see* Fig. 3). In this form, phosphorylase induces the breakdown of glycogen into glucose-1-phosphate. In parallel to this enzymatic process, glycogen synthase, the enzyme that regulates the addition of glucose units to glycogen, is phosphorylated by a cAMP-dependent mechanism and converted from its active form (i) to a less active form (d). Therefore an increase in intracellular cAMP leads to an activation of the enzymatic degradation of glycogen and to an inhibition of the synthesis of the polysaccharide, thus directing all available glucose residues and precursors into the production of phosphate-bound energy (Lehninger, 1982). The intracellular metabolic process described above has been characterized in detail in liver and muscle. In the CNS, similar regulatory processes appear to occur. The presence of glycogen in the brain has been demonstrated biochemically and histochemically (Havet, 1937; Nicholls and Wolfe, 1967; Wolfe and Nicholls, 1967; Sotelo and Palay, 1968; Nahorski and Rogers, 1972; Bruckner and Biesold, 1981; Cammermeyer and Fenton, 1981). Furthermore, most of the enzymes implicated in glucose and glycogen metabolism are present in the brain, and their molecular regulatory processes appear to be similar to those described in liver and muscle (Buell et al., 1958; Breckenridge and Crawford, 1961; Breckenridge and Norman, 1962; 1965; Lowry and Passonneau, 1964; Nelson et al., 1968; Drummond and Bellward, 1970). In addition, several studies have indicated that an increase in intracellular cAMP induces the breakdown of glycogen in several neural tissues (Park and Exton, 1973; Edwards et al., 1974; Nahorski and Rogers, 1975; Nahorski et al., 1975; Wilkening and Makman, 1976, 1977), through the phosphorylation of phosphorylase (b) to (a).

In keeping with these observations, VIP has been demonstrated to promote a concentration-dependent hydrolysis of glycogen in mouse cerebral cortical slices (Magistretti et al., 1981; 1984b). Thus, stimulation of cAMP formation and glycogen breakdown constitute two cellular actions amenable to biochemical analysis that have been demonstrated for VIP. The biochemical approaches used to assess these two cellular actions will be described in the following sections, since they represent simple and valuable means to assess potential actions of newly discovered peptides.

	1	2	3	4	5	6	7	8	9	10	11	12	13	14	15	16	17	18	19	20	21	22	23	24	25	26	27	28	29
VIP	HIS	SER	ASP	ALA	VAL	PHE	THR	ASP	ASN	TYR	THR	ARG	LEU	ARG	LYS	GLN	MET	ALA	VAL	LYS	LYS	TYR	LEU	ASN	SER	ILE	LEU	ASN*	
PHI	HIS	ALA	ASP	GLY	VAL	PHE	THR	SER	ASP	PHE	SER	ARG	LEU	LEU	GLY	GLN	LEU	SER	ALA	LYS	LYS	TYR	LEU	GLU	SER	LEU	ILE*		
h GHRF	TYR	ALA	ASP	ALA	ILE	PHE	THR	ASN	SER	TYR	ARG	LYS	VAL	LEU	GLY	GLN	LEU	SER	ALA	ARG	LYS	LEU	LEU	GLN	ASP	ILE	MET	SER	ARG..
SECRETIN	HIS	SER	ASP	GLY	THR	PHE	THR	SER	GLU	LEU	SER	ARG	LEU	ARG	ASP	SER	ALA	ARG	LEU	GLN	ARG	LEU	LEU	GLN	GLY	LEU	VAL*		
GLUCAGON	HIS	SER	GLN	GLY	THR	PHE	THR	SER	ASP	TYR	SER	LYS	TYR	LEU	ASP	SER	ARG	ARG	ALA	GLN	ASP	PHE	VAL	GLN	TRP	LEU	MET	ASP	THR
HELODERMIN	HIS	SER	ASP	ALA	ILE	PHE	THR	GLN	GLN	TYR	SER	LYS	LEU	LEU	ALA	LYS	LEU	ALA	LEU	GLN	LYS	TYR	LEU	ALA	SER	ILE	LEU	GLY	SER..

AMINO ACID RESIDUES IN THE SAME POSITION AS IN THE VIP MOLECULE ARE UNDERLINED
* DENOTES AMIDATION
ONLY THE FIRST 29 AMINO ACIDS IN THE hGHRF AND HELODERMIN SEQUENCE ARE SHOWN.

Fig. 4.   Amino acid sequence of VIP and related peptides.

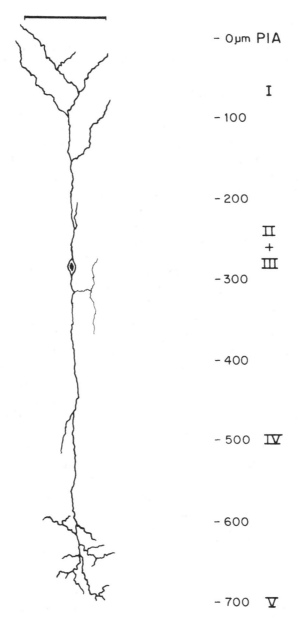

- 0μm PIA

I

- 100

- 200

II
+
III

- 300

- 400

- 500   IV

- 600

- 700   V

Fig. 5.   Diagram showing the shape and dimensions of an idealized VIP-positive bipolar cell in rodent cerebral cortex, based on photomontages of actual cells. The cell is accurately drawn to scale. Bar at top equals 100 μm. Radial depth shown on right. Roman numerals on far right refer to cortical layers (from Morrison et al., 1984).

3.2. Effect on cAMP Levels

Cyclic adenosine 3',5' monophosphate is formed from ATP via an enzymatic reaction catalyzed by the membrane-bound enzyme adenylate cyclase. Since its discovery by Sutherland and collaborators (Rall and Sutherland, 1958; Sutherland et al., 1962), impressive numbers of publications have described and reviewed the methodology for measuring changes in cAMP levels in various cellular systems, including the nervous system (Bloom, 1975; Daly, 1975; Nathanson, 1977; Kebabian and Nathanson, 1982; Drummond, 1983). A commonly used technique for cAMP measurement consists of a competition assay in which a fixed amount of radiolabeled (^{125}I or ^{3}H) cAMP is incubated with (1) unlabeled cAMP standards (usually between 0 and 30 pmol), and (2) a cAMP-binding protein (extracted from bovine adrenal cortex) or an antibody directed against cAMP (Brown et al., 1971; Steiner et al., 1972). cAMP bound to the binding protein or to the antibody is then separated from free cAMP. A standard curve is then generated, allowing for the quantification of cAMP present in the experimental samples assayed in parallel. The sensitivity of this assay is in the picomole/sample range, which is sufficient for most experiments in which nervous tissue is used. It is, however, possible to enhance the sensitivity of the assay by approximately 100-fold with an acetylation step prior to the competition assay. This procedure is of particular value when very limited amounts of tissue are available for testing. Some important points should be borne in mind when designing an experiment in which the effects of a neuropeptide on cAMP levels are examined. If homogenates are used, ATP, the substrate of adenylate cyclase for cAMP formation, should be added in the assay mixture, since during the homogenization process the endogenous stores of ATP are lost. The concentration of ATP added is usually in the millimolar range. A second aspect to be considered when using slices is that the newly formed cAMP is degraded into 5'-AMP and inorganic phosphate by the enzyme phosphodiesterase. The activity of this enzyme can be blocked effectively, for example, by xanthines such as theophylline and isobutyl-methyl-xanthine (IBMX) at millimolar concentrations (Smellie et al., 1979) or by synthetic compounds such as RO-20-1724. These agents can therefore be conveniently added to the incubation buffers in order to prevent the degradation of the newly formed cAMP.

The assessment of the effects of neuroactive substances on

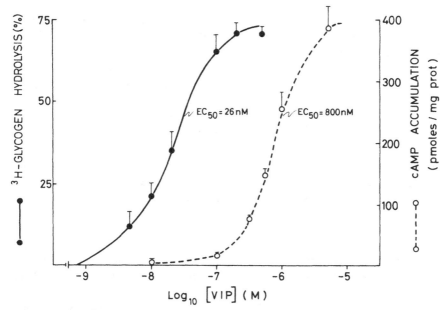

Fig. 6. Concentration–response curves of the stimulation of cAMP formation (○– – –○) and glycogenolysis (●—●) elicited by VIP. Note the difference in the potency (EC_{50}) of VIP to elicit these two actions.

cAMP levels is a simple and extremely well-characterized methodology. It is therefore a particularly suitable biochemical test to assess a cellular action of a newly discovered peptide. To date, however, only few convincing examples of peptide-stimulated cAMP formation in nervous tissue have been demonstrated. Thus, in addition to VIP (Deschodt-Lanckman et al., 1977; Quik et al., 1978; Schorderet et al., 1981; Magistretti and Schorderet, 1984; 1985; Chneiweiss et al., 1985), only substance P and some members of the glucagon-secretin family, such as PHI and secretin, increase cAMP levels in neural tissues (Duffy and Powell, 1975; van Calker et al., 1980; Nathanson, 1981; Huang and Rorstad, 1983; 1984; Roth et al., 1984). The effect of VIP on cAMP levels in mouse cerebral cortical slices is shown in Fig. 6.

3.2.1. Synergism with Noradrenaline

The measurement of cAMP levels, in addition to revealing an action of VIP, has allowed for the examination of interactions between VIP and other neurotransmitters in the cerebral cortex. The monoamine noradrenaline (NA) stimulates cAMP formation via the activation of specific β-adrenergic receptors (Bloom, 1975; Daly, 1975; Nathanson, 1977; Kebabian and Nathanson, 1982; Drummond, 1983). When the combined effects of maximally effective concentrations of VIP and NA (10^{-6} and $10^{-4}M$, respectively) were examined in mouse cerebral cortical slices, a synergistic interaction between the peptide and the monoamine was observed, as shown in Fig. 7. This synergism was antagonized by the specific α-adrenergic antagonist phentolamine, but not by *d,l*-propranolol, a β-adrenergic antagonist (Magistretti and Schorderet, 1984). This observation indicates a modulatory action of NA on a peptide-mediated cellular action, i. e., that NA, by activating specific α-adrenergic receptors, can potentiate the stimulatory effects of VIP on cAMP levels (Magistretti and Schorderet, 1984). A more extensive pharmacological analysis, based on adrenergic agonists' rank-order of potencies, has indicated that the α-adrenergic receptor that mediates the modulatory action of NA is of the α_1 subtype (Magistretti and Schorderet, 1984; 1985; Magistretti et al., 1984a). Recently, these observations have also been reported for guinea pig cerebral cortex (Hollingsworth and Daly, 1985) and for discrete nuclei of the rat hypothalamus (Redgate et al., 1986). Furthermore, this interaction has been confirmed electrophysiologically, since VIP and NA act synergistically to depress spontaneous discharge rate of rat cerebral cortical neurons (Ferron et al., 1985).

3.3. Effect on Glycogen Levels

The major energy substrate of the brain is glucose, which, as a consequence of its hydrolysis, leads to the generation of ATP. Glycogen is the single largest energy reserve in the brain; it can enter the glycolytic pathway after specific enzymatic hydrolysis and may therefore also contribute to the generation of phosphate-bound energy. In peripheral tissues, various mechanisms promoting glycogen hydrolysis exist; they are directly mediated by extracellular signals (neurohumoral transmission), or they are related to the energy state of the cell (see below). The enzyme responsible for glycogen degradation is phosphorylase, which can exist under two forms, phosphorylase a and b. Phosphorylase b is allosterically activated by AMP and inhibited by ATP and glucose-6-phosphate (Lehninger, 1982). Thus, when the energy charge of the cell is low (low concentrations of ATP and high concentrations of AMP), phosphorylase b is activated and promotes the breakdown of glycogen for the generation of high-energy phosphates

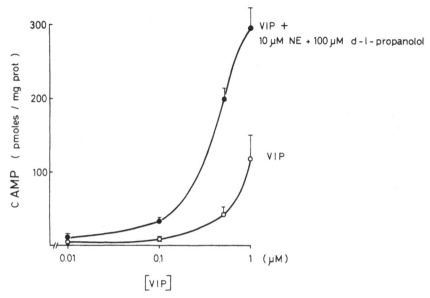

Fig. 7. Concentration–response curve of the stimulatory effect of VIP on cAMP levels in the presence (●—●) or absence (○—○) of 10 µM NA and 100 µM d,l-propranolol, a selective β-adrenergic antagonist. Under these conditions the effect of NA reflects the selective activation of α-adrenergic receptors by the catecholamine (from Magistretti and Schorderet, 1985).

(Lehninger, 1982). Phosphorylase b can also be covalently activated by phosphorylation into phosphorylase a. This phosphorylation occurs as a consequence of an increase in intracellular cAMP and of the subsequent enzymatic steps (Fig. 3; Cohen, 1982). As described in previous sections, cAMP levels in the cell may increase following the interaction of a neurotransmitter (or neurohormone) with specific membrane receptors functionally coupled to an adenylate cyclase, the enzyme catalyzing the synthesis of cAMP from ATP (Sutherland and Robison, 1966; Robison et al., 1970). Increases in intracellular calcium ion concentration, induced by hormones (e. g., NA in adult rat liver; (Exton and Harper, 1975; Exton et al., 1981) or by increased cellular activity (e. g., muscle contraction; Danforth and Helmreich, 1964; Danforth and Lyon, 1964; Heilmeyer et al., 1970; Brostrom et al., 1971) can also activate phosphorylase kinase and eventually lead to glycogen breakdown through the conversion of phosphorylase into its a form. In the brain, similar regulatory mechanisms appear to be operational

(Buell et al., 1958; Breckenridge and Crawford, 1961; Breckenridge and Norman, 1962; 1965; Lowry and Passonneau, 1964; Nelson et al., 1968; Drummond and Bellward, 1970; Passonneau et al., 1971; Edwards et al., 1974; Wilkening and Makman, 1976; 1977).

Thus, the assessment of the effects of a neurotransmitter on glycogen metabolism represents a valuable biochemical approach to study the cellular actions of those neurotransmitters for which a stimulatory action on cAMP formation has been previously demonstrated. This approach has proved useful for the monoamines NA, histamine, and serotonin (5-HT) (Quach et al., 1978; 1980; 1982; Magistretti et al., 1983b), and more recently for VIP (Magistretti et al., 1981; 1984b). The following paragraphs describe in detail the methodology for assessing the glycogenolytic action of neurotransmitters, with particular emphasis on the action of VIP in mouse cerebral cortical slices.

3.3.1. ^{3}H-Glycogen Assay

Cerebral cortical slices are prepared as described in section 2.3 and incubated in the presence of 20 nmol of D-[6-^{3}H]-glucose (specific activity: 500 mCi/mmol). A time-dependent incorporation of ^{3}H-glucose into ^{3}H-glycogen is observed, until a steady-state is reached (usually between 30 and 40 min after the addition of ^{3}H-glucose). The substance under investigation is added at this point for an additional 10 min. The reaction is stopped by rapidly centrifuging the tubes on a table centrifuge (1 min at 9000g). The supernatant is replaced with fresh buffer, and the tissue sonicated for 5 s with a cell disruptor and subsequently placed for 10 min in a waterbath at 95°C. After another centrifugation step (5 min at 9000g), ^{3}H-glycogen is separated from ^{3}H-glucose in the supernatant by ethanol precipitation on filter paper as described below. Proteins are quantified in the pellet as described by Lowry et al. (1951) (Fig. 8).

Tritiated glycogen is determined by ethanol precipitation on filter paper, as described by Sølling and Esmann (1975). Aliquots (150 µL) of the deproteinized supernatants are pipeted onto 31-ET (Whatman) filter paper disks (24 mm in diameter) numbered with a lead pencil to the corresponding tube and placed into an ice-cold solution of ethanol (60%): trichloracetic acid (10%) for 10 min (the final volume of this solution corresponds to 10 mL/filter). They are then placed into a similar volume of ethanol (66%) at room temperature for 10 min. This procedure is repeated six times during the following hour. After the last ethanol (66%) wash, filters are placed for 5 min in acetone, air dried, and placed in vials to which scintilla-

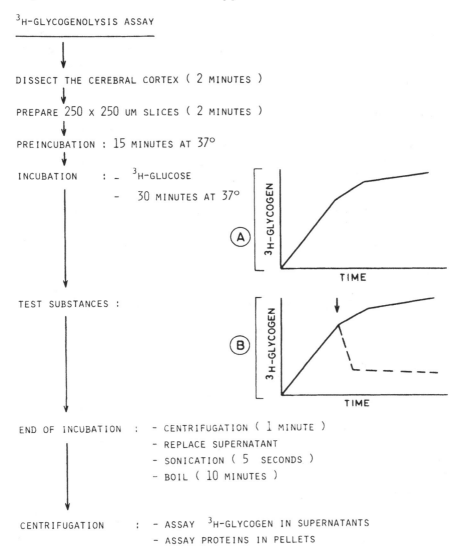

Fig. 8.   The ^{3}H-glycogenolysis assay. A, Time-dependent incorporation of ^{3}H-glucose into ^{3}H-glycogen. B, Decrease in ^{3}H-glycogen levels (hatched line) after application of glycogenolytic agents (arrow).

tion fluid is added and counted for ^{3}H content by liquid scintillation spectrometry.

The contamination by glucose and the recovery of glycogen yielded by this separation technique can be measured. In these experiments, aliquots of 1 mM glycogen (calculated as glycosyl units) or glucose solutions are pipeted on the filter paper disks. One set of filters containing glycogen or glucose is submitted to the ethanol precipitation technique; this procedure is omitted for another identical set. Glycogen and glucose are then determined on each filter using the enzymatic technique originally described by Nahorski and Rogers (1972). The results of this type of control experiment indicate that the recovery of glycogen is greater than 99% and that the contamination by glucose is smaller than 0.5%. The identity of the ^{3}H-labeled product, isolated by the ethanol precipitation technique, with glycogen can also be tested. At the end of a typical experiment, aliquots of the deproteinized supernatants are incubated for 30 min at 37°C in the presence or absence of 4.2 units of 1,4-α-D-glucan glycohydrolase (EC 3.2.1.3), an enzyme liberating glucose from glycogen by hydrolyzing α-(1→6) and α-(1→4) linkages. Samples are then submitted to ethanol precipitation as previously described and counted for tritium content. The radioactivity recovered from the enzymatically treated samples usually represents only 10% of that recovered from the untreated samples.

3.3.2. Glycogenolytic Action of VIP

When applied at the steady-state level of ^{3}H-glycogen synthesis, VIP promotes a rapid decrease in the levels of the newly synthesized ^{3}H-polysaccharide (Fig. 9, Magistretti et al., 1981; 1984b), with 50% of the maximal effect being achieved within 1 min. The glycogenolytic effect of VIP is reversible and, as indicated in Fig. 6, concentration-dependent with an EC_{50} of 26 nM (Magistretti et al., 1981; 1984b). Several other cortical neurotransmitters present in the cerebral cortex do not promote the hydrolysis of glycogen (Table 1; Magistretti et al., 1981; 1984b), thus stressing the specificity of the effect of VIP on cortical carbohydrate metabolism. Peptides structurally related to VIP, such as PHI and secretin, stimulate cortical glycogenolysis with EC_{50} values of 300 and 500 nM, respectively (Fig. 10; Magistretti et al., 1984b). The complete VIP molecule is necessary to elicit a glycogenolytic action, as indicated by the absence of effect of VIP fragments 6-28, 16-28, and 21-28 (Magistretti et al., 1984b).

Noradrenaline has been previously shown to stimulate glycogenolysis in mouse cerebral cortical slices via the activation of specific β-adrenergic receptors. The preparation in which the glycogenolytic action of neurotransmitters is tested consists, as described earlier, of cortical slices 250 μm in section and approx-

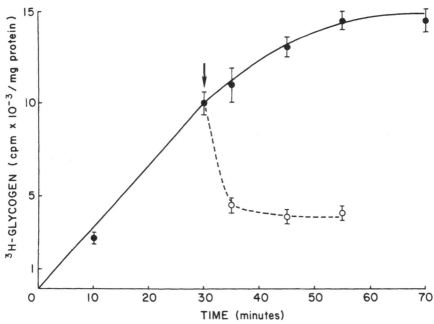

Fig. 9. Time-course of ^{3}H-glycogen synthesis (●——●) and VIP-induced glycogenolysis (○– – –○) in mouse cerebral cortical slices (from Magistretti et al., 1981).

imately 1500 μm in length (depth of the cerebral cortex). Therefore, in these slices a number of cell bodies and processes are intact and some local contacts are maintained. At this level of neuronal organization, an effect observed with one neurotransmitter could be mediated by the release of another. The possibility had therefore to be considered that the glycogenolytic action of VIP could be mediated by the release of NA elicited by the peptide. The observation that *d,l*-propranolol (a specific β-adrenergic antagonist) antagonized the effect of NA but not that of VIP on glycogen levels, however, provided the pharmacological evidence to discard the possibility of a mediation by NA of the effect of VIP. This view was confirmed by lesion studies in which the cerebral cortex was depleted of NA by the selective neurotoxin 6-hydroxydopamine (Malmfors and Thoenen, 1971). The glycogenolytic action of VIP was maintained in the lesioned cortices, thus confirming the independence of the action of VIP from that of NA (Magistretti et al., 1981).

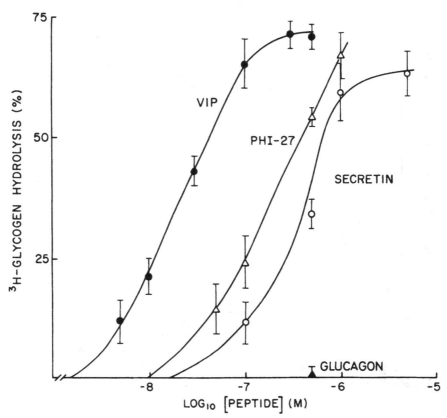

Fig. 10. Comparison of the concentration–response curves of the glycogenolytic effect of VIP, PHI, and secretin. Note that the potency of the peptides is in direct relationship to their degree of sequence homology with VIP. EC_{50}: PHI, 300 nM; secretin, 500 nM (modified from Magistretti et al., 1984b).

3.4. Morphological Correlates

The power of a biochemical approach and the results derived from it can be greatly enhanced by relating the cellular action observed with the morphology of the neuronal system that contains the peptide. Such a correlative analysis has been applied to the actions of VIP in mouse cerebral cortex and to its interactions with the NA-containing system (Magistretti et al., 1981; 1984a,b;

Table 1
Substances Inactive in Promoting ^{3}H-Glycogenolysis in Mouse Cerebral
Cortical Slices

GABA	Leu-Enkephalin
Glutamate	CRF
Aspartate	Vasopressin
Carbamylcholine[a]	Oxytocin
Somatostatin-14	VIP Fragments
Somatostatin-28	6–28
CCK-8	16–28
Met-Enkephalin	21–28

[a]Stable analog of acetylcholine.

Morrison and Magistretti, 1983; Magistretti and Schorderet, 1984; 1985; Morrison et al., 1984; Magistretti and Morrison, 1985).

In rodent cerebral cortex, VIP neurons constitute a rather homogeneous population of intracortical, bipolar cells oriented radially (i. e., in a plane perpendicular to the pial surface) (Loren et al., 1979; McDonald et al., 1982; Connor and Peters, 1984; Morrison et al., 1984). As mentioned earlier, these neurons branch only minimally in the horizontal plane, except in layers I and IV–V, where the dendritic arborization expands horizontally up to 60–100 μm. In the rat visual cortex, the density of these cells is such that on the average there is one VIP-containing neuron per cortical column of 30-μm diameter (Morrison et al., 1984). From this morphological analysis the notion emerges that the cellular actions elicited by the release of VIP molecules from VIP-containing neurons are *locally* restricted, within partially overlapping cortical columns extending from the pial surface to layer V and with a diameter not exceeding 100 μm (Fig. 5; Morrison and Magistretti, 1983; Morrison et al., 1984; Magistretti and Morrison, 1985). The morphology of the NA-containing system within the cerebral cortex is strikingly different. The cell bodies of noradrenergic neurons that innervate the cerebral cortex are located in the nucleus locus ceruleus in the brain stem from where they project diffusely to several regions of the CNS (Dahlstrom and Fuxe, 1964; Fuxe, 1965; Morrison et al., 1978; 1981). In the cerebral cortex, NA is contained in fine fibers organized tangentially, i. e., they run from the frontal to occipital pole of the cortex in a plane parallel to the pial surface and thus orthogonally to the radially oriented VIP-containing intracortical neurons (Fig. 11). Therefore, the noradrenergic cortical system has the capacity to exert its actions *globally*, across a vast rostrocaudal expanse of neocortex simultaneously (Morrison et al., 1978; 1981). These morphological characteristics of the VIP- and NA-containing cortical systems provide the structural background against which to examine the cellular actions of both neurotransmitters and to delineate a possible function of VIP and NA in the cerebral cortex. These considerations are now briefly summarized. VIP and NA promote glycogenolysis in mouse cerebral cortex; this action will result in an increased availability of substrates for the generation of phosphate-bound energy in those cortical cells that are innervated by VIP- and NA-containing neurons. Thus two complementary neuronal systems for the control of cortical energy metabolism appear to exist: (1) *VIP neurons* regulating glycogenolysis *locally*, within cortical columnar modules of 100 μm maximal diameter, and (2) *noradrenergic fibers* exerting their metabolic effect *globally* and synchronously across the neocortex, intersecting a longitudinal array of columns (Magistretti et al., 1981; Morrison and Magistretti, 1983; Magistretti et al., 1984b; Morrison et al., 1984; Magistretti and Morrison, 1985).

Correlations of the synergistic interaction between VIP and NA in stimulating cAMP formation (section 3.2.1) with the structural characteristics of the neuronal systems containing the peptide and the monoamine provide further insights into the functional interplay between these two neurotransmitter systems. Thus the concomitant release of VIP and NA within a discrete volume of the neocortex will generate a "cortical metabolic hot spot" at which cAMP levels are drastically increased in comparison to surrounding regions. The spatial coordinates of such a "hot spot" will be delineated by the intersection of the tangentially organized noradrenergic fibers and a group of activated VIP bipolar intracortical neurons (Fig. 11; Magistretti and Schorderet, 1984; 1985).

These examples illustrate the potential of correlations between the cellular actions of peptides (and other neurotransmitters) with the structural information provided by neurotransmitter-specific neuroanatomical methods. Furthermore these correlations provide a valuable strategy for the extension of biochemical approaches into the domain of the function of neurotransmitters in intercellular communication in the nervous system.

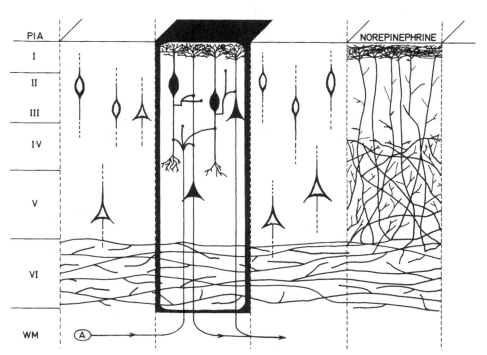

Fig. 11. Diagrammatic representation of a "metabolic hot spot" resulting from the concomitant activation of noradrenergic cortical afferents and a group of VIP-containing neurons. Bottom (layer VI), tangentially organized noradrenergic fibers; Right, detail of the arborization pattern of noradrenergic fibers within the cortex (Morrison et al., 1978; 1981); Ovoid-shaped cells, VIP-intracortical neurons; triangular-shaped cells, pyramidal neurons. Solid symbols refer to activated neurons. Far left, cortical layers; WM, white matter. The concomitant activation of noradrenergic fibers by unexpected sensory stimuli (Aston-Jones and Bloom, 1981a,b) and of a group of VIP-containing intracortical neurons by specific thalamic inputs (A, bottom left) would determine a drastic increase in cAMP levels within a discrete volume of somatosensory cortex (delineated in this drawing by solid, thick lines). *See* section 3.4 for details. For graphic clarity, only the VIP-containing and the pyramidal cells have been represented here. In particular, the principal target cells of the thalamocortical afferents, i. e., the small stellate cells in layer IV, have been omitted. Any cell with the capacity of dendritic reception in layer IV, however, may receive thalamic inputs (White, 1981) (from Magistretti and Schorderet, 1985).

4. Other Experimental Approaches

4.1. Release of Neurotransmitters

The release of neurotransmitters can be elicited in vitro by depolarizing stimuli such as high extracellular K^+ or electrical pulses (*see* also chapter by Dunn and Berridge in this volume). This depolarization-evoked release is generally Ca^{2+}-dependent, as expected from any secretory process. This experimental approach constitutes a valuable tool to establish one of the criteria for a neurotransmitter role for a compound whose presence has been demonstrated in the CNS (*see* section 1.1) (Iversen et al., 1980). Considering the rather nonspecific stimulation represented by high K^+ or electrical pulses, no information as to the synaptic relations of the neuronal system(s) containing the neurotansmitter released can be gathered. In certain instances, however, the release of a given neurotransmitter induced (or inhibited) by another neurotransmitter has been demonstrated (Jessell and Iversen, 1977; Langer, 1977; Starke, 1977; Reubi et al., 1978; Epelbaum et al., 1979; Meyer and Krauss, 1983). Such observations indicate (1) that neurons containing neurotransmitter A (i. e., the neurotransmitter whose release is modulated) possess receptors for neurotransmitter B (i. e., the neurotransmitter that modulates the release of neurotransmitter A), and (2) that the neuronal systems containing neurotransmitters A and B are in synaptic relation. Recently the coexistence of two (usually a "classical" neurotransmitter and a peptide) within the same neuron has been demonstrated in certain neuronal systems (Hökfelt et al., 1980a,b). This observation has prompted the hypothesis that upon activation both neurotransmitters are released into the synaptic cleft and that one of the released neurotransmitters could modulate the release of the other. This hypothesis has found experimental verification in the case of CCK and DA in the cat caudate nucleus, where both neurotransmitters have been shown to coexist within the same terminals (Hökfelt et al., 1980c,d). Thus, CCK in its octapeptide-sulfated form inhibited in vitro the electrically evoked release of ^3H-DA from superfused cat caudate slices at a concentration as low as 10^{-14} M (Markstein and Hökfelt, 1984). This inhibitory effect of CCK was concentration-dependent, between 10^{-14} and 10^{-11} M, and could be observed only in the presence of substances such as bovine serum albumin and bacitracin that protect the peptides from peptidases. In the absence of protective agents, no inhibition by CCK could be observed; rather, the peptide at $10^{-7}M$ markedly enhanced ^{3}H-DA release from cat caudate slices (Markstein and Hökfelt, 1984).

This modulatory effect of CCK on DA release illustrates another biochemical approach that can be used to unravel an action of a neuropeptide. In the following paragraph the methodology currently in use in our laboratory to perform release experiments with nervous tissue is briefly described (for further details and discussion, also *see* the chapter by Dunn and Berridge in this volume).

Slices of the region of interest are prepared as described in section 2.3, and are subsequently placed into superfusion chambers of 500 μL volume. It is important to design chambers of a relatively small volume in order to keep the dead space as limited as possible and prevent excessive dilution of the released neurotransmitter. Oxygenated Krebs-Ringer bicarbonate buffer is superfused through the chambers (maintained at 37°C in a water-bath) at a rate of 800 μL/min. Following 45 min of equilibration, fractions are collected every 2 min and the neurotransmitter released is measured in each fraction. In order to test the potential releasing action of a peptide, the superfusing solution is changed to one containing the peptide under investigation.

4.2. Free Cytosolic Ca^{2+} Concentration

In recent years evidence has emerged pointing to free cytosolic Ca^{2+} as a second messenger for certain transmitter-mediated actions (Cheung, 1980; Rasmussen and Barrett, 1984). The conceptual approach that justifies the assessment of cAMP levels as an index of the action of a given neurotransmitter can be transposed to the measurement of free cytosolic Ca^{2+} concentration. Ca^{2+}-dependent phosphorylations have been demonstrated in nervous tissue and the role of Ca^{2+} in stimulus-secretion coupling has been firmly established in the endocrine system (Cheung, 1980; O'Doherty et al., 1980; Petersen and Maruyama, 1984; Rasmussen and Barrett, 1984). One of the most recent and valuable methodological advances in the assessment of free cytosolic Ca^{2+} concentration has been contributed by Tsien (Tsien, 1980; Tsien et al., 1984; Capponi et al., 1986) with the introduction of the fluorescent probe quin 2. Ca^{2+} is chelated by quin 2, whose fluorescence can be assessed in a fluorescent spectrophotometer at a wavelength of 339 nm for excitation and 492 nm for emission. The fluorescence of quin 2 increases in relation to increases in Ca^{2+} concentration. Quin 2 cannot penetrate into the cell, however, unless it is in the lipophilic acetoxymethyl ester form (quin 2/AM). Once in the cytosol, quin 2/AM is hydrolyzed by cellular esterases to yield hydrophylic free quin 2 to which Ca^{2+} can bind. The apparent dissociation constant of the quin 2–Ca^{2+} complex under the ionic conditions of the cytosol is 115 nM, which correlates well with the concentration of free cytosolic Ca^{2+}. This observation implies that small variations in the concentration of the ion can be effectively monitored with quin 2 spectrophotometry.

This method has been verified extensively in lymphocytes (Tsien et al., 1982; Hesketh et al., 1983) and secretory cells, such as an insulin-secreting cell line (Wollheim and Pozzan, 1984) and clonal pituitary cells (Schlegel and Wollheim, 1984; Schlegel et al., 1984). Very recently, intracellular Ca^{2+} concentrations, $[Ca^{2+}]i$, have been determined in the rabbit vagus nerve in the resting and depolarized state (Pralong et al., 1984), indicating the applicability of the method to nervous tissue. Examples of the effects of peptides on nervous tissue $[Ca^{2+}]i$ assessed by quin 2 fluorescence spectrophotometry are still lacking; this technique seems, however, to constitute a potentially extremely valuable approach for examining the actions of neuropeptides. A limitation of the method, inherent in the mechanism of quin 2 fluorescence, is that it can be applied only to intact cells and preferably to homogeneous cellular systems. The latter consideration should be borne in mind when applying the technique to brain slices, whose heterogeneity in cell types has been discussed earlier (section 2.2). Highly purified primary cultures of defined CNS regions could, however, represent a valuable alternative. In fact, the quin 2 technique has been successfully applied to a cellular system of neuroendocrine origin, such as the pituitary cell line GH3. In this cell line, TRH has been shown to increase and somatostatin-14 to decrease in a concentration-dependent manner $[Ca^{2+}]i$ (Schlegel and Wollheim, 1984; Schlegel et al., 1984).

Recently an improvement in the measurement of free cytosolic Ca^{2+} has been proposed by Tsien and collaborators, with the introduction of fura-2, a new fluorescent Ca^{2+} indicator (Grynkiewicz et al., 1985).

4.3. Protein Phosphorylation

The nervous system and in particular, synaptic specializations, are similar to other tissues in that they contain a large number of protein phosphorylating systems (Cohen, 1982; Rodnight, 1982; Nestler and Greengard, 1983). These systems consist of complexes of (1) protein kinases, which are enzymes capable of transferring phosphate from ATP onto (2) acceptor proteins (which as a consequence become phosphorylated), and (3) protein phosphatases (which dephosphorylate the phosphorylated proteins). Second messengers such as cAMP and Ca^{2+} can activate protein kinases and therefore trigger phosphorylation mechanisms (Cohen, 1982; Nestler and Greengard, 1983). Thus intracellular consequences of neuroactive substances that affect cAMP and Ca^{2+} levels can be monitored biochemically by examining the phosphorylation state of one or more proteins. It should be noted, however, that cAMP- and Ca^{2+}-independent kinase activities have also been observed (Glass and Krebs, 1980; Rodnight, 1982). The general approach to protein phosphorylation assays in nervous tissue consists of the incubation of homogenates or slices in the presence of [gamma-^{32}P]-ATP and the observation of the transfer, catalyzed by protein kinases, of radioactive phosphate onto acceptor proteins. After processing the tissue under appropriate conditions, proteins are separated by discontinuous polyacrylamide gel electrophoresis, and the radioactivity carried by each protein is visualized by autoradiography. A considerable improvement in the resolution of the technique has resulted from the introduction of two-dimensional electrophoretic procedures (Groppi and Browning, 1980; Rodnight, 1982). (For methodological details and alternative procedures, *see* Gispen and Routtenberg, 1982).

This general approach can be employed as an index of a cellular action of neuroactive substances, the expected result being the appearance of a phosphorylated protein following the exposure of the tissue to the agent under investigation. Dopamine- and NA-activated protein phosphorylations have been demonstrated in nervous tissue (Groppi and Browning, 1980; Nestler and Greengard, 1980; Hemmings et al., 1984a,b). The major limitation of this approach may possibly be that the nature and function of the phosphorylated protein are not readily apparent. Correlative analytical biochemical studies or immunohistochemical localization of the protein by light or electron miscroscopy can, however, provide clues as to the function of the phosphorylated protein (Bloom et al., 1979; Dolphin et al., 1983; De Camilli and Navone, 1984; Navone et al., 1984; Ouimet et al., 1984a, b).

4.4. Phosphatidylinositol 4,5-Bisphosphate Metabolism

As outlined in section 1.2, receptor occupation by certain neurotransmitters results in a cellular response that is mediated by events occurring between the receptor supramolecular complex and intracellular effectors. For certain neurotransmitters this link is provided by increases in the intracellular levels of cAMP triggered by receptor-mediated activation of adenylate cyclase. In recent years a number of experimental observations have converged to indicate that phosphatidylinositol 4,5-bisphosphate [PtdIns(4,5)P$_2$], an inositol lipid localized to the inner leaflet of the membrane bilayer, plays a pivotal role in transducing the occupation of the receptor by certain neurotransmitters into an intracellular signal (Berridge and Irvine, 1984; Burgess et al., 1984; Nishizuka, 1984c; Majerus et al., 1985).

The view of the sequence of events in the transduction process that has emerged is the following: [PtdIns(4,5)P$_2$] is hydrolyzed by a Ca^{2+}-dependent phospholipase into diacylglycerol (DG) and inositol-trisphosphate (IP3). DG activates the intramembranous enzyme protein kinase C (Nishizuka, 1984a,b), whereas IP3 mobilizes Ca^{2+} from intracellular stores (Berridge and Irvine, 1984; Burgess et al., 1984; Nishizuka, 1984c; Majerus et al., 1985). Both events (i. e., protein kinase C activation and Ca^{2+} mobilization) will in turn trigger varied cellular responses (Fig. 2). Stimulatory effects on inositol phospholipid turnover have been demonstrated for acetylcholine (via muscarinic receptors), histamine (via H-1 receptors), ACTH$_{1-24}$, and vasopressin (Jolles et al., 1982; Prpić et al., 1982; Fisher et al., 1983; Brown and Masters, 1984; Brown et al., 1984; Kendall and Nahorski, 1984; Masters et al., 1984; Charest et al., 1985). Activation of α_1 adrenergic receptors has similar effects (Ambler et al., 1984; Janowsky et al., 1984). It is likely that the examination of the effects of certain neuropeptides on the levels of the constituents of this signal transduction pathway will constitute a fertile domain of research.

5. Prospects: Toward the Cellular Resolution of Neurochemical Approaches

Most of the experiments in which a biochemical approach has been used to examine the action of a neurotransmitter end, even in the most favorable cases, with a rather frustrating question: In which cell type(s) does the observed effect occur? At first glance this may appear to be an inappropriate question. After all, if an effect of a neurotransmitter has been unmasked and characterized, the original aim of the experiment is achieved. Santiago Ramón y Cajal is always looking over our shoulder, however. One should keep in mind his "neuron theory," in which the anatomical unity of nerve cells was demonstrated, thus introducing the cellular dimension into the study of the nervous system (Ramón y Cajal, 1911; 1954).

To resolve the site of action of a neurotransmitter at the cellular level with biochemical approaches seems, at first examination, a virtually impossible task in view of the heterogeneity of the constitutive elements of the nervous system. A number of strategies can be used, however, to address and potentially answer a question such as: "Which cell type(s) is (are) the target(s) for the action of a given neurotransmitter?"

First, correlations between morphological and functional observations can delineate the spatial constraints within which the action of a neurotransmitter takes place. For example, we know that in the cerebral cortex VIP is localized to local, bipolar interneurons contained, on the average, within cortical columns of approximately 30 μm diameter (Morrison and Magistretti, 1983; Morrison et al., 1984; Magistretti and Morrison, 1985). A column of this dimension is known to contain approximately 100 neurons, 70% of which are pyramidal cells (Powell, 1981). Thus, from a probabilistic point of view, one can infer that VIP molecules released from VIP-containing neurons will diffuse within the column and reach some pyramidal cells. In order for the released VIP molecules to elicit a cellular effect on pyramidal cells, however, receptors for the peptide should be present on this type of neuron. Experimental evidence for this fact is still lacking, however; hence the conclusion that the increases in cAMP and the stimulation of glycogenolysis elicited by VIP occur in pyramidal cells is, at the present time, not warranted. It is clear that from such morphofunctional correlative studies, resolution mainly at the regional (e.g., cerebral cortex vs other CNS regions) or local (e.g., a column of cerebral cortex vs the entire cortex) level can be achieved.

Lesion studies may provide more precise spatial information. If, for example, a population of neurons can selectively be destroyed (e. g., DA-containing neurons in the retina with the selective neurotoxin 6-hydroxydopamine) and the effect of a neurotransmitter disappears after such a manipulation, strong inferences as to the cellular localization of the neurotransmitter's site of action can be derived.

The nature of the cellular action *per se* can provide clues as to the cell type in which the observed effect occurs. Thus, if a neurotransmitter modifies the activity of a key enzyme in the metabolic pathway of another neurotransmitter (e. g., tyrosine hydroxylase in the synthetic pathway of catecholamines), the cell type containing the neurotransmitter whose enzymes are modulated is the likely target. A similar type of reasoning can be applied to release experiments (*see* section 4.1).

Purified preparations of defined cell types constitute another approach in the attempt to achieve the cellular resolution of the effect observed with a given neurotransmitter. Bulk-isolated glia (astrocytes and oligodendrocytes) and neurons can be prepared (Norton and Poduslo, 1970; Raine et al., 1971; Poduslo and Norton, 1972; Sellinger and Azcurra, 1974; Henn, 1980; Norton et al., 1983; Snyder et al., 1983), thus providing a certain degree of cellular specificity. Ideally, single classes of neurons could be used. Recently procedures have been described for the isolation of purified populations of retinal cells (Schaeffer et al., 1980; Van Buskirk and Dowling, 1981; Trachtenberg and Packey, 1983). This approach, if applied to other CNS regions, could provide an extremely powerful tool to resolve the actions of neurotransmitters at the cellular level. Cerebral microvessel preparations have also been described and the effects of various peptides, including VIP and parathyroid hormone, on cAMP levels in such preparations have been observed (Huang and Drummond, 1979; Huang and Rorstad, 1983; 1984).

Primary cultures of purified astroglia, oligodendroglia, and neurons are available in the armamentarium of the biochemical pharmacologist in search of cellular specificity (Fischbach and Nelson, 1977; Adler et al., 1979; Manthorpe et al., 1979; Kimhi, 1981; Kriegstein and Dichter, 1983). Thus the observation that VIP promotes glycogenolysis in cultured astrocytes suggests that VIP-containing neurons may regulate the availability of energy substrates by hydrolyzing glycogen stores in the astrocytes (Fig. 12; Magistretti et al., 1983a). Furthermore this observation supports the notion of the existence of a neuron–glia interaction aimed at maintaining homeostasis in response to local increases in neuronal activity.

The most satisfying approach in attempting to resolve the actions of a neurotransmitter at the cellular level would clearly be the direct visualization of the chemical modification(s) produced by the neurotransmitter in an identified cell. Quin-2 (or fura-2) fluorescence spectrophotometry as an indicator of changes in in-

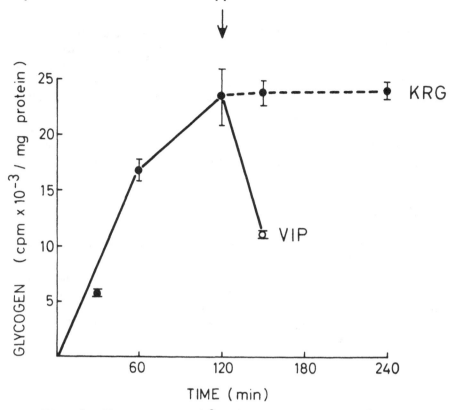

Fig. 12. Time course of ³H-glycogen synthesis (●) and VIP-induced glycogenolysis (○) in cultured astroglia from neonatal rat brain. Concentration of VIP equals 500 nM (from Magistretti et al., 1983a).

tracellular free Ca^{2+} concentration could be of great potential if routinely applicable to CNS preparations.

Hybridoma technology now provides the possibility of preparing monoclonal antibodies (Mab) highly specific for small epitopes (Kennett et al., 1980). If, for example, the active form of an enzyme could be distinguished from the non-active form by Mabs, the target cells of a neurotransmitter which modifies the activation state of that enzyme could be visualized by indirect immunohistochemistry. In this vein, Greengard and collaborators have visualized the target cells of DA-containing neurons in the corpus striatum, by localizing DARPP (dopamine- and adenosine 3',5'-monophosphate-regulated phosphoprotein-32,000), a protein selectively present in dopaminoceptive cells (Ouimet et al., 1984b; Walaas and Greengard, 1984).

In conclusion, I would like to note how important it is for the biochemical pharmacologist to know not only *what* a neurotransmitter does, but also *where* and *when* it does it. The focal point of any strategy in examining the mechanisms of nervous system function is the neuron, viewed both in its individuality and in its dynamic interplay with the other constitutive elements of the nervous system. Thus the action of a neurotransmitter acquires its unique qualitative features and functional impact by occurring at a precise point in time and space in the active nervous system.

Acknowledgments

Research in Dr. Pierre J. Magistretti's laboratory is supported by a grant from Fonds National Suisse de la Recherche Scientifique (3.423.-0.83). The author also wishes to thank Drs. W. Pralong, W. Schlegel, and M. Schorderet for their critical comments and for providing valuable references, Ms. Sylvianne Bonnet for her outstanding skill in preparing the manuscript, and Mr. Fred Pillonel for the graphical work.

References

Adler R., Manthrope M., and Varon S. (1979) Separation of neuronal and nonneuronal cells in monolayer cultures from chick embryo optic lobe. *Dev. Biol.* **69,** 424–435.

Ahren B., Alumets J., Ericson M., Fahrenkrug J., Fahrenkrug L., Hakanson R., Hedner P., Loren I., Melander A., Rerup C., and Sundler F. (1980) VIP occurs in intrathyroidal nerves and stimulates thyroid hormone secretion. *Nature* **287,** 343–345.

Ambler S. K., Brown R. D., and Taylor P. (1984) The relationship between phosphatidylinositol metabolism and mobilization of intracellular calcium elicited by alpha$_1$-adrenergic receptor stimulation in BC3H-1 muscle cells. *Mol. Pharmacol.* **26,** 405–413.

Amoss M., Burgus R., Blackwell R., Vale W., Fellows R., and Guillemin R. (1971) Purification, amino acid composition and N-terminus of the hypothalamic luteinizing hormone releasing factor (LRF) of ovine origin. *Biochem. Biophys. Res. Commun.* **44,** 205.

Aston-Jones G. and Bloom F. E. (1981a) Activity of norepinephrine-containing locus coeruleus neurons in behaving rats anticipates fluctuations in the sleep-walking cycle. *J. Neurosci.* **1,** 876–886.

Aston-Jones G. and Bloom F. E. (1981b) Norepinephrine-containing locus coeruleus neurons in behaving rats exhibit pronounced responses to non-noxious environmental stimuli. *J. Neurosci.* **1,** 887–900.

Baudry M. and Lynch G. (1980) Regulation of hippocampal glutamate receptors: Evidence for the involvement of a calcium-activated protease. *Proc. Natl. Acad. Sci. USA* **77,** 2298–2302.

Baudry M., Evans J., and Lynch G. (1986) Excitatory amino acids inhibit stimulation of phosphatidylinositol metabolism by aminergic agonists in hippocampus. *Nature* **319,** 329–331.

Baudry M., Bundman M., Smith E., and Lynch G. (1981) Micromolar levels of calcium stimulate proteolytic activity and glutamate receptor binding in rat brain synaptic membranes. *Science* **212,** 937–938.

Berridge M. J. and Irvine R. F. (1984) Inositol trisphosphate, a novel second messenger in cellular signal transduction. *Nature* **312,** 315–321.

Bloom F. E. (1975) The role of cyclic nucleotides in central synaptic transmission. *Rev. Physiol. Biochem. Pharmacol.* **74,** 1–103.

Bloom F. E. (1981) Neuropeptides. *Sci. Amer.* **245,** 114–125.

Bloom F. E. (1983) Chemical Communication in the CNS: Neurotransmitters and Their Function, in *Molecular and Cellular Interactions Underlying Higher Brain Functions, Progress in Brain Research* vol. 58 (Changeux J.-P., Glowinski J., Imbert M., and Bloom F. E., eds.), Elsevier, Amsterdam.

Bloom F. E. (1984) The functional significance of neurotransmitter diversity. *Am. J. Physiol.* **246,** C184–C194.

Bloom F. E. and McGinty J. F. (1981) Cellular Distribution and Function of Endorphins, in *Endogenous Peptides and Learning and Memory Processes* (Martinez J. L., Jensen R. A., Messing R. B., Righter H., and McGaugh J. L., eds.), Academic, New York.

Bloom F. E., Ueda T., Battenberg E., and Greengard P. (1979) Immunocytochemical localization, in synapses, of protein I, an endogenous substrate for protein kinases in mammalian brain. *Proc. Natl. Acad. Sci. USA* **76,** 5982–5986.

Boler J., Enzman F., Folkers J., Bowers C. Y., and Schally A. V. (1969) The identity of chemical and hormone properties of the thyrotropin-releasing hormone and pyroglutamyl-histidyl-prolineamide. *Biochem. Biophys. Res. Commun.* **37,** 705–710.

Brazeau P., Vale W., Burgus R., Ling N., Butcher M., Rivier J., and Guillemin R. (1973) Hypothalamic polypeptide that inhibits the secretion of immunoreactive pituitary growth hormone. *Science* **179,** 77–79.

Breckenridge B. M. and Crawford E. J. (1961) The quantitative histochemistry of the brain. Enzymes of glycogen metabolism. *J. Neurochem.* **7,** 234–240.

Breckenridge B. M. and Norman J. H. (1962) Glycogen phosphorylase in brain. *J. Neurochem.* **9,** 383–392.

Breckenridge B. M. and Norman J. H. (1965) The conversion of phosphorylase b to phosphorylase a in brain. *J. Neurochem.* **12,** 51–57.

Brostrom C. O., Hunkeler F. L., and Krebs E. J. (1971) The regulation of skeletal muscle phosphorylase kinase by Ca^{2+}. *J. Biol. Chem.* **246,** 1961–1967.

Brown J. H. and Masters S. B. (1984) Muscarinic regulation of phosphatidylinositol turnover and cyclic nucleotide metabolism in the heart. *Fed. Proc.* **43,** 2613–2617.

Brown B. L., Albano J. D. M., Ekins R. P., Sgherzi A. M., and Tampion W. (1971) A simple and sensitive saturation assay method for the measurement of adenosine 3′:5′-cyclic monophosphate. *Biochem. J.* **121,** 561–562.

Brown E., Kendall D. A., and Nahorski S. R. (1984) Inositol phospholipid hydrolysis in rat cerebral cortical slices. I. Receptor characterisation. *J. Neurochem.* **42,** 1379–1387.

Bruckner G. and Biesold D. (1981) Histochemistry of glycogen deposition in perinatal rat brain: Importance of radial glia cells. *J. Neurocytol.* **10,** 749–757.

Buchel L. and McIlwain H. (1950) Narcotics and the inorganic and creatine phosphates of mammalian brain. *Br. J. Pharmacol.* **5,** 465–473.

Buell M. V., Lowry O. H., Roberts N. R., Chang M. L. W., and Kapphahn J. I. (1958) The quantitative histochemistry of the brain. V. Enzymes of glucose metabolism. *J. Biol. Chem.* **232,** 979–993.

Burgess G. M., Godfrey P. P., McKinney J. S., Berridge M. J., Irvine R. F., and Putney, Jr., J. W. (1984) The second messenger linking receptor activation to internal Ca release in liver. *Nature* **309,** 63–66.

Cammermeyer J. and Fenton I. M. (1981) Improved preservation of neuronal glycogen by fixation with iodoacetic acid-containing solutions. *Exp. Neurol.* **72,** 429–445.

Capponi A. M., Lew P. D., Schlegel W., and Pozzan T. (1986) Use of Intracellular Calcium and Membrane Potential Fluorescent Indicators in Neuro-endocrine Cells, in *Methods in Enzymology* Academic, New York.

Carraway R. and Leeman S. E. (1976) Characterization of radioimmunoassayable neurotensin in the rat. Its differential distribution in the central nervous system, small intestine, and stomach. *J. Biol. Chem.* **251,** 7045–7052.

Charest R., Prpić V., Exton J. H., and Blackmore P. F. (1985) Stimulation of inositol trisphosphate formation in hepatocytes by vasopressin, adrenaline and angiotensin II and its relationship to changes in cytosolic free Ca^{2+}. *Biochem. J.* **227,** 79–90.

Cheung W. Y. (1980) Calmodulin plays a pivotal role in cellular regulation. *Science* **207,** 19–27.

Chneiweiss H., Glowinski J., and Prémont J. (1985) Vasoactive intestinal polypeptide receptors linked to an adenylate cyclase, and their relationship with biogenic amine- and somatostatin-sensitive adenylate cyclases on central neuronal and glial cells in primary cultures. *J. Neurochem.* **44,** 779–786.

Cohen P. (1982) The role of protein phosphorylation in neural and hormonal control of cellular activity. *Nature* **296,** 613–620.

Connor J. R. and Peters A. (1984) Vasoactive intestinal polypeptide-immunoreactive neurons in rat visual cortex. *Neuroscience* **12,** 1027–1044.

Cooper J. R., Bloom F. E., and Roth R. H., eds. (1982) *The Biochemical Basis of Neuropharmacology.* Oxford University, New York.

Costa M., Furness J. B., Buffa R., and Said S. I. (1980) Distribution of enteric nerve cells bodies and axons showing immunoreactivity for vasoactive intestinal polypeptide in the guinea pig intestine. *Neuroscience* **5,** 587–596.

Cotman C. W. (1974) Isolation of Synaptosomal and Synaptic Plasma Membrane Fractions, in *Methods in Enzymology* vol. 31, part A (Fleischer S. and Packer L., eds.), Academic, New York.

Dahlstrom A. and Fuxe K. (1964) Evidence for the existence of monoamine-containing neurons in the central nervous system. I. Demonstration of monoamines in the cell bodies of brain stem neurons. *Acta Physiol. Scand.* **62,** (suppl. 232), 1–55.

Daly J. (1975) Role of Cyclic Nucleotides in the Nervous System, in *Handbook of Psychopharmacology* (Iversen L. L., Iversen S. D., and Snyder S. H., eds.), Plenum, New York.

Danforth W. H. and Helmreich E. (1964) Regulation of glycolysis in muscle. The conversion of phosphorylase b to phosphorylase a in frog sartorius muscle. *J. Biol. Chem.* **239,** 3133–3138.

Danforth W. H. and Lyon J. B. (1964) Glycogenolysis during tetanic contraction of isolated mouse muscles in the presence and absence of phosphorylase a. *J. Biol. Chem.* **239,** 4047–4050.

Davis G. R., Morawski S. G., Santa Ana C. A., and Fordtran J. D. (1980) Mechanism of vasoactive intestinal polypeptide (VIP) action on jejunal water and electrolyte transport in man. *Clin. Res.* **28,** 764A.

De Camilli P. and Navone F. (1984) Immunocytochemistry as a tool in the study of neurotransmitter actions. *Trends Pharmacol. Sci.* **5,** 300–303.

deDuve C. (1965) The separation and characterization of subcellular particles. *Harvey Lectures* **59,** 49–87, Academic, New York.

deRobertis E., deIraldi A. P., Rodriguez de Lores Arnaiz G., and Gomez C. (1961) On the isolation of nerve endings and synaptic vesicles. *J. Biophys. Biochem. Cytol.* **9,** 229–235.

Deschodt-Lanckman M., Robberecht P., and Christophe J. (1977) Characterization of VIP-sensitive adenylate cyclase in guinea pig brain. *FEBS Lett.* **83,** 76–80.

Dingledine R., ed. (1984) *Brain Slices.* Plenum, New York.

Dockray G. J. (1980) Cholecystokinins in rat cerebral cortex: Identification, purification and characterization by immunochemical methods. *Brain Res.* **188,** 155–165.

Dolphin A. C., Detre J. A., Schlichter D. J., Nairn A. C., Yeh H. H., Woodward D. J., and Greengard P. (1983) Cyclic nucleotide-dependent protein kinases and some major substrates in the rat cerebellum after neonatal X-irradiation. *J. Neurochem.* **40,** 577–581.

Douglas W. W. (1980) Autacoids, in *The Pharmacological Basis of Therapeutics* (Goodman Gilman A., Goodman L. S., and Gilman, A., eds.), Macmillan, New York.

Drummond G. I. (1983) Cyclic Nucleotides in the Nervous System, in *Advances in Cyclic Nucleotides Research* vol. 15 (Greengard P. and Robison G. A., eds.), Raven, New York.

Drummond G. I. and Bellward G. (1970) Studies on phosphorylase b kinase from neural tissues. *J. Neurochem.* **17,** 475–482.

Duffy M. J. and Powell D. (1975) Stimulation of brain adenylate cyclase activity by the undecapeptide substance P and its modulation by the calcium ion. *Biochem. Biophys. Acta* **385,** 275–280.

duVigneaud V. (1956) Hormones of the posterior pituitary gland: Oxytocin and vasopressin. *Harvey Lectures* **50,** 1–26.

duVigneaud V., Ressler C., Swan J. M., Roberts C. W., Katsoyannis P. G., and Gordon S. (1953) The synthesis of an octapeptide amide with the hormonal activity of oxytocin. *J. Am. Chem. Soc.* **75,** 4879–4880.

duVigneaud V., Gish D. T., and Katsoyannis P. G. (1954) A synthetic preparation possessing biological properties associated with arginine vasopressin. *J. Am. Chem. Soc.* **76,** 4751–4752.

Eckenstein F. and Baughman R. W. (1984) Two types of cholinergic innervation in cortex, one co-localized with vasoactive intestinal polypeptide. *Nature* **309,** 153–155.

Edwards C., Nahorski S. R., and Rogers K. J. (1974) In vivo changes of cerebral cyclic adenosine 3',5'-monophosphate induced by biogenic amines: Association with phosphorylase activation. *J. Neurochem.* **22,** 565–572.

Epelbaum J., Tapia-Arancibia L., Besson J., Rotsztejn W. H., and Kordon C. (1979) Vasoactive intestinal peptide inhibits release of somatostatin from hypothalamus in vitro. *Eur. J. Pharmacol.* **58,** 493–495.

Exton J. H. and Harper S. C. (1975) Role of Cyclic AMP in the Actions of Catecholamines on Hepatic Carbohydrate Metabolism, in *Advances in Cyclic Nucleotide Research* vol. 5 (Drummond G. I., Greengard P., and Robison G. A., eds.), Raven, New York.

Exton J. H., Cherrington A. D., Blackmore P. F., Dehaye J. P., Strickland W. G., Jordan J. E., and Chrisman T. D. (1981) Hormonal Regulation of Liver Glycogen Metabolism, in *Cold Spring Harbor Conference on Cell Proliferation* vol. 8, Cold Spring Harbor Press.

Ferron A., Siggins G. R., and Bloom F. E. (1985) Vasoactive intestinal polypeptide acts synergistically with norepinephrine to depress spontaneous discharge rate in cerebral cortical neurons. *Proc. Natl. Acad. Sci. USA* **82,** 8810–8812.

Finley J. C. W., Maderdrut J. L., Roger L. J., and Petrusz P. (1981) The immunocytochemical localization of somatostatin-containing neurons in the rat central nervous system. *Neuroscience* **6,** 2173–2192.

Fischbach G. D. and Nelson P. G. (1977) Cell Culture in Neurobiology, in *Handbook of Physiology* section I: *The Nervous System,* vol. I: *Cellular Biology of Neurons,* part 2 (Kandel E., ed.), American Physiology Society, Bethesda.

Fisher S. K., Klinger P. D., and Agranoff B. W. (1983) Muscarinic agonist binding and phospholipid turnover in brain. *J. Biol. Chem.* **258,** 7358–7363.

Fuxe K. (1965) Evidence for the existence of monoamine neurons in the central nervous system. IV. Distribution of monoamine nerve terminals in the central nervous system. *Acta Physiol. Scand.* **64** (suppl. 247), 36–85.

Gispen W. H. and Routtenberg A., eds. (1982) *Brain Phosphoproteins. Characterization and Function, Progress in Brain Research* vol. 56, Elsevier, Amsterdam.

Glass D. B. and Krebs E. G. (1980) Protein phosphorylation catalyzed by cyclic AMP-dependent and cyclic GMP-dependent protein kinases. *Ann. Rev. Pharmacol. Toxicol.* **20,** 363–388.

Gray E. G. and Whittaker V. P. (1962) The isolation of nerve endings from brain: An electron microscopic study of the cell fragments of homogenization and centrifugation. *J. Anat.* **96,** 79–88.

Groppi Jr., V. E. and Browning E. T. (1980) Norepinephrine-dependent protein phosphorylation in intact C-6 glioma cells. Analysis by two-dimensional gel electrophoresis. *Mol. Pharmacol.* **18,** 427–437.

Grossman H. I. (1981) General Concepts, in *Gut Peptides* (Bloom S. R. and Polak J. H., eds.), Churchill Livingstone, Edinburgh.

Grynkiewicz G., Poenie M., and Tsien R. Y. (1985) A new generation of Ca^{2+} indicators with greatly improved fluorescence properties. *J. Biol. Chem.* **260**, 3440–3450.

Guillemin R. (1978a) Peptides in the brain: The new endocrinology of the neuron. *Science* **202**, 390–402.

Guillemin R. (1978b) Biochemical and Physiological Correlates of Hypothalamic Peptides. The New Endocrinology of the Neuron, in *The Hypothalamus* (Reichlin S., Baldessarini R. J., and Martin J. B., eds.) Raven, New York.

Guillemin R., Brazeau P., Böhlen P., Esch F., Ling N., and Wehrenberg W. B. (1982) Growth hormone-releasing factor from a human pancreatic tumor that caused acromegaly. *Science* **218**, 585–587.

Havet J. (1937) Le glycogène dans les centres nerveux. *Cellule* **46**, 179–182.

Heilmeyer L. M. G., Meyer F., Haschke R. H., and Fischer E. H. (1970) Control of phosphorylase activity in a muscle glycogen particle. II. Activation by calcium. *J. Biol. Chem.* **245**, 6649–6656.

Hemmings, Jr., H. C., Greengard P., Lim Tung H. Y., and Cohen P. (1984a) DARPP-32, a dopamine-regulated neuronal phosphoprotein, is a potent inhibitor of protein phosphatase-1. *Nature* **310**, 503–505.

Hemmings, Jr., H. C., Nairn A. C., Aswad D. W., and Greengard P. (1984b) DARPP-32, a dopamine- and adenosine 3':5'-monophosphate-regulated phosphoprotein enriched in dopamine-innervated brain regions. II. Purification and characterization of the phosphoprotein from bovine caudate nucleus. *J. Neurosci.* **4**, 99–110.

Hendry S. H. C., Jones E. G., and Beinfeld M. C. (1983) Cholecystokinin-immunoreactive neurons in rat and monkey cerebral cortex make symmetric synapses and have intimate associations with blood vessels. *Proc. Natl. Acad. Sci. USA* **80**, 2400–2404.

Henn F. A. (1980) Separation of Neuronal and Glial Cells and Subcellular Constituents, in *Advances in Cellular Neurobiology* vol. 1 (Fedoroff S. and Hertz L., eds.), Academic, New York.

Hesketh T. R., Smith G. A., Moore J. P., Taylor M. V., and Metcalfe J. C. (1983) Free cytoplasmic calcium concentration and the mitogenic stimulation of lymphocytes. *J. Biol. Chem.* **258**, 4876–4882.

Hökfelt T., Ljungdahl A., Steinbusch H., Verhofstad A., Nilsson G., Brodin E., Pernow B., and Goldstein M. (1978) Immunohistochemical evidence of substance P-like immunoreactivity in some 5-hydroxy-tryptamine-containing neurons in the rat central nervous system. *Neuroscience* **3**, 517–538.

Hökfelt T., Johansson O., Ljungdahl A., Lundberg J. M., and Schultzberg M. (1980a) Peptidergic neurones. *Nature* **284**, 515–521.

Hökfelt T., Lundberg J. M., Schultzberg M., Johansson O., Skirboll L., Angard A., Fredholm B., Hamberger B., Pernow B., Rehfeld J., and Goldstein M. (1980b) Cellular localization of peptides in neural structures. *Proc. Roy. Soc. Lond* **B210**, 63–77.

Hökfelt T., Rehfeld J. F., Skirboll L. R., Ivemark B., Goldstein M., and Markey K. (1980c) Evidence for coexistence of dopamine and CCK in mesolimbic neurons. *Nature* **285**, 476–478.

Hökfelt T., Skirboll L., Rehfeld J. F., Goldstein M., Markey K., and Dann O. (1980d) A subpopulation of mesencephalic dopamine neurons projecting to limbic areas contains a cholecystokinin-like peptide: Evidence from immunohistochemistry combined with retrograde tracing. *Neuroscience* **5**, 2093–2124.

Hollingsworth E. B. and Daly J. W. (1985) Accumulation of inositol phosphates and cyclic AMP in guinea-pig cerebral cortical preparations. Effects of norepinephrine, histamine, carbamylcholine and 2-chloroadenosine. *Biochem. Biophys. Acta* **847**, 207–216.

Hoshino M., Yanaihara C., Hong Y.-M., Kishida S., Katsumaru Y., Vandermeers A., Vandermeers-Piret M.-C., Robberecht P., Christophe J., and Yanaihara N. (1984) Primary structure of helodermin, a VIP-secretin-like peptide isolated from Gila monster venom. *FEBS Lett.* **178**, 233–239.

Huang M. and Drummond G. I. (1979) Adenylate cyclase in cerebral microvessels: Action of guanine nucleotides, adenosine and other agonists. *Mol. Pharmacol.* **16**, 462–472.

Huang M. and Rorstad O. P. (1983) Effects of vasoactive intestinal polypeptide, monoamines, prostaglandins, and 2-chloroadenosine on adenylate cyclase in rat cerebral microvessels. *J. Neurochem.* **40**, 719–726.

Huang M. and Rorstad O. P. (1984) Cerebral vascular adenylate cyclase: Evidence for coupling to receptors for vasoactive intestinal peptide and parathyroid hormone. *J. Neurochem.* **43**, 849–856.

Hughes J., Smith T. W., Kosterlitz H., Fothergill L., Morgan B., and Morris H. (1975) Identification of two related pentapeptides from the brain with potent opiate agonist activity. *Nature* **258**, 577–579.

Iversen L. L. (1984) Amino acids and peptides: Fast and slow chemical signals in the nervous system? *Proc. Roy. Soc. Lond.* **B221**, 245–260.

Iversen L. L., Lee C. M., Gilbert R. F., Hunt S., and Emson P. C. (1980) Regulation of neuropeptide release. *Proc. Roy. Soc. Lond.* **B210**, 91–111.

Janowsky A., Labarca R., and Paul S. M. (1984) Characterization of neurotransmitter receptor-mediated phosphatidylinositol hydrolysis in the rat hippocampus. *Life Sci.* **35**, 1953–1961.

Jessell T. M. and Iversen L. L. (1977) Opiate analgesics inhibit substance P release from rat trigeminal nucleus. *Nature* **268**, 549–551.

Johansson O., Hökfelt T., and Elde R. P. (1984) Immunohistochemical distribution of somatostatin-like immunoreactivity in the central nervous system of the adult rat. *Neuroscience* **13**, 265–339.

Jolles J., van Dongen C. J., ten Haaf J., and Gispen W. N. (1982) Polyphosphoinositide metabolism in rat brain: Effects of neuropeptides, neurotransmitters and cyclic nucleotides. *Peptides* **3**, 709–714.

Kebabian J. W. and Nathanson J. H., eds. (1982) *Handbook of Experimental Pharmacology* 58/II. Springer Verlag, Berlin.

Kendall D. A. and Nahorski S. R. (1984) Inositol phospholipid hydrolysis in rat cerebral cortical slices. II. Calcium requirement. *J. Neurochem.* **42**, 1388–1394.

Kennett R. H., McKearn T. J., and Bechtol K. B., eds. (1980) *Monoclonal antibodies, Hybridomas: A New Dimension in Biological Analyses* Plenum, New York.

Kerins C. and Said S. I. (1973) Hyperglycemic and glycogenolytic effects of vasoactive intestinal polypeptide. *Proc. Soc. Exp. Biol. Med.* **142**, 1012–1014.

Kimhi Y. (1981) Clonal Systems, in *Excitable Cells in Tissue Culture* (Nelson P. G. and Lieberman M., eds.), Plenum, New York.

Krejs G. J., Barkley R. M., Read N. W., and Fordtran J. S. (1978) Intestinal secretion induced by vasoactive intestinal polypeptide. *J. Clin. Invest.* **61**, 1337–1345.

Krieger D. T. (1983) Brain peptides: What, where and why? *Science* **222**, 975–985.

Krieger D. T. and Martin J. B. (1981a) I. Brain peptides. *New Engl. J. Med.* **304**, 876–885.

Krieger D. T. and Martin J. B. (1981b) II. Brain peptides. *New Engl. J. Med.* **304**, 944–951.

Krieger D. T., Brownstein M. J., and Martin J. B., (1983) *Brain Peptides.* John Wiley, New York.

Kriegstein A. R. and Dichter M. A. (1983) Morphological classification of rat cortical neurons in cell culture. *J. Neurosci.* **3**, 1634–1647.

Langer S. Z. (1977) Presynaptic receptors and their role in the regulation of transmitter release. *Br. J. Pharmacol.* **60**, 481–497.

Larsson L. I. and Rehfeld J. F. (1979) Localization and molecular heterogeneity of cholecystokinin in the central and peripheral nervous system. *Brain Res.* **165**, 201–218.

Larsson L. I., Fahrenkrug J., and Schaffalitzky de Muckadell O. B. (1977) Vasoactive intestinal polypeptide occurs in nerves of the female genitourinary tract. *Science* **197**, 1374–1375.

Larsson L. I. Fahrenkrug J., Holst J. J., and Schaffalitzky de Muckadell O. B. (1978) Innervation of the pancreas by vasoactive intestinal polypeptide (VIP) immunoreactive nerves. *Life Sci.* 22, 773–780.

LeBaron F. N. (1955) The resynthesis of glycogen by guinea pig cerebral-cortex slices. *Biochemistry* 61, 80–85.

Lehninger A. L., ed. (1982) *Principles of Biochemistry.* Worth, New York.

Li C. L. and McIlwain H. (1957) Maintenance of resting membrane potentials in slices of mammalian cerebral cortex and other tissues in vitro. *J. Physiol.* 139, 178–190.

Lorén I., Emson P. C., Fahrenkrug J., Björklund A., Alumets J., Hakanson R., and Sundler F. (1979) Distribution of vasoactive intestinal polypeptide in the rat and mouse brain. *Neuroscience* 4, 1953–1976.

Lowry O. H. and Passonneau J. V. (1964) The relationship between substrates and enzymes of glycolysis in brain. *J. Biol. Chem.* 239, 31–42.

Lowry O. H., Rosebrough N. J., Farr A. L., and Randall R. J. (1951) Protein measurements with the folin phenol reagent. *J. Biol. Chem.* 193, 265–275.

Magistretti P. J. and Morrison J. H. (1985) VIP neurons in the neocortex. *Trends Neurosci.* 8, 7–8.

Magistretti P. J. and Schorderet M. (1984) VIP and noradrenaline act synergistically to increase cyclic AMP in cerebral cortex. *Nature* 308, 280–282.

Magistretti P. J. and Schorderet M. (1985) Norepinephrine and histamine potentiate the increases in cyclic adenosine 3':5'-monophosphate elicited by vasoactive intestinal polypeptide in mouse cerebral cortical slices: mediation by α_1-adrenergic and H_1-histaminergic receptors. *J. Neuroscience* 5, 362–368.

Magistretti P. J., Morrison J. H., Shoemaker W. J., Sapin V., and Bloom F. E. (1981) Vasoactive intestinal polypeptide induces glycogenolysis in mouse cortical slices: A possible regulatory mechanism for the local control of energy metabolism. *Proc. Natl. Acad. Sci. USA* 78, 6535–6539.

Magistretti P. J., Manthorpe M., Bloom F. E., and Varon S. (1983a) Functional receptors for vasoactive intestinal polypeptide in cultured astroglia from neonatal rat brain. *Regul. Pept.* 6, 71–80.

Magistretti P. J., Morrison J. H., Shoemaker W. J., and Bloom F. E. (1983b) Effect of 6-hydroxydopamine lesions on norepinephrine-induced ^{3}H-glycogen hydrolysis in mouse cortical slices. *Brain Res.* 261, 159–162.

Magistretti P. J., Hof P., and Schorderet M. (1984a) The increase in cyclic-AMP levels elicited by vasoactive intestinal peptide (VIP) in mouse cerebral cortical slices is potentiated by ergot alkaloids. *Neurochem. Int.* **6,** 751–753.

Magistretti P. J., Morrison J. H., Shoemaker W. J., and Bloom F. E. (1984b) Morphological and functional correlates of VIP neurons in cerebral cortex. *Peptides* **5,** 213–218.

Majerus P. W., Wilson D. B., Connolly T. M., Bross T. E., and Neufeld E. J. (1985) Phosphoinositide turnover provides a link in stimulus-response coupling. *Trends Biochem. Sci.* **10,** 168–171.

Malmfors T. and Thoenen H., eds. (1971) *6-Hydroxydopamine and Catecholamine Neurons.* Elsevier, Amsterdam.

Manthorpe M., Adler R., and Varon S. (1979) Development, reactivity and GFA immunofluorescence of astroglia-containing monolayer cultures from rat cerebrum. *J. Neurocytol.* **8,** 605–621.

Markstein R. and Hökfelt T. (1984) Effect of cholecystokinin-octapeptide on dopamine release from slices of cat caudate nucleus. *J. Neurosci.* **4,** 570–575.

Masters S. B., Harden T. K., and Brown J. H. (1984) Relationships between phosphoinositide and calcium responses to muscarinic agonists in astrocytoma cells. *Mol. Pharmacol.* **26,** 149–155.

Matsuzaki Y., Hamasaki Y., and Said S. I. (1980) Vasoactive intestinal peptide: A possible transmitter of noradrenergic relaxation of guinea pig airways. *Science* **210,** 1252–1253.

McDonald J. K., Parnavelas J. G., Karamanlidis A. N., and Brecha N. (1982) The morphology and distribution of peptide-containing neurons in the adult and developing visual cortex of the rat. II. Vasoactive intestinal polypeptide. *J. Neurocytol.* **11,** 825–837.

McIlwain H. (1951) Metabolic response in vitro to electrical stimulation of sections of mammalian brain. *Biochem. J.* **49,** 382–393.

McIlwain H. (1952) Phosphates and nucleotides of the central nervous system. *Biochem. Soc. Symp.* **8,** 27–43.

McIlwain H. (1953) Substances which support respiration and metabolic response to electrical impulses in human cerebral tissues. *J. Neurol. Neurosurg. Psychiatry* **16,** 257–266.

McIlwain H. (1984) Introduction: Cerebral Subsystems as Biological Entities, in *Brain Slices* (Dingledine R., ed.), Plenum, New York.

McIlwain H. and Tresize M. A. (1956) The glucose, glycogen and aerobic glycolysis of isolated cerebral tissues. *Biochem. J.* **63,** 250–257.

McIlwain H., Ayres P. J. W., and Forda O. (1952) Metabolic response to electrical stimulation in separated portions of human cerebral tissues. *J. Met. Sci.* **98,** 265–272.

Meyer D. K. and Krauss J. (1983) Dopamine modulates cholecystokinin release in neostriatum. *Nature* **301,** 338–340.

Morrison J. H. and Magistretti P. J. (1983) Monoamines and peptides in cerebral cortex. *Trends Neurosci.* **6,** 146–151.

Morrison J. H., Grzanna R., Molliver M., and Coyle J. T. (1978) The distribution and orientation of noradrenergic fibers in neocortex of the rat: An immunofluorescence study. *J. Comp. Neurol.* **181,** 17–40.

Morrison J. H., Molliver M. E., Grzanna R., and Coyle J. T. (1981) The intra-cortical trajectory of the coeruleo-cortical projection in the rat: A tangentially organized cortical afferent. *Neuroscience* **6,** 139–158.

Morrison J. H., Benoit R., Magistretti P. J., and Bloom F. E. (1983) Immunohistochemical distribution of pro-somatostatin-related peptides in cerebral cortex. *Brain Res.* **262,** 344–351.

Morrison J. H., Magistretti P. J., Benoit R., and Bloom F. E. (1984) The distribution and morphological characteristics of the intracortical VIP-positive cell: An immunohistochemical analysis. *Brain Res.* **292,** 269–282.

Nahorski S. R. and Rogers K. J. (1972) An enzymatic fluorometric micromethod for determination of glycogen. *Anal. Biochem.* **49,** 492–497.

Nahorski S. R. and Rogers K. J. (1975) The role of catecholamines in the action of amphetamine and L-dopa on cerebral energy metabolism. *Neuropharmacol.* **14,** 283–290.

Nahorski S. R., Rogers K. J., and Edwards C. (1975) Cerebral glycogenolysis and stimulation of α-adreno-receptors and histamine H_2 receptors. *Brain Res.* **92,** 529–533.

Nathanson J. A. (1977) Cyclic nucleotides and nervous system function. *Physiol. Rev.* **57,** 157–256.

Nathanson J. A. (1981) Cellular Interactions of Biogenic Amines, Peptides, and Cyclic Nucleotides, in *Neurosecretion and Brain Peptides* (Martin J. B., Reichlin S., and Bick K. L., eds.), Raven, New York.

Navone F., Greengard P., and De Camilli P. (1984) Synapsin I in nerve terminals: Selective association with small synaptic vesicles. *Science* **226,** 1209–1211.

Nelson S. R., Schulz D. W., Passonneau J. V., and Lowry O. H. (1968) Control of glycogen levels in brain. *J. Neurochem.* **15,** 1271–1279.

Nestler E. J. and Greengard P. (1980) Dopamine and depolarizing agents regulate the state of phosphorylation of protein I in the mammalian superior cervical sympathetic ganglion. *Proc. Natl. Acad. Sci. USA* **77,** 7479–7483.

Nestler E. J. and Greengard P. (1983) Protein phosphorylation in the brain. *Nature* **305,** 583–588.

Nicholls J. G. and Wolfe D. E. (1967) Distribution of ^{14}C-labelled sucrose, insulin, and dextran in extracellular spaces and in cells of the leech central nervous system. *J. Neurophysiol.* **30,** 1574–1592.

Nishizuka Y. (1984a) Protein kinases in signal transduction. *Trends Biochem. Sci.* **9**, 163–166.

Nishizuka Y. (1984b) The role of protein kinase C in cell surface signal transduction and tumour promotion. *Nature* **308**, 693–698.

Nishizuka Y. (1984c) Turnover of inositol phospholipids and signal transduction. *Science* **225**, 1365–1370.

Norton W. T. and Poduslo S. E. (1970) Neuronal soma and whole neuroglia of rat brain: A new isolation technique. *Science* **167**, 1143–1145.

Norton W. T., Farooq M., Fields K. L., and Raine C. S. (1983) The long term culture of bulk-isolated bovine oligodendroglia from adult brain. *Brain Res.* **270**, 295–310.

O'Doherty J., Youmans S. J., McD. Armstrong W., and Stark R. J. (1980) Calcium regulation during stimulus-secretion coupling continuous measurement of intracellular calcium activities. *Science* **209**, 510–513.

Ouimet C. C., McGuinness T. L., and Greengard P. (1984a) Immunocytochemical localization of calcium/calmodulin-dependent protein kinase II in rat brain. *Proc. Natl. Acad. Sci. USA* **81**, 5604–5608.

Ouimet C. C., Miller P. E., Hemmings, Jr., H. C., Walaas S. I., and Greengard P. (1984b) DARPP-32, a dopamine- and adenosine 3' : 5'-monophosphate-regulated phosphoprotein enriched in dopamine-innervated brain regions. III. Immunocytochemical localization. *J. Neurosci.* **4**, 111–124.

Park C. R. and Exton J. H. (1973) Glucagon and the Metabolism of Glucose, in *Glucagon Molecular Physiology, Clinical and Therapeutic Implications* (Lefebvre P. J. and Unger R. H., eds.), Pergamon, New York.

Passonneau J. V., Brunner E. A., Molstad C., and Passonneau R. (1971) The effects of altered endocrine states and of ether anesthesia on mouse brain. *J. Neurochem.* **18**, 2317–2328.

Peikin S. R., Rottman A. J., Batzri S., and Gardner J. D. (1978) VIP effects on exocrine pancreas. *Am. J. Physiol.* **235**, E743–E749.

Peters A., Miller M., and Kimerer L. M. (1983) Cholecystokinin-like immunoreactive neurons in rat cerebral cortex. *Neuroscience* **8**, 431–448.

Petersen O. H. and Maruyama Y. (1984) Calcium-activated potassium channels and their role in secretion. *Nature* **307**, 693–696.

Piper P. J., Said S. I., and Vane J. R. (1970) Effects of smooth muscle preparations of unidentified vasoactive peptides from intestine and lung. *Nature* **225**, 1144–1145.

Poduslo S. E. and Norton W. T. (1972) Isolation and some chemical properties of oligodendroglia from calf brain. *J. Neurochem.* **19**, 727–736.

Powell T. P. S. (1981) Certain Aspects of the Intrinsic Organisation of the

Cerebral Cortex, in *Brain Mechanisms and Perceptual Awareness* (Pompeiano O. and Ajmone Marsan C., eds.), Raven, New York.

Pralong W. F., Jirounek P., and Straub R. W. (1984) Free calcium in mammalian nerve axons measured by quin-2. *Neurosci. Lett.* (suppl. 18) S338.

Prpić V., Blackmore P. F., and Exton J. H. (1982) Phosphatidylinositol breakdown induced by vasopressin and epinephrine in hepatocytes is calcium-dependent. *J. Biol. Chem.* **257**, 11323–11331.

Quach T. T., Rose C., and Schwartz J. C. (1978) (^{3}H)-Glycogen hydrolysis in brain slices: Responses to neurotransmitters and modulation of noradrenaline receptors. *J. Neurochem.* **30**, 1335–1341.

Quach T. T., Duchemin A. M., Rose C., and Schwartz J. C. (1980) (^{3}H)-Glycogen hydrolysis elicited by histamine in mouse brain slices: Selective involvement of H_1-receptors. *Mol. Pharmacol.* **17**, 301–308.

Quach T. T., Rose C., Duchemin A. M., and Schwartz J. C. (1982) Glycogenolysis induced by serotonin in brain: Identification of a new class of receptor. *Nature* **298**, 373–375.

Quik M., Iversen L. L., and Bloom S. R. (1978) Effect of vasoactive intestinal peptide (VIP) and other peptides on cAMP accumulation in rat brain. *Biochem. Pharmacol.* **27**, 2209–2213.

Raine C. S., Poduslo S. E., and Norton W. T. (1971) The ultrastructure of purified preparations of neurons and glial cells. *Brain Res.* **27**, 11–24.

Rall T. W. and Sutherland E. W. (1958) Formation of cyclic adenine ribonucleotide by tissue particles. *J. Biol. Chem.* **232**, 1065–1076.

Ramón y Cajal S. (1911) *Histologie du système nerveux de l'homme et des vertébrés* vol. 2, Azoulay L., transl. Maloine, Paris, 1911.

Ramón y Cajal S. (1954) *Neuron theory or reticular theory? Objective evidence of the anatomical unity of nerve cells.* English translation, Consejo Superior de Investigaciones Scientificas, Instituto Ramón y Cajal, Madrid, by Ubeda Purkiss M. and Fox C. A.

Rasmussen H. and Barrett P. Q. (1984) Calcium messenger system: An integrated view. *Physiol. Rev.* **64**, 938–984.

Redgate E. S., Depree J. D., and Axelrod J. (1986) Interaction of neuropeptides and biogenic amines on cyclic adenosine monophosphate accumulation in hypothalamic nuclei. *Brain Res.* **365**, 61–69.

Rehfeld J. F. (1980) Cholecystokinin. *Trends Neurosci.* **3**, 65–67.

Rehfeld J. F. (1978) Immunochemical studies on cholecystokinin. II. Distribution and molecular heterogeneity in the central nervous system and small intestine of man and hog. *J. Biol. Chem.* **253**, 4022–4030.

Renaud L. P., Pittman Q. J., Blume H. W., Lamour Y., and Arnauld E. (1979) Effects of Peptides on Central Neuronal Excitability, in *Central Nervous System Effects of Hypothalamic Hormones and other Peptides* (Collu R., Barbeau A., and Ducharme J. R., eds.), Raven, New York.

Reubi J. C., Emson P. C., Jessell T. M., and Iversen L. L. (1978) Effects of GABA, dopamine and substance P on the release of newly synthesized ³H-5-HT from rat substantia nigra in vitro. *Naunyn Schmiedeberg's Arch. Pharmacol.* **304,** 271–275.

Rivier J., Spiess J., Thorner M., and Vale W. (1982) Characterization of a growth hormone-releasing factor from a human pancreatic islet tumour. *Nature* **300,** 276–278.

Rivier J., Spiess J., and Vale W. (1983) Characterization of rat hypothalamic corticotropin-releasing factor. *Proc. Natl. Acad. Sci. USA* **80,** 4851–4855.

Robberecht P., Waelbroeck M., Dehaye J.-P., Winand J., Vandermeers A., Vandermeers-Piret M-C., and Christophe J. (1984) Evidence that helodermin, a newly extracted peptide from Gila monster venom, is a member of the secretin/VIP/PHI family of peptides with an original pattern of biological properties. *FEBS Lett.* **166,** 277–282.

Robison G. A., Schmidt M. J., and Sutherland E. W. (1970) On the Development and Properties of the Brain Adenyl Cyclase System, in *Role of Cyclic AMP in Cell Functions, Advances in Biochemical Psychopharmacology* vol. 3 (Greengard P. and Costa A., eds.), Raven, New York.

Rodnight R. (1982) Aspects of Protein Phosphorylation in the Nervous System with Particular Reference to Synaptic Transmission, in *Brain Phosphoproteins. Characterization and Function. Progress in Brain Research* vol. 56 (Gispen W. H. and Routtenberg A., eds.), Elsevier, Amsterdam.

Rose S. P. R. and Sinha A. K. (1969) Some properties of isolated neuronal cell fractions. *J. Neurochem.* **16,** 1319–1328.

Roth B. L., Beinfeld M. C., and Howlett A. C. (1984) Secretin receptors on neuroblastoma cell membranes: Characterization of ¹²⁵I-labeled secretin binding and association with adenylate cyclase. *J. Neurochem.* **42,** 1145–1152.

Said S. I., ed. (1982) *Advances in Peptide Hormone Research Series, Vasoactive Intestinal Peptide.* Raven, New York.

Said S. I. and Mutt V. (1970a) Potent peripheral and splanchic vasodilator peptide from normal gut. *Nature* **225,** 863–864.

Said S. I. and Mutt V. (1970b) Polypeptide with broad biological activity: Isolation from small intestine. *Science* **169,** 1217–1218.

Schaeffer J. M., Schmeckel D. E., Conn P. M., and Brownstein M. J. (1980) A simple and rapid method to isolate rat retinal cells for biochemical analysis. *Neuropeptides* **1,** 39–45.

Schally A. V., Arimura A., Baba Y., Nair R. M. G., Matsuo H., Redding T. W., and Debeljuk L. (1971) Isolation and properties of the FSH and LH-releasing hormone. *Biochem. Biophys. Res. Commun.* **43,** 393–399.

Schebalin M., Said S. I., and Makhlouf G. M. (1977) Stimulation of insulin and glucagon secretion by vasoactive intestinal peptide. *Am. J. Physiol.* **232**, E197–E200.

Schlegel W. and Wollheim C. B. (1984) Thyrotropin-releasing hormone increases cytosolic free Ca^{2+} in clonal pituitary cells (GH_3 cells): Direct evidence for the mobilization of cellular calcium. *J. Cell Biol.* **99**, 83–87.

Schlegel W., Wuarin F., Wollheim C. B., and Zahnd G. R. (1984) Somatostatin lowers the cytosolic free Ca^{2+} concentration in clonal rat pituitary cells (GH_3 cells). *Cell Calcium* **5**, 223–236.

Schorderet M., Sovilla J. Y., and Magistretti P. J. (1981) VIP- and glucagon-induced formation of cyclic AMP in intact retinae in vitro. *Eur. J. Pharmacol.* **71**, 131–133.

Sellinger O. Z. and Azcurra J. M. (1974) Bulk Separation of Neuronal Cell Bodies and Glial Cells in the Absence of Added Digestive Enzymes, in *Research Methods in Neurochemistry* vol. II (Marks N. and Rodnight R., eds.), Plenum, New York.

Siegel G. J., Albers R. W., Agranoff B. W., and Katzman R., eds. (1981) *Basic Neurochemistry*. Little, Brown, Boston.

Siggins G. R. (1981) Catecholamines and Endorphins as Neurotransmitters and Neuromodulators, in *Regulatory Mechanisms of Synaptic Transmission* (Tapia R. and Cotman C. W., eds.), Plenum, New York.

Siggins G. R. and Bloom F. E. (1981) Modulation of Unit Activity by Chemically Coded Neurons, in *Brain Mechanisms and Perceptual Awareness* (Pompeiano O. and Ajmone Marsan C., eds.), Raven, New York.

Siman R., Baudry M., and Lynch G. (1984) Brain fodrin: Substrate for calpain I, an endogenous calcium-activated protease. *Proc. Natl. Acad. Sci. USA* **81**, 3572–3576.

Sladeczek F., Pin J.-P., Récasens M., Bockaert J., and Weiss S. (1985) Glutamate stimulates inositol phosphate formation in striatal neurones. *Nature* **317**, 717–719.

Smellie F. W., Davis C. W., Daly J. W., and Wells J. N. (1979) Alkylxanthines: Inhibition of adenosine-elicited accumulation of cyclic AMP in brain slices and of brain phosphodiesterase activity. *Life Sci.* **24**, 2475–2482.

Smitherman T. C., Sakio H., Geumei A. M., Yoshida T., Oyamada M., and Said S. I. (1982) Coronary Vasodilator Action of VIP, in *Vasoactive Intestinal Peptide* (Said S. I., ed.), Raven, New York.

Snyder S. H. (1980) Brain peptides as neurotransmitters. *Science* **209**, 976–983.

Snyder D. S., Zimmerman, Jr., T. R., Farooq M., Norton W. T., and Cammer W. (1983) Carbonic anhydrase, 5'-nucleotidase, and 2',3'-

cyclic nucleotide-3'-phosphodiesterase activities in oligodendrocytes, astrocytes, and neurons isolated from the brains of developing rats. *J. Neurochem.* **40,** 120–127.

Sølling H. and Esmann V. (1975) A sensitive method of glycogen determination in the presence of interfering substances utilizing the filter-paper technique. *Anal. Biochem.* **68,** 664–668.

Sotelo C. and Palay S. L. (1968) The fine structure of the lateral vestibular nucleus in the rat. I. Neurons and neuroglial cells. *J. Cell. Biol.* **36,** 151–179.

Spiess J., Rivier J., and Vale W. (1983) Characterization of rat hypothalamic growth hormone-releasing factor. *Nature* **303,** 532–535.

Starke K. (1977) Regulation of noradrenaline release by presynaptic receptor systems. *Rev. Physiol. Biochem.* **77,** 1–124.

Staun-Olsen P., Ottesen B., Bartels P. D., Nielsen M. H., Gammeltoft S., and Fahrenkrug J. (1982) Receptors for vasoactive intestinal polypeptide on isolated synaptosomes from rat cerebral cortex. Heterogeneity of binding and desensitization of receptors. *J. Neurochem.* **39,** 1242–1251.

Steiner A. L., Pagliera A. S., Chase L. R., and Kipnis D. M. (1972) Radioimmunoassay for cyclic nucleotides. II. Adenosine 3',5'-monophosphate and guanosine 3',5'-monophosphate in mammalian tissues and body fluids. *J. Biol. Chem.* **247,** 1114–1120.

Sutherland E. W. and Robison G. A. (1966) The role of cyclic 3',5'-AMP in responses to catecholamines and other hormones. *Pharmacol. Rev.* **18,** 145–161.

Sutherland E. W., Rall T. W., and Menon T. (1962) Adenyl cyclase. I. Distribution, preparation and properties. *J. Biol. Chem.* **237,** 1220–1227.

Taylor D. P. and Pert C. B. (1979) Vasoactive intestinal polypeptide: Specific binding to rat brain membranes. *Proc. Natl. Acad. Sci. USA* **76,** 660–664.

Thulin L. and Samnegard H. (1978) Circulatory effect of gastrointestinal hormone and related peptides. *Acta Chir. Scand.* **482,** 73–74.

Trachtenberg M. C. and Packey D. J. (1983) Rapid isolation of mammalian Muller cells. *Brain Res.* **261,** 43–52.

Tsien R. Y. (1980) New calcium indicators and buffers with high selectivity against magnesium and protons: Design, synthesis and properties of prototype structures. *Biochemistry* **19,** 2396–2404.

Tsien R. Y., Pozzan T., and Rink T. J. (1982) Calcium homeostasis in intact lymphocytes: Cytoplasmic free calcium monitored with a new, intracellular trapped fluorescent indicator. *J. Cell Biol.* **94,** 325–334.

Tsien R. Y., Pozzan T., and Rink T. J. (1984) Measuring and manipulating cytosolic Ca^{2+} with trapped indicators. *Trends Biochem. Sci.* **9**, 263–265.

Uhl G. R. and Snyder S. H. (1976) Regional and subcellular distributions of brain neurotensin. *Life Sci.* **19**, 1827–1832.

Vale W., Spiess J., Rivier C., and Rivier J. (1981) Characterization of a 41-residue ovine hypothalamic peptide that stimulates secretion of corticotropin and β-endorphin. *Science* **213**, 1394–1397.

Van Buskirk R. and Dowling J. E. (1981) Isolated horizontal cells from carp retina demonstrate dopamine-dependent accumulation of cyclic AMP. *Proc. Natl. Acad. Sci. USA* **78**, 7825–7829.

van Calker D., Muller M., and Hamprecht B. (1980) Regulation by secretin, vasoactive intestinal peptide and somatostatin of cyclic AMP accumulation in cultured brain cells. *Proc. Natl. Acad. Sci. USA* **77**, 6907–6911.

Walaas S. I. and Greengard P. (1984) DARPP-32, a dopamine- and adenosine 3':5'-monophosphate-regulated phosphoprotein enriched in dopamine-innervated brain regions. I. Regional and cellular distribution in the rat brain. *J. Neurosci.* **4**, 84–98.

Wei E. P., Kontos H. A., and Said S. I. (1980) Mechanism of action of vasoactive intestinal polypeptide on cerebral arterioles. *Am. J. Physiol.* **239**, H765–H768.

White E. L. (1981) Thalamocortical Synaptic Relations, in *The Organization of the Cerebral Cortex* (Schmitt F. O., Worden F. G., Adelman G., and Dennis S. G., eds.), MIT Press, Cambridge, Massachusetts.

Wilkening D. and Makman M. H. (1976) Stimulation of glycogenolysis in rat caudate nucleus slices by *l*-isopropylnorepinephrine, dibutyryl cyclic AMP and 2-chloroadenosine. *J. Neurochem.* **26**, 923–928.

Wilkening D. and Makman M. H. (1977) Activation of glycogen phosphorylase in rat caudate nucleus slices by *l*-isopropylnorepinephrine and dibutyryl cyclic AMP. *J. Neurochem.* **28**, 1001–1007.

Wolfe D. E. and Nicholls J. G. (1967) Uptake of radioactive glucose and its conversion to glycogen by neurons and glial cells in the leech central nervous system. *J. Neurophysiol.* **30**, 1593–1609.

Wollheim C. B. and Pozzan T. (1984) Correlation between cytosolic free Ca^{2+} and insulin release in an insulin secreting cell line. *J. Biol. Chem.* **259**, 2262–2267.

Wood C. L. and Blum J. J. (1982) Effect of vasoactive intestinal polypeptide on glycogen metabolism in rat hepatocytes. *Am. J. Physiol.* **242**, E262–E272.

Zieglgansberger W. (1980) Peptides in the Regulation of Neuronal Function, in *Peptides: Integrators of Cell and Tissue Function* (Bloom F. E., ed.), Raven, New York.

Behavioral Tests

Their Interpretation and Significance in the Study of Peptide Action

Adrian J. Dunn and Craig W. Berridge

1. General Considerations

Neuropeptides (i.e., peptides found in the nervous system) have received wide general acceptance as regulators of neuronal activity (e.g., Krieger and Liotta, 1979; Hökfelt, 1980). In many cases, neuropeptides were previously known as peptide hormones or had been first isolated from other tissues, frequently the gut. In other cases, hypothalamic peptides known to serve as releasing hormones (i.e., molecules released from the hypothalamus that could either stimulate or inhibit the release of hormones from the pituitary) were subsequently found to have an independent existence elsewhere in the central nervous system (CNS). In some cases, peptides were first discovered in extracts of brain tissue, and, in a very few cases so far, deduced from mRNA sequences found in only a relatively small population of neurons (Sutcliffe et al., 1983). In all cases, neuropeptides appear to have powerful activities as regulators of nervous system function.

Whether neuropeptides should be considered as neurotransmitters or neuromodulators is a semantic question. We prefer to use the term neurotransmitter in the broadest sense of the term, to mean a substance released by a neuron that has a specific effect on the activity of another cell. The term neuromodulator has been used in many different senses, e.g., to describe substances that have slow electrophysiological effects, alter the response to other neurotransmitters, act at remote (nonsynaptic) sites, and so on, and we do not believe it useful to define terms to cover all the possible activities of neuroactive secretions (*see* Dismukes, 1979). It is important that the terms neurotransmitter and neuromodulator describe functions, not chemical compounds, which may act differently in different anatomical locations. Regardless of definitions, it is clear that neuropeptides have the ability to affect the activity of neurons, thus altering the output of the nervous system, i.e., in behavior.

From a biochemical point of view, the special features of neuropeptides that distinguish them from other neurotransmitters are twofold. (1) they are synthesized as large protein precursors in the perikarya of neurons, processed by proteolytic cleavage(s) and/or glycosylation or acylation, and transported axoplasmically to the neuronal terminals packaged in vesicles, (2) after Ca^{2+}-dependent release, peptides must be degraded to their constituent amino acids before further metabolism, because evidence is lacking for cellular uptake of intact peptides that might permit recycling of active molecules. These facets contrast with those of the "classical neurotransmitters," which are largely synthesized in nerve terminals from appropriate precursors and can be taken back up into presynaptic terminals by specific uptake systems, permitting reuse (recycling) of active chemicals.

For these reasons, the secretion of neuropeptides is metabolically expensive compared to that of the longer-established neurotransmitters. This is compensated for by an increased potency of peptides, reflected in a higher affinity for their receptors [dissociation constants (K_d values) typically 10^{-12} to $10^{-9}M$, as compared to 10^{-7} to $10^{-5}M$ for the biogenic amines]. This increased potency allows for effective actions from the release of smaller quantities of the neuropeptides and/or the potential for actions at remote sites, i.e., the released peptides can be active after transport over significant distances, e.g., in extracellular fluid (ECF), cerebrospinal fluid (CSF), or blood.

Therefore, extraordinarily sensitive assays are necessary to determine biologically significant quantities of peptides. In practice the only assays sensitive enough are radioimmunoassays (RIAs), which are somewhat imprecise procedures and rather sensitive to interference. Because of the potential problems with cross-reactivity, peptide chemists prefer to precede the RIA with a powerful purification procedure, such as high-performance liquid chromatography (HPLC). This makes the procedures for neuropeptide assay very cumbersome because of the very large number of HPLC fractions that must be analyzed by RIA. In some cases, bioassays may be even more sensitive than RIAs, which are typically useful down to the low picomole range or better, depending on the antibody and the specific activity of the ligand.

Neuropeptides exhibit high specificity for their receptors, but an unexpected property is the high specificity of their actions. Whereas classical neurotransmitters such as acetylcholine or norepinephrine (noradrenaline) have little effect after injection of moderate doses into animals, neuropeptides in small doses may have remarkably specific effects. The analgesic properties of the endogenous opioid peptides (endorphins) are well known, but other examples are angiotensin, which administered intracerebrally (ic) in tiny doses elicits drinking in rodents (Phillips, 1984); adrenocorticotropin (ACTH), which when administered similarly elicits grooming behavior (Gispen et al., 1975) (*see* section 2.1), and bombesin, which elicits scratching (Gmerek and Cowan, 1983). There is a long list of examples, some of which may explain the results of the studies done previously on the transfer of memory (Ungar, 1973). The important aspects are that the doses necessary can be very low [in the low femtomole (10^{-15}) range] and the specificity of action can be great.

In this chapter we intend an overview of the important principles that have been established for the study of peptide-induced behaviors. We do not attempt to be comprehensive, but by way of illustration will discuss the current status of work on three different behavioral models: ACTH-induced grooming (section 2.1), vasopressin's effects on learning and memory (section 2.2), and cholecystokinin-induced cessation of feeding (section 2.3).

1.1. The Question of Dose

1.1.1. How Much Is Too Much: How Much Is Too Little?

There is probably no issue as problematic as that of the appropriate dose of a neuropeptide. It is frequently argued that the dose of a peptide administered is too high for a resultant effect to be "physiological." Yet rarely do we know what the physiologically effective concentration of a neuropeptide is, nor do we know what concentration of the administered peptide reaches the active site. Both factors represent a significant problem in neuropeptide research.

At the neuromuscular junction it has been estimated that the concentration of acetylcholine from one released quantum is 0.3

mM (Kuffler and Yoshikami, 1975), far above the K_d for the muscle acetylcholine receptor. Because the K_d values for neuropeptide receptors are so much lower than this, we can anticipate much lower concentrations of peptides at their sites of action. In general, concentrations in the range of the K_d of the receptor should be effective. Thus, depending on the affinity of the particular receptor, an effective concentration of 10^{-12} to $10^{-9}M$ is reasonable. This corresponds to a dose of about 1 pmol to 1 nmol distributed over the entire rat brain. For a peptide with a molecular weight of 1000 (e.g., an octapeptide), this would require a dose of between 1 ng and 1 μg, well in the active range for many peptides given intracerebroventricularly (icv; i.e., into the cerebral ventricles).

The second aspect may be more important; namely how much intact peptide actually reaches the active site. Peptides are very labile to biological degradation in some circumstances, and remarkably stable in others. In plasma, peptide hormones can have considerable half-lives, e.g., 10–60 min for ACTH (Liotta et al., 1978; Tanaka et al., 1978). Also, intact peptides are frequently isolated from CSF. Nevertheless, peptides can be broken down very rapidly in tissue and intracellularly. Although several proteases are present in brain, it is not known whether these have access to extracellular peptides. It is likely, however, that membrane-bound peptidases inactivate peptides in extracellular fluid, by analogy with the membrane-bound cholinesterases believed to hydrolyze acetylcholine released in the brain.

Degradation of the peptide may not be the only problem, however. When the peptide is administered icv, it may have to diffuse a considerable distance through the tissue or ECF to reach its site of action. Although there is no actual barrier to such diffusion, rates can be quite low, as indicated by the slowness with which dyes or radiolabeled compounds penetrate the tissue. The concentration of peptide reaching the active site will be determined by the difference between the rate of diffusion and the rate of removal by degradation or other mechanisms. Neither of these is measurable by current empirical methods.

1.1.2. The Importance of Dose–Response Relationships

Dose–response curves enable the determination of effective concentrations of the peptide, and the dose required for a half-maximal effect (ED_{50}) should be approximately equivalent to the K_d for the receptor concerned. Good dose–response relationships also allow for the use of pharmacological tools (e.g., antagonists) to determine specificity for the peptide actions.

Knowledge of the dose–response relationship is also necessary for trivial reasons. Clearly one needs to know how much peptide must be administered to elicit an effect; the lack of any observable effect in a test may occur because the administered dose is too low. Conversely, the dose administered may produce a maximal (ceiling) effect, so that decreasing the dose may not significantly decrease the behavioral response. Alternatively, known antagonists may be ineffective because of the supramaximal dose of "agonist" peptide used.

Control injections (zero dose) must of course be given, but the nature of the appropriate control solutions is arguable. Normally, administered peptides are dissolved in physiological saline, but occasionally acidification is necessary to dissolve the peptide. An appropriate control is an injection of the same volume of the vehicle used as the solvent for the peptide. Some have argued that the best control is a solution either of the hydrolyzed peptide or a mixture of its constituent amino acids in appropriate concentrations. One should be aware that icv amino acids are not without activity, e.g., proline (Cherkin et al., 1976) and D-phenylalanine (Ehrenpreis et al., 1978), which may, of course, be relevant after injected peptides are hydrolyzed. Administration of an "inactive" related peptide may be preferable. Probably the best control will become apparent when a number of different peptide analogs have been studied (*see* section 1.2).

To generate a proper dose–response curve can require a very large number of animals, which is especially difficult if precise ic cannulation is necessary. In some cases it is possible to improve the situation by generating an entire dose–response curve in one animal. This can be done only if it is established that there are no changes in the response (sensitization or desensitization) with repeated doses of the peptide. The best design involves a complete Latin square in which each animal receives each dose of the peptide on a random basis (*see*, for example, Dunn and Vigle, 1985). It is normally best to perform the injections a day or more apart, but some paradigms do permit repeated injections on the same day or even in the same session.

1.1.3. Timing and Repeated Doses

When the behavioral response being studied is a simple emitted one, the timing is not particularly critical (except with respect to time of day, season, or sexual cycle; *see* section 1.4.5). The behavioral sequence merely follows the administration of the peptide. When the behavior under study is dependent on other factors, however, timing can obviously be critical. In examples discussed below on the effect of ACTH and vasopressin on learning and memory (*see* section 2.2.2), whether or not one observes a behavioral effect may depend on the time of peptide administration. This may be less critical when the effect of the peptide is long lasting (e.g., in the case of vasopressin), as compared to ACTH whose activity is only evident when injected shortly before testing.

In other cases repeated injections may be necessary. For example, when studying the reversal of performance deficits following hypophysectomy, it is hardly surprising that repeated (so-called subacute) or even chronic peptide administration can be required (de Wied, 1965). There are several available procedures for chronic administration of a peptide. The oldest relies on the special properties of zinc hydroxide (de Wied, 1966). Peptides are mixed with a solution of a zinc salt, then coprecipitated by raising the pH with alkali. The gelatinous precipitate containing the peptide is then injected subcutaneously (sc). At the sc pH (about 5), the complex zinc hydroxide precipitate slowly dissolves, releasing the active peptide as it does so. Another procedure is to implant pieces of silastic tubing, containing solutions of the peptide, which slowly leak into the circulation. In either case, experimentation is necessary to achieve the desired rate of delivery of the peptide. The most convenient method uses osmotic minipumps, commercially available devices that are implanted under the skin. Influx of sc fluid caused by osmotic pressure causes collapse of a small plastic bag, slowly (over several days) expelling its contents, which can be a concentrated solution of the peptide.

1.2. Utility of Peptide Analogs

1.2.1. Degradation-Resistant Analogs

One powerful advantage of studying peptides over other pharmacological agents active on the CNS is the opportunity to create structural analogs by substituting different amino acids in the primary sequence. A major goal of such studies is to create peptides that are less biologically labile and resist degradation by introducing chemical bonds that are not susceptible to cleavage by endogenous enzymes. An example is the use of nonphysiological analogs of amino acids (e.g., D-analogs) close to the active site.

[Met^{4}SO$_2$,D-Lys^8,Phe^9]ACTH$_{4-9}$ (ORG 2766) is an analog of the core sequence of ACTH that has approximately 1000 times the potency of natural ACTH on avoidance behavior, largely because of its resistance to degradation (Witter et al., 1975).

1.2.2. Potent Agonists and Antagonists

An important aim of substitutions in peptide sequences is to produce potent agonists and antagonists. Potency can be manipulated not only by increasing the resistance to biological degradation, but also by seeking structures that better fit the receptors (van Nispen et al., 1977). For experimental purposes, antagonists can be even more important to establish physiological roles for the peptides (*see* section 1.3.2). A number of such analogs has been synthesized for vasopressin (Sawyer and Manning, 1985), and a few are available for angiotensin (Rioux et al., 1973), corticotropin-releasing factor (CRF) (Rivier et al., 1984), and so on.

Sometimes substitutions of amino acids in the parent peptide can produce unexpected results. For example, [D-Phe^7]ACTH$_{4-10}$ is not only degradation resistant, but has effects on conditioned avoidance behavior opposite to those of the natural ACTH fragment [L-Phe^7]ACTH$_{4-10}$ (Greven and de Wied, 1977). Even more surprisingly, the [D-Phe^7]ACTH$_{4-10}$ is active in inducing grooming, whereas ACTH$_{4-10}$ itself is not (see section 2.1.1).

1.2.3. Analogs With Diverse Actions

Another use for such manipulations is to produce analogs that lack some of the activities of the natural peptide. Natural ACTH contains 39 amino acids, but the sequence ACTH$_{1-24}$ contains full adrenocortical activity. A much shorter sequence is sufficient to produce effects on avoidance behavior, however, with ACTH$_{4-10}$ being quite sufficient (de Wied et al., 1975). The importance of this is that ACTH$_{4-10}$ has essentially no adrenocortical activity. Thus, behavioral effects of ACTH$_{4-10}$ cannot be attributed to its normal endocrine activity on the adrenal cortex. Similarly desglycinamide-lysine vasopressin (DG-LVP; i.e., LVP$_{2-9}$) has little of the classical pressor or antidiuretic activities of vasopressin itself, yet has activity in some behavioral tests (*see* section 2.2.4). The use of selective analogs can be a powerful means of revealing the existence of different kinds of receptors. Such procedures have clearly established the existence of two different receptors for vasopressin (V$_1$ and V$_2$), and presumptive evidence exists that the cerebral receptor is different from either of these (Cornett and Dorsa, 1985)

In a few cases it has been possible to make nonpeptide analogs of peptides. The classic example is the opioid peptides for which a long list of alkaloid analogs exists, including a natural analog, morphine itself.

1.3. Sites of Action of the Peptide

One of the most important pieces of information sought by the neuroscientist interested in the actions of a neuropeptide is the site of such action. This problem can be approached in a variety of ways, but the paucity of peptides for which we know the precise site of action attests to the experimental difficulties.

1.3.1. The Role of the Blood–Brain Barrier

Do peptides cross or do they not cross, that is the question. The extensive literature on this subject has frequently been misleading and often plain wrong. The ignorance stems from the failure to understand the true nature of the blood–brain barrier. The barrier is a *selective*, not an absolute, one, and does not have a single anatomical counterpart (Davson, 1972). It should be obvious to any biological scientist that there is no such thing as an absolute barrier, and that any compound administered to an animal will penetrate every organ to some extent. The question then is how much. The physical counterparts of the blood–brain barrier are the cellular membranes of the glial and neuronal cells and of the endothelial cells forming the cerebral capillaries. The special features of the capillaries in most parts of the brain are that the walls are not fenestrated (i.e., have no windows or pores) as are those of other tissues and that there are tight junctions between the endothelial cells so that large molecules such as proteins cannot pass. Thus, in common with all membranes, the blood–brain barrier restricts access to nonlipophilic materials. There is also a size factor, so that there is no effective blood–brain barrier to small lipophilic molecules such as alcohol or chloroform, but as size and polarity increase the barrier becomes more restrictive. Negatively charged (acidic) compounds are more restricted than those that are positively charged because of the negative polarity of the phospholipid bilayer.

The blood–brain barrier does not exist in all regions of the brain, however. A number of extremely important brain regions do not exhibit any barrier at all. These include not only such nonneural structures as the pineal gland and the anterior pituitary (adenohypophysis), but also true neural structures such as the posterior pituitary (neurohypophysis), the intermediate lobe of the pituitary in those animals that possess one, the infundibular stalk, the median eminence region of the hypothalamus, the organum vasculosum laminae terminalis (OVLT), the subfornical organ, the area postrema on the floor of the fourth ventricle in the brainstem, and other midline periventricular structures (sometimes known collectively as the circumventricular organs, CVOs) (Weindl, 1973). The absence of a blood–brain barrier in these regions is paralleled by the presence of fenestrations and the absence of tight junctions in their capillary endothelial cells (Davson, 1972).

The significance of this is that peptides circulating in the plasma have access to neurons in these special areas. For example, circulating peptide hormones can affect neurons in the hypothalamic median eminence to provide feedback on the secretion of releasing factors. So much for the blanket statement that peptides do not penetrate the brain.

Another way in which compounds can avoid the blood–brain barrier is by being selected by one or another of the specific uptake systems that function in the uptake of glucose, amino acids, or other essential nutrients. This avenue has long been used by pharmacologists to provide access to the brain for drugs, and it is likely that some peptides, especially small lipophilic peptides can use this route. It may explain the large variability in uptake measured with different peptides.

Some authors have used techniques designed to measure gross uptake of substances into the brain such as the Oldendorf technique (e.g., Cornford et al., 1978) and, finding no significant uptake of labeled peptides, have erroneously concluded that there is no penetration of the peptide concerned. Unfortunately these techniques are unreliable for substances with 10% of the permeability of water and could not detect penetrance of those with permeabilities of 1%. Yet 1% of a 1 µg/mL solution of a peptide can easily provide a dose of peptide physiologically significant in a specific brain area. Other more sensitive techniques have detected significant uptake of a range of peptides (*see*, for example, Rapoport et al., 1980; Banks and Kastin, 1985, for a review), although many of the published studies are flawed. For example, when using a labeled peptide as a tracer to determine uptake, it is essential to demonstrate unequivocally that the uptake of the tracer substance is located in intact molecules of the type administered, and is not merely in degraded radiolabeled catabolites. This is rarely true for uptake of ^{125}I, favored because of the relative ease of labeling peptides and because of its high specific radioactivity. It is also necessary to correct adequately for nonspecific absorbance of the labeled peptides to the tissue and for their presence in contaminating fluids such as blood.

In considering access of peptides to the brain, it is also important that there is no effective barrier between the CSF and the brain, so that peptides injected into the ventricles or the subarachnoid space are limited in their access to the tissue only by diffusion. It should also be understood that the blood–brain barrier is bidirectional. The same kinds of compound that encounter difficulty entering brain tissue will also have problems leaving the brain. (Substances injected into the ventricular system will drain with CSF back into venous blood. The time taken for this to occur will vary with the site and volume of the injection, but could take up to 30 min. A recent report demonstrated that cholecystokinin injected into the lateral ventricle of a rabbit appeared rapidly (5 min) in the plasma by a noncarrier-mediated mechanism (Passaro et al., 1982). In this experiment, however, the peptide was injected as a bolus in a volume (200 µL) large enough to cause rapid mixing, and the results may not necessarily hold for injections of more reasonable size.)

1.3.2. Peripheral vs Central Sites of Action

For the neuroscientist, it is vital first to know whether or not the neuropeptide acts directly on the brain or through some peripheral organ. In the simplest cases, the administered peptide may merely act on some endocrine organ, eliciting the release of other hormones or factors that can affect behavior. Obvious examples are pituitary peptides such as the gonadotrophins, which can cause secondary release of steroids from the gonads, or ACTH, which causes release of glucocorticosteroids from the adrenal cortex. There may, however, be more subtle examples—insulin decreasing plasma glucose or vasopressin increasing blood pressure (see section 2.2).

In the absence of other information, how does one decide whether the behavioral actions of a peptide are central or peripheral? The first line of experimentation should be to compare administration directly into the brain (ic), as contrasted with peripheral administration sc. Subcutaneous administration is preferable to intraperitoneal (ip), because degradation is more likely to occur if the peptide is absorbed into the gut. Because any compound administered peripherally may reach a brain site in some concentration (see above), dose–response curves are extremely important. If the effective dose of a peptide administered ic is far below that effective peripherally, this can be taken as prima facie evidence for a CNS action. But it does not constitute proof.

An alternative approach has been used in the study of the effects of vasopressin on learning and memory (*see* section 2.2). Doses of vasopressin that are behaviorally active centrally are much lower than those effective peripherally. Peripheral administration of a vasopressin antagonist, however, prevents the behavioral effects of peripherally administered vasopressin (Le Moal et al., 1981). Because the antagonist is not thought to cross the blood–brain barrier, it was argued that the site of action of vasopressin is peripheral. This is a weak argument for all the reasons discussed above. Small concentrations even of the antagonist may penetrate the brain and the crucial active sites may not have a barrier anyway. Furthermore, the peripherally administered antagonist is very likely to inhibit the cerebral uptake of the vasopressin itself by competition at its uptake sites.

Similarly, specific antibodies to the peptide can be used as antagonists. Because penetration of the brain by antibodies from the periphery is likely to be very limited, the ability of a peripherally administered antibody to inhibit the actions of a peripherally administered peptide has been taken as evidence that the site of action is peripheral. It should be remembered, however, that the antibody may sequester the peptide circulating peripherally and thus decrease the pool available for cerebral uptake. This argument is therefore fallacious. Nevertheless, because the blood–brain barrier is bidirectional, the ability of ic injected antibodies to antagonize the behavioral actions of a peptide (either peripherally or centrally administered) can provide good evidence that the site of action is intracerebral, provided that peripherally administered antibody is ineffective against centrally administered peptide.

It is worthwhile to point out several limitations of the use of antibodies as antagonists. First, even though a good antibody binds the peptide antigen specifically, the binding equilibrium may not be such that all the available peptide is sequestered. Many commercially available antibodies were selected for use in radioimmunoassays in which equilibrium conditions, not irreversible binding of the antigen, are important. Also be aware that unpurified or partially purified antisera may already be at least partially saturated with antigen. Furthermore, if the antigenic site on the peptide is not the same as its active site, binding to the antibody may not inactivate the peptide, although of course it may prevent it entering the brain or limit its penetration of tissue. Also, just as diffusion through extracellular space can limit penetration of icv-administered peptides into brain tissue, the penetration of icv-injected antibodies into brain tissue is very limited and slow. Thus to be effective, antibodies may have to be administered in high concentrations and will work best if the site(s) of action of the peptide are on the ventricular surface or close to it.

A major advantage of the use of antagonists (peptide analogs or antibodies) is their ability to antagonize actions of the endogenous peptides. Thus the ability of antagonists to inhibit an ongoing behavior can provide evidence supporting an endogenous (physiological) role of the peptide in the behavior. Moreover, the effective routes of administration for the antagonist can provide information on the location of the active sites of the peptide. Examples of this exist for the effects of vasopressin, oxytocin, ACTH, and α-melanocyte-stimulating hormone (MSH) on memory (van Wimersma Greidanus et al., 1975, 1978; Bohus et al., 1978) (*see* section 2.2), angiotensin-induced drinking (Fitzsimons et al., 1977), ACTH-induced grooming (Dunn et al., 1979) (*see* section 2.1), and cholecystokinin (Della-Fera et al., 1981) (*see* section 2.3).

1.3.3. Intracerebral Sites of Action

A surprising number of peptides are active when injected icv, usually into the lateral or third cerebral ventricle. In other cases, peptides may need to be injected directly into precise brain areas. Determining the location of such areas can be a tedious process. The obvious method is to implant cannulae into a variety of different brain areas and inject the peptide in all of them (*see*, for example, van Wimersma Greidanus and de Wied, 1971). The problem is that not all cannulae work equally and diffusion may occur over variable distances. Thus it is important to generate dose–response curves from a particular location from multiple animals.

There are ancillary techniques to assist in finding the sites of action, including use of brain lesions to determine brain structures vital for the peptide-induced behavior (*see* examples in section 2.1.4). The problem is that an almost limitless number of different lesions is possible. Another approach is the pharmacological one. Appropriate drugs can be used to determine chemical systems (especially well-characterized neurotransmitter systems, such as the catecholamines) essential for the expression of the peptide-induced behavior. Good examples can again be found from the work on ACTH-induced grooming behavior (*see* section 2.1.3). The problem with both lesion and pharmacological approaches is that although they may be useful for determining brain structures or pathways involved in the expression of the behavior, they will rarely reveal the primary site of action of the peptide.

A more direct approach is to seek receptors for the peptide. This can be done most conveniently by binding studies using an appropriate labeled ligand. Peptide ligands are more tricky to use than those for other binding sites because they are often more labile and peptide receptors may be present in very low concentrations, but have been used successfully in several cases. A most convenient way is to use so-called in vitro autoradiography (Kuhar and Yamamura, 1975), in which slices of tissue are exposed to the labeled ligand so that subsequent autoradiograms reveal the anatomical distribution of the binding sites, which are presumed to be receptors for the neuropeptide action (e.g., corticotropin-releasing factor; CRF) (De Souza et al., 1984). A variation of this procedure uses fluorescently labeled peptides as ligands, and then identifies active sites using fluorescence microscopy on frozen tissue sections (Landas et al., 1980). For common peptides, there may be multiple sites, perhaps distributed in diverse regions of the brain. The problem then is to determine which are important for the particular response under study. This can be done by local cannulation or lesion studies.

If a suitable ligand is not available, it is possible to seek receptors in other ways. For example, on the assumption that the receptor can be coupled to adenylate cyclase, it would be possible to study the adenylate cyclase activity of membrane preparations from different regions of brain for stimulation by the peptide. This approach could also be applicable to receptors known to be coupled to phosphoinositide metabolism (Farese, 1983; Berridge, 1985).

Another approach would be to use autoradiographic or other techniques to determine changes in cerebral blood flow (CBF) following peptide administration. This approach has been used for several peptides, including ACTH (Goldman et al., 1979). Similarly, the 2-deoxyglucose procedure for determining glucose utilization (Sokoloff, 1981) could be used. This could be accessible to PET (positron-emitting tomographic) scanning procedures in man or other animals with large brains. This approach has been used for ACTH and vasopressin (Dunn et al., 1980). Alternatively a whole host of other biochemical responses could be studied, most usefully if a response is already known or characterized. In common with the lesion and pharmacological procedures, however, the results would not necessarily reveal the primary site(s) of action.

An interesting technique for studying the site of ic action of peptides uses the injection of viscous compounds to block the flow of CSF within the ventricular system, or to coat and insulate the surface of the ventricles. The technique, pioneered by Herz et al. (1970), requires pre-implantation of stainless steel cannulae through which either cold-cream (Nivea) or silicone grease is injected under pressure into specific brain sites. For example, it is relatively easy to plug the cerebral aqueduct (between the third and fourth ventricles). Then, a subsequent third ventricle injection of the peptide will no longer be effective if its site of action was downstream in the fourth ventricle. To control for nonspecific damage, a fourth ventricle injection should still be effective. Whereas if the active site were in the third ventricle, injections there, but not in the fourth ventricle, would be effective. A similar approach can be used by plugging the interventricular foramen (between the lateral and third ventricles), to distinguish a lateral from a third ventricle site. The technique can be extended to use smears of the cream on the walls of the ventricles. Verification of the location of the blockading substances can be obtained by sectioning the frozen brain once the behavioral experiments have been completed. The white cold-cream or dyed silicone grease is clearly visible in such sections, and, if desired, the extent of diffusion of the injected peptide solution can be determined by inclusion of a suitable dye in the injectate. This technique has been used to determine the site of action of angiotensin II on drinking (Hoffman and Phillips, 1976), bradykinin on blood pressure (Lewis and Phillips, 1984), and ACTH on grooming (Dunn and Hurd, 1986). In all cases the crucial site appears to be in the anteroventral quadrant of the third ventricle, in or close to the OVLT.

1.3.4. Cerebroventricular Transport of Neuropeptides: A Physiological Route?

There is some reason to believe that transport of peptides in the CSF may be physiological (Dunn, 1978; Rodriguez, 1984). The major function of CSF is apparently to cushion the brain against damage from blows to the head. There is little evidence for a role in the supply of nutrients or in the removal of waste products (Davson, 1972). For the shock-absorbing function the ventricles serve no purpose. It is also to be noted that the CSF, whether it arises from the choroid plexus or from extracellular fluid, is produced largely within the ventricular system and flows rapidly in a consistent pathway through the ventricles and the subarachnoid space. The time for CSF to flow through this system is remarkably similar in all species. Kety (1970) has pointed out that many of the cells of the brain appear to be striving to send processes to the cerebral surface, either to the ependyma or the pia. He suggested that this might be to detect, and hence to respond to, factors in the CSF, both in the ventricular and the subarachnoid space. It is thus conceivable that the function of ventricular flow of CSF is intracerebral communication.

Axons of neurons containing vasopressin and oxytocin appear to innervate the ependyma and choroid plexus, and some even appear to terminate in the ventricles (Brownfield and Koslowski, 1977; Scott et al., 1977). Thus these pathways could constitute routes for the secretion of peptides into CSF, which could than be detected by neurons in structures surrounding the cerebral aqueduct and the fourth ventricle. Although CSF as normally isolated does contain proteases (this may be in part contamination from the withdrawal procedure), intact peptides are commonly isolated from CSF. A suggestion that ependymal cells may play a role in peptide metabolism is their high concentration of gamma-glutamyl transferase, an enzyme that has high affinity for peptides (Prusiner et al., 1976) and may play a role in peptide transport.

Such a neurohumoral transport system might explain the presence of relatively high concentrations of neuropeptides in CSF and the potent behavioral activity of peptides administered icv. It would also explain the remarkable ability of icv-injected antisera to block physiological actions of neuropeptides (*see* section 1.3.2). The concept of an intracerebral endocrine system is consistent with ideas of neurohumoral control of affect, since large populations of cells can be bathed with the secretions. It is notable that all the important ascending projection systems for the biogenic amines have cell bodies very close to the ventricles, ideally located for responding to chemical messengers in the CSF. Scheibel and Scheibel (1974) showed that dendrites of raphe neurons penetrate the ependymal layer and terminate in the ventricles. Furthermore, ependymal cells on the floor of the fourth ventricle send processes deep into the brain stem, including structures such as the locus ceruleus. It is also interesting that if the CSF mediates communication in a descending direction, the biogenic amine systems, notably those for norepinephrine (NE), dopamine (DA), and serotonin (5-hydroxytryptamine, 5-HT) send ascending signals to almost every region of the brain. Thus a wide-ranging, two-way neurohumoral communication is achieved.

1.4. Behavioral Considerations

1.4.1. Choice of a Behavior

In choosing a behavioral assay, one is obviously at the mercy of prior knowledge. If a chosen peptide is already known to have a behavioral activity, then clearly that should be the behavioral assay of choice; unless, of course, it is desired to establish a behavioral profile for the peptide, or discover a new assay. Behavioral screening procedures are not exceptionally comprehensive. One experimental psychologist would probably recommend first studying open-field behavior, and another might suggest an instrumental conditioning task. The truth is that each and every kind of test has severe limitations and none can reveal all the behavioral effects of each administered compound. Open-field behavior (i.e., placing the animal in a confined space new to the animal) may reveal effects of the peptide on general activity (i.e., moving, locomotion, rearing) and may identify soporific properties, or scratching or grooming behavior, but would not, for example, reveal the ability of angiotensin to elicit drinking (unless of course water were present), the ability of oxytocin to elicit maternal behavior (in the absence of pups), or the ability of cholecystokinin to inhibit eating (unless food were present and the animal was eating). Likewise, effects on learning and memory would not be revealed unless such effects were specifically sought. Clearly the screening task can be formidable, but let us assume that some data are available in the literature indicating behavioral responses to administration of the peptide.

1.4.2. Need for Quantification

The highest priority should be for a readily quantifiable behavior. This is because most induced behaviors are not all-or-none and precise quantification in behavioral research is as important as it is in any other science. Good quantification enables the generation of good dose–response curves, which are useful for determining specificity, sites of action, and so on (*see* sections 1.1.2. and 1.3.2) and can provide a bioassay for the peptide or its analogs.

1.4.3. Emitted vs Stimulus-Controlled Behaviors

Two fundamentally distinct forms of behavioral tests are possible: emitted and external stimulus-controlled. In the former, the peptide is administered to the animals and their behaviors observed in an environment selected according to the kind of behavioral response anticipated. In the latter, the effect of the peptide treatment on the animals' performance of a predetermined task, such as pressing a bar to receive a food pellet, is measured. Effects of peptides may be evident in one or both kinds of task. In such tasks, different kinds of effects may be found, depending on how the experiment is performed. For example, in a learning task the peptide treatment may be used to determine whether the animal learns the task more rapidly (an effect on *acquisition*) or whether it performs the task better when it is retested at a subsequent time (an effect on memory function). A difference in interpretation may also concern the time of administration over which the peptide is effective. Thus in the above example, if the peptide is effective when administered before, during, or shortly after the training session, it may be considered to improve the *formation (consolidation)* of memory. If it is only effective when given during the retention test itself, however, the effect may be considered to be on the *retrieval* of the memory or on its *performance*.

1.4.4. Species, Strain, Sex, and Age

The choice of species, strain, sex, and age may be critical in observing reliable behavioral effects of peptides. To a great extent one may be limited by the particular peptide-behavior combination being studied. The natural choice is for readily available laboratory animals such as rats and mice, but a large number of other possibilities exist, limited only by economic and practical factors. Strain and even the supplier of an animal can be critical. For example, the behavior of mice varies markedly among different strains. Certain strains are very excitable and others much more docile. In practice, certain strains may not even be able to learn certain tasks. For example, jump-up avoidance tasks are rapidly learned by C57s, but very slowly if at all by DBAs.

Sex is clearly a major factor in behaviors such as mating, and in sexually specific behaviors, such as maternal behavior (as, for example, induced by oxytocin; Pedersen and Prange, 1979), food hoarding, and nest-building. Sex differences may also appear in other peptide-induced behaviors, however. If female animals are used, the particular phase of the menstrual cycle during the testing may be of paramount importance. This is why male animals are used for so many behavioral tests. Cycling can also occur in males housed together with cycling females.

Most behaviors are not age-specific, but animals of a certain age may be more sensitive to the peptide-induced effects. Moreover, older animals can be more expensive and more difficult to replace.

The most conservative approach is to choose animals of a species, strain, sex, and age that are consistently available from a reliable source.

1.4.5. Environmental Factors

Animal behavior is as sensitive to environmental factors as is human behavior. The need for a controlled environment in terms of temperature and light cycle is obvious. Seasonal factors may also be of great importance, not only in sexually related behaviors. Effects may be related to environmental temperature, humidity, and so on. Nevertheless, it is well documented that behavioral activities may be related to season *per se* (e.g., Shashoua, 1973).

However, more subtle factors are known to cause variations in behavior. For example, the nature of the bedding material can matter; different results may be obtained when animals are bedded with cedar wood shavings as compared to pine. The size and nature of the cages may also be important. Experimental results using rats raised in the celebrated *environmentally enriched* condition, in which they are housed several to a cage in large cages with a variety of (children's) toys to play with (*see* Greenough, 1976), have revealed clear anatomical, neurochemical, and behavioral differences from rats housed under standard laboratory conditions. It is, of course, likely that an enriched environment is more normal compared to a natural one, whereas normal laboratory housing is "impoverished."

The number of animals per cage can also be critical. If different sexes are housed together the consequences will become obvious. But even in a unisexual environment a dominance hierarchy is established (just as in humans). All animals in a group cage are not equal and may not yield similar behavioral responses, with or without peptide administration. Even if this were not so, one should be aware that removal (and indeed replacement) of one animal from a group-housed cage is stressful for the remaining subjects. This can be documented by progressive increases in plasma corticosterone concentrations in the remaining animals as animals are removed one by one. However, it may be equally dangerous to house animals individually. It is well established that isolation of mice induces aggressive behavior; mice that have been isolated for several weeks or less will attack other mice as soon as they encounter them. Housing different species in the same room should also be avoided for reasons aside from disease control. Unlike domestic cats and dogs raised together, different species do not always cohabit well. Rats of many strains will naturally attack and kill mice when given the opportunity (muricidal behavior), and it is unlikely that mice kept in the sight and smell of rats will behave like those that are not.

1.4.6. Automated Systems vs Human Observers

Regardless of whether behavioral data are collected by an automated system or a human observer, it is vital that the experiment be performed "blind," which is to say that the observer or whoever is handling the animals is unaware of the specific treatment each subject receives. In the simplest case this means that when comparing a peptide injection with placebo, the solutions should be coded and that the code not be known by the experimenter who injects the animals. The code should not be broken until all the behavioral data have been analyzed and summarized. This is because it is relatively easy to introduce conscious or unconscious bias, not only in the way one interprets and records a behavioral response, but also in the handling of the animals and their placement in the behavioral apparatus. Animals can be sensitive to subtle cues. Experimenters who train animals have been known to boast that they can predetermine which way a rat will turn in a T-maze by altering the way they place the animal in the starting area. This is useful for winning bets, but is not in the best scientific interests of the experiment.

Not all behaviors are readily adaptable to automatic data collection. When they are, it is clearly a very useful way of increasing the ease of data collection and running large numbers of animals. It can also substantially reduce observer bias. The problem is that data collected automatically can be misleading; the subjects can fool the investigator, if you will. It is imperative that human observation be used in the early stages of the experiment and whenever any substantial change in the procedure is introduced. For example, a common automated procedure is to use devices designed to count vibrations indicative of motor movements. Sensitive detectors can in many cases distinguish between body movements involved in grooming, for example, as opposed to those involved in locomotion (walking about the cage). Violent movements not involving locomotion may be misinterpreted, however. Similar problems occur when photodetector systems are used to detect movement. The breaking of light beams may be interpreted as locomotion, but an animal that decides to scratch itself repeatedly in the right part of the cage can generate a large apparent mileage! More sophisticated systems can be used to automate avoidance training and instrumental conditioning, for example, pigeons pecking bars in certain sequences or at certain rates to receive food reward. Such experiments would be much more difficult and tedious without automation.

Despite the problems of automation, automated methods are far less susceptible to investigator bias than are human data collection systems. If the latter are necessary, then it is important that obvious precautions are taken. Rodents are extremely sensitive to odors, and there are many anecdotal reports of the perfume or after-shave of animal experimenters producing unusual results. To avoid this and other problems, animals should be observed from another room using a one-way glass window, or a video-camera system. If these are not available, observers should be trained to sit quietly in the room with the animals, preferably for a considerable period before data collection is begun. Loud and sudden noises are to be avoided, and a white-noise generator can be very useful to block out ambient noise common in most buildings. A convenient economical white-noise generator can be made by detuning an FM radio receiver.

The most important rule is to maintain consistent environmental conditions for data collection and to ensure that animals receive the various experimental treatments randomly throughout the observation period and are randomly distributed throughout the room when multiple animals are observed simultaneously.

2. Specific Examples of Peptide-Induced Behaviors

The number of tests available for the study of the effects of peptides on behavior is at least as great as the number of behaviors that can be exhibited in the laboratory. Included among these are those for the study of learning, ingestion and satiety, locomotor activity, pain, and sexual behavior. The use of these tests has been described in much detail in the literature and to do so here would be impossible. Rather, we wish to supplement our general comments with a discussion of the evidence concerning the role of a few specific peptides in mediating specific behaviors. These brief reviews will serve as examples of the value and limitations of studies of the behavioral actions of peptides and the manner in which research of this nature can be approached.

2.1. ACTH-Induced Grooming Behavior

Ferrari et al. (1963) first noted increased grooming in rats and rabbits before the onset of the stretching and yawning syndrome (SYS), which they studied after icv administration of ACTH. But grooming was first studied systematically by Gispen et al. (1975). Much lower doses of ACTH were necessary to elicit excessive grooming than to elicit SYS, with the effective dose of $ACTH_{1-24}$ being between 10 and 100 ng per rat. The grooming response is considered to be elicited by a direct action on the brain because of its short latency following icv injection and because it is not elicited by peripherally administered $ACTH_{1-24}$ at doses of up to 5 mg/kg (Gispen et al., 1975). Aside from rodents, ACTH-induced grooming is also observed in pigeons (Deviche and Delius, 1981) and chickens (Williams and Scampoli, 1984).

2.1.1. Structure–Function Relationships

ACTH-induced excessive grooming shows structure–function relationships in rats similar to those of SYS, namely that $ACTH_{1-13}$-NH_2 and $ACTH_{5-16}$-NH_2 elicit it, whereas $ACTH_{1-10}$, $ACTH_{4-10}$, and $[Met^4SO_2,D\text{-}Lys^8,Phe^9]ACTH_{4-9}$ (ORG 2766) do not (Gispen et al., 1975). Curiously, $[D\text{-}Phe^7]ACTH_{4-10}$, an analog containing an amino acid in the center of its core sequence in the D- as opposed to the L-configuration, does induce grooming. Also, although $ACTH_{4-10}$, $ACTH_{4-9}$, $ACTH_{4-8}$, and $ACTH_{4-6}$ are inactive, $ACTH_{4-7}$ does promote grooming, but $ACTH_{5-7}$ does not (Wiegant and Gispen, 1977). Rees et al. (1976) confirmed the ability of $ACTH_{1-24}$ and $[D\text{-}Phe^7]ACTH_{4-10}$ to elicit grooming in mice, along with the lack of activity of $ACTH_{4-10}$.

These structure–function relationships indicate that whatever receptor is involved in grooming is quite distinct from that involved in modulating avoidance behavior. In particular, $ACTH_{4-10}$, which retards extinction of avoidance behavior, does not elicit grooming, whereas $[D\text{-}Phe^7]ACTH_{4-10}$, which accelerates extinction of avoidance behavior, does elicit grooming (Greven and de Wied, 1977). Furthermore, the "grooming receptor" is distinct from the adrenocortical ACTH-receptor, because although longer sequences of ACTH (such as $ACTH_{1-24}$) are necessary in both cases, $ACTH_{1-16}$, which is not steroidogenic, does elicit grooming (Wiegant et al., 1979). This is an excellent example of the use of structural analogs to distinguish receptor types and functions.

2.1.2. The Specificity of ACTH-Induced Grooming

ACTH administered icv elicits grooming in both male and female rats and also in hypophysectomized, adrenalectomized, and castrated male rats (Gispen et al., 1975), indicating that the effect is not dependent on secondary secretions from the pituitary, adrenals, or gonads.

The behavioral response is apparently specific to grooming, which competes with other ongoing behaviors, such as locomotion or exploration (Isaacson and Green, 1978). ACTH-induced grooming is not stereotyped, and it closely resembles naturally occurring grooming. The proportion of time spent on various parts of the body and the grooming sequence is not distinguishable from natural grooming (Gispen and Isaacson, 1981).

Many other peptides have been reported to elicit grooming following icv injection. These include vasopressin (Delanoy et al., 1978), angiotensin (Phillips, 1984), bombesin (Katz, 1980; Gmerek and Cowan, 1983), eledoisin (Katz, 1980), substance P (Katz, 1979), prolactin (Drago et al., 1980), CRF (Morley and Levine, 1982), dynorphin (Zwiers et al., 1981), and β-endorphin, but not α-endorphin, β-LPH, nor the enkephalins (Gispen et al., 1976; see below). None of these peptides is capable of eliciting grooming to the extent that ACTH can. One μg of $ACTH_{1-24}$ injected into the lateral ventricle of a rat can cause the animal to groom for more than 90% of the time over the next hour (Gispen and Isaacson, 1981; Dunn et al., 1984). In other cases, notably vasopressin (Delanoy et al., 1978; Meisenberg and Simmons, 1983) and bombesin (van Wimersma Greidanus et al., 1985), the peptide-induced grooming does not resemble natural grooming, but is more of a compulsive or stereotyped scratching. In yet other cases, the effect of the peptides may be explained by their ACTH-releasing properties (e.g., CRF, vasopressin, angiotensin). Thus ACTH may be the only endogenous peptide to induce a grooming behavior resembling natural grooming.

2.1.3. Pharmacology of ACTH-Induced Grooming

ACTH-induced grooming is prevented by prior treatment with the specific opiate antagonists naloxone and naltrexone in low doses (Gispen and Wiegant, 1976), suggesting the involvement of an opiate receptor. Subsequently, it was found that low doses of morphine administered icv (50–400 μg) (Gispen and Wiegant, 1976) elicited grooming, as did β-endorphin (Gispen et al., 1976). The ability of ACTH analogs to elicit grooming closely parallels their ability to counteract morphine-induced analgesia and to compete for dihydromorphine binding to presumed opiate receptors (Wiegant et al., 1977a). Moreover, Dunn et al. (1981b) have shown that naloxazone, an irreversible opiate antagonist, prevents ACTH-induced grooming with a time course of recovery that parallels its effects on morphine-induced antinociception. It should not, however, be presumed that the "grooming" receptor is an opiate receptor, because the potency of opiates and endorphins to elicit grooming is lower than that of $ACTH_{1-24}$ (Dunn, 1987). Moreover, the maximum grooming induced by the opiates is never as high as with ACTH. It seems more likely that an opiate receptor is involved somewhere in the sequence of expression of the grooming behavior, rather than as the primary site of action of the peptides. This may explain some of the rather complex structure–function relationships described above.

ACTH-induced grooming exhibits other opiate-like characteristics. Although there is no loss of grooming potency with daily injections of ACTH (Colbern et al., 1978), the potency is diminished for up to 18 h after a single injection (Jolles et al., 1978). There is complete cross-reactivity between $ACTH_{1-24}$, [D-Phe^7]$ACTH_{4-10}$, β-endorphin, and morphine (Jolles et al., 1978). Prevention of the grooming with naloxone also blocks the "tolerance," however (Jolles et al., 1978). Furthermore, the peptide injection may elicit a stress response because sequences of ACTH that elicit grooming when injected icv also increase plasma concentrations of corticosterone, even though those same sequences (e.g., $ACTH_{1-16}$, and [D-Phe^7]$ACTH_{4-10}$) have no adrenocortical activity when injected systematically (Wiegant et al., 1979).

Prior studies had suggested that ACTH administration activated brain catecholaminergic systems (see reviews by Versteeg, 1980; Dunn, 1984), but a direct involvement of brain dopaminergic systems is suggested because ACTH-induced grooming is inhibited by haloperidol and other DA-receptor antagonists at low doses (Wiegant et al., 1977b; Guild and Dunn, 1982). Also, $ACTH_{1-24}$ elicits grooming when injected into the substantia nigra, but not the striatum, whereas haloperidol injected into the striatum, but not the substantia nigra, inhibits grooming elicited by icv-administered $ACTH_{1-24}$ (Wiegant et al., 1977b). Cholinergic systems may also be involved. ACTH-induced grooming is very sensitive to muscarinic cholinergic antagonists such as atropine and scopolamine (Dunn and Vigle, 1985). The muscarinic receptors concerned appear to be located in the CNS, because the methyl bromide derivatives of atropine and scopolamine that cross the blood–brain barrier poorly do not inhibit the grooming following peripheral administration, but are effective centrally. Also, small (microgram) icv doses of hemicholinium, an inhibitor of choline uptake that is known to deplete brain acetylcholine, also inhibits ACTH-induced grooming (Dunn and Vigle, 1985). The involvement of adrenergic systems is unlikely. Neither propanolol (β-antagonist) nor phenoxybenzamine (α-antagonist) affected the grooming behavior, but phentolamine (α-antagonist) did (Dunn, 1987). None of this implies any direct action of ACTH on receptors for these neurotransmitters, but it does imply that they are involved in the expression of the grooming behavior.

2.1.4. Sites on Which ACTH Acts To Elicit Grooming

The site on which ACTH acts to elicit grooming is not known. Indirect data exist from studies of glucose utilization using the deoxyglucose procedure during ACTH-induced grooming (Dunn et al., 1980). Regionally specific changes occurred in several brain regions, and a decrease of glucose utilization in the pyriform cortex and an increase in the cerebellum were statistically correlated with the expression of the grooming behavior. Electrolytic lesions of the septum, preoptic area, mammillary bodies, amygdala, posterior thalamus, and dorsal or ventral hippocampus did not affect $ACTH_{1-24}$-induced grooming, but aspiration of the hippocampus did (Colbern et al., 1977). Amygdaloid or hippocampal lesions enhanced SYS.

As indicated above, small doses of $ACTH_{1-24}$ injected into the substantia nigra elicited grooming, whereas large doses in the striatum did not (Wiegant et al., 1977b). Surprisingly, Ryan and Isaacson (1983) reported that ACTH injected into the nucleus accumbens induced grooming. Grooming is also elicited shortly after third ventricle application of ACTH, whereas SYS is considerably delayed, and Gessa et al. (1967) considered it to be a fourth ventricle phenomenon. Recent data show that $ACTH_{1-24}$ injected into the lateral or third ventricle effectively induces grooming in rats in which the cerebral aqueduct has been blocked by a small injection of cold-cream. Blockade of the interventricular foramen prevented grooming from ACTH injections into the corresponding lateral ventricle, but not from injections into the third ventricle. These results suggest a third ventricle site of action for ACTH. Using cold-cream injections to block potential active sites in the third ventricle, the crucial areas were found to be in the anteroventral quadrant of the third ventricle, in the region of the OVLT (Dunn and Hurd, 1986).

2.1.5. Significance of ACTH-Induced Grooming

Some interesting observations suggest that grooming behavior may be elicited by endogenous release of ACTH. Grooming is elicited in rats by mild stress, such as transfer to a novel environment (Colbern et al., 1978; Jolles et al., 1979) or "nontraumatic" noise–light stress (Katz and Roth, 1979). This increased grooming is impaired by low doses of naloxone (0.2 mg/kg) (Green et al., 1979) that do not alter the baseline (home cage) grooming activity (Gispen and Wiegant, 1976; Green et al., 1979) or locomotor or exploratory activities in a hole-board apparatus (an open field with holes in the floor for the rats to investigate (Green et al., 1979). Novelty-induced grooming is also prevented by haloperidol at doses (0.2 mg/kg sc) that did not alter locomotor or exploratory activities (Green et al., 1979). Thus, pharmacologically novelty-induced grooming resembles ACTH-induced grooming.

In one report, hypophysectomy severely inhibited novel-cage-induced grooming (Dunn et al., 1979), whereas in another it did not (Jolles et al., 1979). In the latter study, however, hypophysectomized rats showed higher home-cage grooming. In an extension of the Utrecht studies, Gispen et al. (1980) were unable to determine the reason for the discrepancy. Most important, icv administration of antiserum to ACTH inhibited novel-cage-induced grooming (Dunn et al., 1979). The antiserum used was specific to ACTH and did not interact with $ACTH_{1-10}$, α-MSH, β-LPH, β-endorphin, or the enkephalins. It was not effective when given sc even in much larger amounts. If novelty-induced grooming is caused by stress-induced secretion of ACTH, then treatments that reduce the stress should decrease the grooming. Consistent with this, benzodiazepines (diazepam or chlordiazepoxide), at doses that did not affect ACTH- or beta-endorphin-induced grooming in mice, reduced novelty-induced grooming (Dunn et al., 1981a). Neonatal monosodium glutamate (MSG) treatment, however, which largely destroys hypothalamic ACTH-containing cells, did not alter novelty-induced grooming or that in response to $ACTH_{1-24}$ in rats (Dunn et al., 1985).

The significance of ACTH-induced grooming is obscure. Nonessential grooming is considered by many ethologists to be a displacement behavior, by which they mean that it is not functionally directed and has neutral value as perceived by other animals. One could speculate that displacement behaviors have biological value during stress. Thus stress may elicit grooming unless some more important behavior, such as fighting or fleeing, overrides it. This would have the advantage of not signaling weakness or indecision to a real or potential adversary. If this speculation is correct, it is not obvious why the behavior should be ACTH- or β-endorphin-mediated rather than a direct neural response.

2.2. Vasopressin-Induced Effects on Learning and Memory

2.2.1. The Pituitary and Memory

Removal of the posterior and intermediate lobe of the pituitary from rats does not affect their rate of acquisition of a conditioned avoidance response, but does increase the rate of its extinction (de Wied, 1965). In these experiments, the rat is trained to respond to a conditioned stimulus (CS), typically a buzzer or light, in order to avoid the unconditioned stimulus (US), electric shock. Two commonly used conditioned avoidance response chambers are the shuttle box, in which the animal crosses a barrier separating two compartments of the conditioning box in response to the CS (de Wied, 1964), and the pole-jump, in which the animal jumps onto a pole placed in the middle of the conditioning box to escape shock (de Wied, 1966). Generally, training consists of a number of trials given daily for up to 14 consecutive days. After this acquisition period, extinction sessions are run in which the CS is presented for 5 s without a subsequent US, and the rate of extinction of the response is followed. Subcutaneous administration of pitressin, a crude extract of the pituitary, or ACTH attenuates this increased rate of extinction in posterior lobectomized animals. From these observations, it was suggested that pituitary peptides are involved in memory function.

A number of factors need to be studied carefully before such a hypothesis should be considered seriously. First, it must be determined whether the observed behavioral effect results from a central or peripheral action of the peptides, which also have potent endocrine activities. Second, it must be determined whether the behavioral activity is distinct from those endocrine activities. Third, it must be shown that the behavior is affected by a specific interaction of the peptides with memory, rather than some nonspecific effect.

Over the past 20 years, these questions have been studied using a variety of behavioral and pharmacological techniques. In the following, we shall examine the various approaches used, the information they have yielded, and the questions still remaining. A recent review (Strupp and Levitsky, 1985) analyzes the behavioral issues in detail.

2.2.2. ACTH vs Vasopressin

Although both ACTH and vasopressin can delay extinction of the avoidance response in hypophysectomized rats, they appear to produce this effect by different mechanisms. This is evident from the differences in the time course of their action. A single dose of lysine vasopressin (LVP)* facilitates the retention of the avoidance response when given either during acquisition or extinction. By contrast, ACTH is effective only when administered during extinction and for only a relatively short period after its administration (de Wied and Bohus, 1966; Ader et al., 1972). Because a direct effect on memory consolidation would be expected to be apparent for more than 1 d after peptide administration, these results suggest that ACTH does not directly influence memory processes. Its effects on behavioral performance have been suggested to be associated with arousal or selective attentional mechanisms (*see*, for example, LaHoste et al., 1980; Sandman and Kastin, 1981). Vasopressin's more permanent influence on behavioral responses, may, however, indicate a more direct effect on memory processes.

2.2.3. Active vs Passive Avoidance Behavior

During active avoidance behavior a number of cerebral systems are activated, including sensory and motor systems, as well as those involved in the general stress response. It is possible that

---

*Vasopressin exists in two different forms, lysine vasopressin (LVP) and arginine vasopressin (AVP), that may differ only in the N-terminal amino acid residue (lysine or arginine, respectively). Different species synthesize the different molecules (e.g., LVP in the pig; AVP in the mouse, rat, and man), but there is no evidence for any differences in the biological activities of AVP and LVP.

vasopressin exerts a long-lasting effect on one of these systems, which in turn facilitates the behavioral response. This question can be approached behaviorally by examining the effects of the peptide in tests of learning other than active avoidance behavior. If vasopressin interacts with systems specifically involved in memory, then one would expect to observe memory-enhancing effects in other behavioral tests, assuming that the basic memory mechanisms are the same for these learned responses, an assumption that may or may not be true.

A useful test for these purposes is the one-trial passive (inhibitory) avoidance test (Ader et al., 1972). Briefly, this test consists of placing a rat or mouse on a brightly lit platform connected to a darkened chamber. When the animal enters the preferred darkened chamber, it receives a single brief electric footshock. The animal is then replaced on the platform after an appropriate intervening period (typically 24 or 48 h), and the latency to enter the darkened chamber is interpreted as an indication of the strength of the memory, or retention, of the training session. One advantage of the one-trial test is that the time at which learning occurs is known with some precision, because learning is confined to that one trial. Obviously, any treatment that impairs motor function will give the false impression that the animal retained the knowledge obtained in the previous session. By contrast, motor impairment in active avoidance behavior would appear as poor learning. Therefore, concordant results in both passive and active avoidance behavioral tasks would suggest effects on learning *per se*, rather than motoric effects.

LVP, given sc either immediately after the passive avoidance training trial or immediately before the retention test, 24 or 48 h after training, improved responding during the test session (Ader and de Wied, 1972). When the animals were tested 24 h after the learning trial, AVP was most effective when administered either immediately after training or just prior to testing, whereas it had only a slight effect when given 3 h, and no effect at all when administered 6 h after training (Bohus et al., 1978). These results suggest an effect of AVP on both the formation of memory and its retrieval.

2.2.4. Central vs Peripheral Actions

The dose of AVP effective in prolonging retention of the passive avoidance response administered peripherally was found to be 20 ng/rat sc (Bohus, 1977). Intracerebroventricularly, AVP is effective at a dose of 1 ng per rat (Bohus et al., 1978), indicating a far greater potency of the peptide intracerebrally.

In support of a central action of vasopressin on passive avoidance behavior, antiserum to AVP administered icv either shortly before or after training significantly decreased subsequent performance of the response compared to control animals (van Wimersma Greidanus et al., 1975). Moreover, systemic administration of 100 times the amount of antiserum, which removed all circulating AVP as measured by the analysis of urinary vasopressin, did not affect the retention of the behavior.

Nevertheless, results obtained with vasopressin antagonists have raised the possibility that actions outside the CNS are important for the behavioral effect of vasopressin. Le Moal et al. (1981) replicated the observations of the Utrecht group that low doses of vasopressin administered peripherally (1–6 μg AVP sc) retarded extinction of the pole-jump avoidance response. They found, however, that this effect of AVP could be prevented by peripheral administration of the vasopressin antagonist 1-deaminopenicillamine-2-(*O*-methyl)-tyrosine AVP (dPTyr(Me)AVP). Because this compound was considered to cross the blood–brain barrier poorly, Le Moal et al. argued that their results suggested a peripheral site for the action of peripherally administered AVP on the extinction of avoidance behavior. As discussed above, this is a weak argument until the site of action of vasopressin and the properties of dPTyr-(Me)AVP are fully understood.

Koob et al. (1981) also found that the dose of icv AVP (1 ng) effective in delaying the extinction of pole-jump avoidance responding was considerably lower than that effective peripherally (1 μg). However, dPTyr(Me)AVP, which facilitated extinction when injected sc, was effective icv only at doses similar to those effective sc. Therefore, Koob et al. argued that endogenous AVP may act on a peripheral site to produce the prolonged extinction effect. They did note, however, that a direct action within the CNS cannot be ruled out just because the antagonist was behaviorally active icv only when administered at a high dose. dPTyr(Me)AVP is more hydrophobic than AVP, so that it might enter the circulation more readily than AVP and become distributed throughout the body as though systemically administered. Alternatively, it is possible that even if vasopressin is directly involved in memory, other systems might also affect learning or performance of the task. Koob et al. (1981) used a high shock current to aid the detection of facilitation of the extinction response. At this higher level of shock, other systems not involving vasopressin (which would therefore not be sensitive to a vasopressin antagonists) might be affected, increasing the strength of the memory. Ettenberg (1984) also reported that a high dose of dPTyr(Me)AVP icv is necessary to prevent the behavioral effects of sc AVP (in this case, in a food finding task, see below).

The problem of the results obtained with vasopressin antagonists was addressed extensively by de Wied et al. (1984). Using a passive avoidance paradigm, they found that both AVP, and the more potent catabolite of vasopressin [pGlu,Cyt]AVP$_{4-8}$ (AVP$_{4-8}$), administered sc facilitated passive avoidance behavior. Both AVP and AVP$_{4-8}$ were much more potent behaviorally when injected icv than when systematically injected (440 and 23,000 times, respectively). Two different vasopressin antagonists, dPTyr(Me)AVP and d(CH$_3$)$_5$Tyr(Me)AVP, administered sc antagonized the behavioral effects of peripherally injected AVP and AVP$_{4-8}$. Moreover, icv injection of d(CH$_3$)$_5$Tyr(Me)AVP prevented the facilitatory effects of icv AVP and AVP$_{4-8}$, and of sc AVP. This same antagonist injected sc also prevented the effect of icv AVP. These results suggest that the pressor effects of vasopressin are exerted in the periphery and can be blocked by either antagonist. They also suggest that vasopressin can elicit behavioral effects by a central action, which can be differentiated from its peripheral pressor effects, based not only on the location of the receptors, but also on their structural requirements.

These results contrast with those of Koob et al. (1981), Ettenberg (1984), and Lebrun et al. (1985), who found dPTyr(Me)AVP administered icv to be relatively ineffective in preventing the behavioral responses to peripheral AVP (de Wied et al. did not mention experiments using this antagonist icv, although they presented data using it sc). The conflict between the Californian and Utrecht groups amounts to different results with two different antagonists in two different tasks (identical experiments have not been reported). A possible resolution of this conflict could occur if the behavioral effects of sc and icv vasopressin derived from different mechanisms.

2.2.5. Endocrine Factors: Direct vs Indirect Effects

It is possible that the action of vasopressin on behavior stems from a nonspecific effect, secondary to its endocrine activity. In early work, de Wied et al. (1972) addressed this problem with the use of DG-LVP, which is practically devoid of the pressor, antidiuretic, oxytocic, and corticotropin-releasing activities of the intact molecule. Despite dramatically reduced classical endocrine activities, DG-LVP actively prolonged the retention of the pole-jump avoidance response in intact rats when injected immediately after the last acquisition trial. Moreover, the AVP catabolite AVP_{4-8}, which is about three times as potent behaviorally as AVP sc, and 300 times as potent icv, had no significant effect on blood pressure or heart rate at behaviorally active doses (de Wied et al., 1984). These results are reminiscent of the effect of ACTH on pole-jump avoidance extinction, which although distinct from that of LVP, is also unrelated to its adrenocortical actions because the fragment $ACTH_{4-10}$, which lacks adrenocorticotropic activity, can restore the normal extinction profile in posterior lobectomized animals (Bohus et al., 1973).

But, not all authors have found DG-AVP to be behaviorally potent (Ettenberg et al., 1983b; *see* Strupp and Levitsky, 1985, for a review). And, Koob et al. (1981) and Le Moal et al. (1984) consider that the behavioral activity of peripherally injected AVP is related to its pressor activity. In support of their hypothesis are the data that peripherally administered vasopressin antagonists effectively prevent the behavioral effects of peripherally administered AVP. Indeed, they showed that behaviorally active doses of AVP also produce changes in blood pressure, which are reversed by dPTyr-(Me)AVP along with the behavior. By contrast, de Wied et al. (1984) reported that AVP_{4-8} is behaviorally active at doses far below those that affect blood pressure or heart rate. But, Lebrun et al. (1985) found that icv injections of dPTyr(Me)AVP antagonized the behavioral effects of peripherally administered AVP only at doses that antagonized its pressor effects. Moreover, Koob et al. (1985) recently reported that osmotic stress, produced by injecting concentrated solutions of sodium chloride, could mimic the behavioral effect of sc AVP; and that the behavioral effect of the osmotic stress could also be antagonized by dPTyr(Me)AVP.

These latter results certainly provide strong support for the hypothesis that the behavioral activity of peripherally administered vasopressin is related to its pressor activity exerted peripherally. But, they do not exclude central behavioral activity for vasopressin.

2.2.6. Behavioral Specificity: Specific vs Nonspecific Actions

So far, we have discussed results exclusively from studies using aversively motivated tasks. To strengthen the argument for a function of vasopressin in memory, it is important to show that vasopressin acts similarly in positively (i.e., reward) motivated behavioral tasks. Unfortunately, there have only been a limited number of studies examining this question, and these have produced mixed results.

Using sexually motivated approach behavior, Bohus (1977) found that a higher percentage of male rats administered DG-LVP chose the correct arm of a T-maze to reach a receptive female. This effect was dependent on the rat receiving copulation reward. Similarly, in a black–white discriminative T-maze, in which one group was rewarded for choosing the black arm, whereas a second group was rewarded for choosing the white arm, rats showed a prolonged period of extinction of the learned response when they received LVP sc during extinction, but only when choosing the black arm was reinforced. The inhibition of extinction of the behavioral response was, however, dependent on vasopressin administration throughout the extinction period; administration limited to the acquisition period was ineffective in inhibiting extinction of the response (Hostetter et al., 1977). Thus the temporal aspects of the action of vasopressin in this task are quite different from those seen in aversively motivated behavioral tasks, and not what would be predicted from a direct action of vasopressin on memory.

Thus although vasopressin apparently affects performance in at least some positively motivated tasks, the characteristics differ from those seen in aversively motivated tasks. In fact, in one of these tasks the time course of action of vasopressin closely resembles that observed with ACTH in aversively motivated tasks. The information concerning vasopressin's effects on positively motivated tasks is far from sufficient to allow for any definitive conclusions. Future studies should examine the temporal requirements of vasopressin in a variety of positively motivated tasks.

Ettenberg et al. (1983a) studied the effects of vasopressin in one-trial food- or water-finding paradigms. In these experiments, a nondeprived animal was exposed to a novel open field containing food or water, removed, immediately injected with vasopressin, and tested subsequently in the same open field after 48 h food or water deprivation. Ettenberg et al. (1983a) reported that sc AVP (but not DG-AVP) facilitated the water-finding task. Further, sc AVP decreased the latency of 48 h food-deprived rats to find food (Ettenberg, 1984). In this study, icv dPTyr(Me)AVP prevented the effect of AVP only when administered in high doses, suggesting the possibility of a peripheral site of action for the antagonist.

One possible reason for the activity of vasopressin in aversive-. ly motivated tasks is that the administration of the peptide itself is aversive. Consistent with this idea, peripherally administered AVP has been shown to generate both taste and place avoidance responses (Ettenberg et al., 1983b). This is important because it suggests that one way in which vasopressin might improve retention of a behavioral response is by arousing the animal, perhaps because of some aversive properties. Increasing arousal could in turn facilitate retention of the task. If such a mechanism occurs, retention would be affected specifically, but the effect would not be related to a direct action on memory systems. In support of this idea, it was shown that LiCl (which possesses aversive properties) mimicked the effect of AVP in the above-mentioned water-finding task (Ettenberg 1983b).

This concept is important because an arousal mechanism could explain the behavioral activity of vasopressin in positively motivated as well as aversive tasks.

2.2.7. Conclusions on the Actions of Vasopressin

Vasopressin and its derivatives can apparently facilitate the retention of certain learned responses in a time-dependent manner. Peripherally injected AVP can display aversive properties, which may be associated with its pressor effects. The behavioral and pressor activities can be prevented by peripheral, and perhaps icv, administration of vasopressin antagonists. This activity enables vasopressin to improve the retention of animals in aversively motivated, and perhaps other, tasks.

This peripheral activity of vasopressin and its behavioral consequences do not, however, appear to explain the potency of the peptide administered icv, and probably do not explain the ability of icv administered antagonists to reverse vasopressin's behavioral effects. It seems very likely that there is also a direct effect of vasopressin in the CNS, perhaps even from peripherally injected vasopressin. This CNS action can be differentiated from its known endocrine activities with the use of analogs and selective antagonists. It is apparent in both aversive and positively reinforced tasks. There is no good evidence that icv vasopressin at behaviorally effective doses exerts peripheral pressor effects.

These results are consistent with a direct effect of icv vasopressin on memory, but they are not sufficient. The possibility remains that vasopressin acts nonspecifically in the CNS to enhance memory by, for example, altering blood flow or metabolism in critical regions of the brain. Vasopressin may thus alter the arousal or attentional state of the animal, facilitating retention of the task. It cannot yet be excluded that icv vasopressin is aversive, although the data are not consistent with such a simple interpretation. Nevertheless, the possibility remains that the behavioral effects of vasopressin both peripherally and centrally occur because it arouses the animal.

2.3. Effects of Cholecystokinin on Feeding Behavior

Cholecystokinin (CCK), one of a number of peptides first isolated from the gut, is released in response to food entering the duodenum and upper jejunum and stimulates gallbladder contraction and pancreatic enzyme secretion. As with so many other gut peptides, CCK-like peptides have been found outside the gastrointestinal tract, within various regions of the brain, including hypothalamic nuclei, limbic structures, and the cerebral cortex. The functional role of this cerebral CCK is presently unknown. Cholecystokinin is of interest for this discussion because it has been postulated to play a role in the termination of feeding.

As was the case with vasopressin, we must examine the action of this peptide on feeding carefully and objectively. We wish to know whether it affects feeding mechanisms specifically and whether the evidence supports a physiological role of endogenous CCK in the termination of a meal. Although termination of a meal is a relatively simple behavior to observe and quantify, contradictions are present in the literature, and controversy over the subject continues.

2.3.1. Feeding Behavior: Specific vs Nonspecific Effects

Maclagan (1937) first reported the possibility that an intestinal product was involved in the initiation of satiety, noting that sc injections of a crude extract of canine intestine, enterogastrone, inhibited feeding in rabbits. Subsequent studies showed that secretin and pancreatic glucagon, two peptides found in enterogastrone, lacked this ability to inhibit food intake (Schally et al., 1967). Proof that this effect of enterogastrone was caused specifically by CCK was obtained when it was shown that the synthetic C-terminal octapeptide of CCK (CCK-8), which retains all the known biological activity of the endogenous CCK, inhibited food intake in a manner similar to that of enterogastrone (Gibbs et al., 1973). This inhibition of feeding by peripherally administered CCK has been observed in the dog, rat, and monkey (Johnson and Grossman, 1970; Gibbs et al., 1976).

To test the specificity of this action, the effect of CCK on water intake and intake of a liquid diet was examined. CCK did not suppress water intake, whereas it did inhibit ingestion of the liquid diet, suggesting that CCK specifically influences feeding mechanisms (Gibbs et al., 1973). In addition, because intake of water and the liquid diet involves similar mouth and tongue movements, this result suggests that CCK does not produce a general inhibition of activity that might result from toxic side effects. Further evidence against the inhibition produced by CCK being an aversive effect derives from the observation that the peptide does not produce a taste aversion in the bait-shyness test (Gibbs et al., 1973; Holt et al., 1974).

Although CCK appears to have the ability to inhibit feeding, this does not prove that it directly influences satiety mechanisms. Cholecystokinin has many known functions, including stimulation of the secretion of glucagon, insulin, amylase, and other pancreatic proteins, in addition to the stimulation of bile flow. Most of these responses to CCK have been examined for their ability to induce satiety. Removal of the gallbladder has no effect on CCK's ability to inhibit feeding. Although glucagon has some satiety-eliciting properties, CCK appears to act independently of glucagon to inhibit feeding, because glutaramic acid inhibited satiety induced by administration of CCK-8, but had no effect on the satiety induced by glucagon administration (Collins et al., 1983). Thus, most of the known actions of CCK have been shown not to be responsible for its satiety effect.

The effectiveness of CCK in inducing satiety can be demonstrated nicely using a sham feeding system. In this experimental procedure, animals are fitted with gastric cannulae that allow the immediate removal of a liquid diet directly from the stomach. Using this preparation, all the oropharyngeal stimulation is present during feeding, although most of the signals arising from the intestine are eliminated. Animals deprived of food for 17 h will eat continuously for periods exceeding 7 h (Young et al., 1974). Thus, oropharyngeal-related information alone is not sufficient to signal satiety. It had been noted previously that rats display a specific behavioral sequence on termination of a meal, with resting occurring shortly after (Antin et al., 1975). Partially purified CCK injected ip into food-deprived, sham feeding animals is capable of eliciting the same behavioral pattern in normal satiated rats (Antin et al., 1975).

To test further whether CCK inhibits feeding through any general toxic effect, sham feeding animals were given quinine-adulterated food or an anorectic dose of amphetamine. Whereas feeding is terminated using either procedure, the animals do not exhibit the behavioral sequence characteristic of satiety (Antin et al., 1975). Therefore, termination of the meal is not a sufficient condition for the appearance of the behavioral sequence.

Additional support for CCK's role as an endogenous satiety factor is derived from its synergistic interaction with pregastric food stimulation in producing satiety (Antin et al., 1978). In sham feeding animals, after 17 h of food deprivation, CCK was more effective when given 6 min before or after feeding began than when administered 12 min before. It was most effective when given 12 min after feeding began. This interaction between CCK and pregastric stimulation is very similar to that seen with the satiety-eliciting duodenal infusions of liquid food in sham feeding rats, a procedure shown to stimulate the release of endogenous CCK.

2.3.2. Peripheral Actions

That exogenous CCK influences satiety through peripheral receptors in the rat is supported by the observation that bilateral abdominal vagotomy abolishes the inhibitory action of CCK-8 on feeding (Smith et al., 1981; Morley et al., 1982; Lorenz and Goldman, 1982). Specifically, the gastric branch of the vagus nerve is necessary; transections of the hepatic or celiac branches have no effect on CCK's activity (Smith et al., 1981). Based on pharmacological and selective lesion studies, the afferent vagal fibers are the necessary components of the gastric vagal nerve for the action of CCK (Smith et al., 1981; 1983; Kadar et al., 1984).

Recently, it was reported that ip administration of anti-CCK antibodies in either lean or obese rats increased food consumption during the 30-min period following the injection. Furthermore, the production of endogenous anti-CCK antibodies resulted in an increase in both food intake and weight gain over a 3 mo period in lean, but not in obese, rats (McLaughlin et al., 1985). These results, if confirmed, provide strong support for a peripheral role of CCK in mediating satiety.

If the action of CCK in the rat is mediated through signals sent from the periphery, it is unclear where this information is processed in the brain. Midbrain transection blocks the satiety effect of CCK, suggesting that areas rostral to the midbrain are involved (Crawley et al., 1984). One area to which such signals might be expected to be relayed is the hypothalamus, considered to be a major structure for both the initiation and termination of feeding. Evidence for the involvement of this structure in the effect of CCK on feeding is far from conclusive. Intraperitoneal injection of CCK has been reported to stimulate the release of NE from the anterior hypothalamic area and the ventromedial nucleus of the hypothalamus (VMH) (Myers and McCaleb, 1981). In a more comprehensive examination of the neurochemical changes associated with peripherally administered CCK, the NE and DA content of the hypothalamus were decreased (Kadar et al., 1984). Thus, catecholamine systems in the hypothalamus may be involved in the satiety effect of CCK. Lesions of the VMH, however, the area typically believed to play a major role in the mediation of satiety, did not eliminate the ability of ip CCK to inhibit feeding in the rat (Kulkosky and Breckenridge, 1976; Smith et al., 1981). In contrast, another study showed that bilateral lesions of the VMH abolished the inhibition of food intake by cerulein, a peptide similar to CCK in structure and gastrointestinal activity (Anika et al., 1977).

2.3.3. Central Actions

Although the evidence for a peripheral site of action of CCK on feeding is fairly convincing, at least in some species, there is also evidence suggesting that centrally administered CCK can inhibit feeding. In the rat, the species in which most studies concerning CCK and feeding have been conducted, icv-injected CCK inhibits tail-pinch-induced eating (Antelman and Szechtma, 1975). Whereas in one study a fairly large dose of CCK-8 (3 μg) was required to obtain this result (Nemeroff et al., 1978), subsequent studies have replicated the results using much lower doses (e.g., 250 ng) (Morley, 1982). Also, low doses (75–150 ng) of CCK-8 injected directly into hypothalamic structures inhibited eating stimulated by NE microinjected into the hypothalamus (Myers and McCaleb, 1981).

A problem with studies of this nature is that the feeding is stimulated by nonphysiological mechanisms. Thus it is difficult to determine how relevant the subsequent inhibition of feeding is to normal feeding. Tail-pinch can be considered a stressor, therefore CCK might attenuate feeding by influencing systems related to the stress perception or response, rather than those systems specifically involved in feeding. The fact that tail-pinch-induced chewing was not attenuated by icv CCK, however, unlike the amount of food injested, suggests that the effect of CCK was specific for feeding (Nemeroff et al., 1978).

Another difficulty in interpreting these reports is that adequate descriptions of the animals' behavioral responses were not given. If CCK elicits a normal state of satiety, than the animal would be expected to exhibit behaviors normally associated with satiety. Tail-pinch, however, activates multiple systems so that the animal exhibits a number of responses, some of which might compete with the behaviors associated with satiety. This limits the use of such experimental paradigms for a behavioral analysis of the role of CCK in feeding.

Another study examined the effect of CCK (ip or icv) on food intake as measured by an operant response (Maddison, 1977). Cholecystokinin administered icv produced a dose-related decrease in food-rewarded lever pressing. This effect was observed longer after injection than with ip administration, although this might be explained by the time taken for the peptide to diffuse to its site of action. Again, no mention of satiety-associated behaviors was made, making interpretation of these results difficult. In the one study that used a normal feeding paradigm, extremely large doses of CCK were needed to significantly inhibit the food intake in 17-h food-deprived rats (Lorenz and Goldman, 1982). Future studies will need to examine more closely whether intrahypothalamic injections of reasonable doses of CCK can elicit satiety in a normally feeding rat before the question of whether CCK has any central effects in the rat can be answered more definitively.

Currently available data suggest that CCK is most effective in the rat when given peripherally, but there appear to be species differences in this respect. In the sheep, continuous administration of picomole quantities of CCK-8 into the cerebral ventricles inhibits feeding (Della-Fera and Baile, 1979). Further, icv injection of antibodies to CCK stimulates feeding in sheep, but not in rats (Della-Fera et al., 1981). Administration of similar quantities of icv CCK-8 into the ventricles of pigs also inhibited feeding in 17-h food-deprived animals, without affecting drinking in water-deprived animals. In contrast, when the highest dose of CCK used in this study was injected peripherally, there was no effect on food intake (Parrot and Baldwin, 1981). Finally, inhibition of feeding using picomolar quantities of CCK-8 administered icv was also observed in the chick (Denbow and Myers, 1982).

2.3.4. Conclusions on the Actions of CCK

It is clear that exogenous CCK possesses satiety-eliciting properties distinct from its gastrointestinal activities. The site of action appears to differ depending on the species. Thus, in rats, CCK is most effective when administered peripherally, where it probably stimulates gastric vagal afferents. Whereas in the sheep, pig, and chick the peptide is apparently capable of acting centrally.

Although exogenous CCK can induce satiety, this does not prove that endogenous CCK regulates satiety in the normally feeding animal. Present difficulties in the accurate measurement of circulating CCK preclude answering this question. Some studies have shown that agents known to stimulate the secretion of CCK also inhibit feeding, suggesting that endogenous CCK can act similarly to exogenous CCK. This result, however, is insufficient to indicate whether CCK is part of the normal mechanism of termination of a meal. One would like to show that the removal of circulating CCK results in a deficiency of the animals' ability to terminate feeding and that its replacement can reinstate the behavior. As mentioned above, antibodies to CCK increased food intake in rats when injected peripherally, and in sheep when injected into the lateral ventricles. Thus these latter results provide support for a role of endogenous CCK in the regulation of feeding behavior, but they are not yet conclusive.

Acknowledgments

The authors' research cited in this chapter was supported by the US National Institute of Mental Health (1R01-MH-25486). We gratefully acknowledge the clerical assistance of Phyllis Stambaugh.

References

Ader R. and de Wied D. (1972) Effects of lysine vasopressin on passive avoidance learning. *Psychonom. Sci.* **29,** 46–48.

Ader R., Weijnen J. A. W. M., and Moleman P. (1972) Retention of a passive avoidance response as a function of the intensity and duration of electric shock. *Psychonom. Sci.* **26,** 125–128.

Anika S. M., Houpt T. R., and Houpt K. A. (1977) Satiety elicited by cholecystokinin in intact and vagotomized rats. *Physiol. Behav.* **19,** 761–766.

Antelman S. M. and Szechtman H. (1975) Tail pinch induces eating in sated rats which appears to depend on nigrostriatal dopamine. *Science* **189,** 731–733.

Antin J., Gibbs J., Holt J., Young R. C., and Smith G. P. (1975) Cholecystokinin elicits the complete behavioral sequence of satiety in rats. *J. Comp. Physiol. Psychol.* **89,** 784–790.

Antin J., Gibbs J., and Smith G. P. (1978) Cholecystokinin interacts with pregastric food stimulation to elicit satiety in the rat. *Physiol. Behav.* **20,** 67–70.

Banks W. A. and Kastin A. J. (1985) Permeability of the blood–brain barrier to neuropeptides: The case for penetration. *Psychoneuroendocrinology* **10,** 385–399.

Berridge M. J. (1985) The molecular basis of communications within the cell. *Sci. Am.* **253,** 142–152.

Bohus B. (1977) Effect of desglycinamide-lysine vasopressin (DG-LVP) on sexually motivated T-maze behavior of the male rat. *Horm. Behav.* **8,** 52–61.

Bohus B., Gispen W. H., and de Wied D. (1973) Effect of lysine vasopressin and $ACTH_{4-10}$ on conditioned avoidance behavior of hypophysectomized rats. *Neuroendocrinology* **11,** 137–143.

Bohus B., Kovacs G. L., and de Wied D. (1978) Oxytocin, vasopressin and memory: opposite effects on consolidation and retrieval processes. *Brain Res.* **157,** 414–417.

Brownfield M. S. and Koslowski G. P. (1977) The hypothalamo-choroidal tract. I. Immunohistochemical demonstration of neurophysin pathways to telencephalic choroid plexuses and cerebrospinal fluid. *Cell Tissue Res.* **178,** 111–127.

Cherkin A., Eckardt M. J., and Gerbrandt L. K. (1976) Memory: Proline induces retrograde amnesia in chicks. *Science* **193,** 242–244.

Colbern D., Isaacson R. L., Bohus B., and Gispen W. H. (1977) Limbic-midbrain lesions and ACTH-induced excessive grooming. *Life Sci.* **21,** 393–402.

Colbern D. L., Isaacson R. L., Green E. J., and Gispen W. H. (1978) Repeated intraventricular injections of ACTH1-24: The effects of home or novel environments on excessive grooming. *Behav. Biol.* **23,** 381–387.

Collins S., Walker D., Forsyth P., and Belbeck L. (1983) The effects of proglumide on cholecystokinin-, bombesin-, and glucagon-induced satiety in the rat. *Life Sci.* **32,** 2223–2229.

Cornett L. E. and Dorsa D. M. (1985) Vasopressin receptor subtypes in dorsal hindbrain and renal medulla. *Peptides* 6, 85–89.

Cornford E. M., Braun L. D., Crane P. D., and Oldendorf W. H. (1978) Blood–brain barrier restriction of peptides and the low uptake of enkephalins. *Endocrinology* 103, 1297–1303.

Crawley J. N., Kiss J. Z., and Mezey E. (1984) Bilateral midbrain transections block the behavioral effects of cholecystokinin on feeding and exploration in rats. *Brain Res.* 322, 316–321.

Davson H. (1972) The Blood–Brain Barrier, in *Structure and Function of Nervous Tissue* vol. 4 (Bourne G. H., ed.), Academic, New York.

Delanoy R. L., Dunn A. J., and Tintner R. (1978) Behavioral responses to intracerebroventricularly administered neurohypophyseal peptides in mice. *Horm. Behav.* 11, 348–362.

Della-Fera M. A. and Baile C. A. (1979) Cholecystokinin octapeptide: Continuous picomole injections into the cerebral ventricles of sheep suppress feeding. *Science* 206, 471–473.

Della-Fera M. A., Baile C. A., Schneider B. S., and Grinker J. A. (1981) Cholecystokinin antibody injected in cerebral ventricles stimulates feeding in sheep. *Science* 212, 687–689.

Denbow D. M. and Myers R. D. (1982) Eating, drinking and temperature responses to intracerebroventricular cholecystokinin in the chick. *Peptides* 3, 739–743.

De Souza E. B., Perrin M. H., Insel T. R., Rivier J., Vale W. W., and Kuhar M. J. (1984) Corticotropin-releasing factor receptors in rat forebrain: Autoradiographic identification. *Science* 224, 1449–1451.

Deviche P. and Delius J. D. (1981) Short-term modulation of domestic pigeon *(Columba livia L.)* behaviour induced by intraventricular administration of ACTH. *Z. Tierpsychol.* 55, 335–342.

de Wied D. (1964) Influence of anterior pituitary on avoidance learning and escape behavior. *Am. J. Physiol.* 207, 255–259.

de Wied D. (1965) The influence of the posterior and intermediate lobe of the pituitary and pituitary peptides on the maintenance of a conditioned avoidance response in rats. *Int. J. Neuropharmacol.* 4, 157–167.

de Wied D. (1966) Inhibitory effect of ACTH and related peptides on extinction of conditioned avoidance behavior in rats. *Proc. Soc. Exp. Biol. Med.* 122, 28–31.

de Wied D. and Bohus B. (1966) Long term and short term effects on retention of a conditioned avoidance response in rats by treatment with long acting pitressin and α-MSH. *Nature* 212, 1484–1486.

de Wied D., Greven H. M., Lande S., and Witter A. (1972) Dissociation of the behavioral and endocrine effects of lysine vasopressin by tryptic digestion. *Br. J. Pharmacol.* 45, 118–122.

de Wied D., Witter A., and Greven H. M. (1975) Behaviorally active ACTH analogues. *Biochem. Pharmacol.* **24**, 1463–1468.

de Wied D., Gaffori O., van Ree J. M., and de Jong W. (1984) Vasopressin antagonists block peripheral as well as central vasopressin receptors. *Pharmacol. Biochem. Behav.* **21**, 393–400.

Dismukes R. K. (1979) New concepts of molecular communication among neurons. *Brain Behav. Sci.* **2**, 409–448.

Drago F., Canonico P. L., Bitetti R., and Scapagnini U. (1980) Systemic and intraventricular prolactin induces excessive grooming. *Eur. J. Pharmacol.* **65**, 457–458.

Dunn A. J. (1978) Peptides and behavior: A critical analysis of research strategies. *Neurosci. Res. Progr. Bull.* **16**, 554–555.

Dunn A. J. (1984) Effects of ACTH, β-Lipotropin and Related Peptides on the Central Nervous System, in *Peptides, Hormones and Behavior: Molecular and Behavioral Neuroendocrinology* (Nemeroff C. B. and Dunn A. J., eds.), Spectrum, New York.

Dunn A. J. (1987) Studies on the neurochemical mechanisms and significance of ACTH-induced grooming. *Ann. NY Acad. Sci.*, in press.

Dunn A. J., and Hurd R. W. (1986) ACTH acts via an anterior third ventricular site to elicit grooming behavior. *Peptides* **7**, 651–657.

Dunn, A. J. (1978) Peptides and behavior: A critical analysis of research strategies. *Neurosci. Res. Prog. Bull.* **16**, 554–555.

Dunn A. J. and Vigle G. (1985) ACTH-induced grooming behavior involves cerebral cholinergic neurons and muscarinic receptors. *Neuropharmacology* **24**, 329–331.

Dunn A. J., Green E. J., and Isaacson R. L. (1979) Intracerebral adrenocorticotropic hormone mediates novelty-induced grooming in the rat. *Science* **203**, 281–283.

Dunn A. J., Steelman S., and Delanoy R. (1980) Intraventricular ACTH and vasopressin cause regionally specific changes in cerebral deoxyglucose uptake. *J. Neurosci. Res.* **5**, 485–495.

Dunn A. J., Guild A. L., Kramarcy N. R., and Ware M. D. (1981a) Benzodiazepines decrease grooming in response to novelty but not ACTH or β-endorphin. *Pharmacol. Biochem. Behav.* **15**, 605–608.

Dunn A. J., Childers S. R., Kramarcy N. R., and Villiger J. W. (1981b) ACTH-induced grooming involves high-affinity opiate receptors. *Behav. Neural Biol.* **31**, 105–109.

Dunn A. J., Alpert J. E., and Iversen S. D. (1984) Dopamine denervation of frontal cortex or nucleus accumbens does not affect ACTH-induced grooming behaviour. *Behav. Brain Res.* **12**, 307–315.

Dunn A. J., Webster E. W., and Nemeroff C. B. (1985) Neonatal treatment with monosodium glutamate does not alter grooming behavior induced by novelty or adrenocorticotropic hormone. *Behav. Neural Biol.* **44,** 80–89.

Ehrenpreis S., Comaty J. E., and Myles S. B. (1978) Naloxone reversible analgesia produced by D-phenylalanine in mice. *Soc. Neurosci. Abstr.* **4,** 459.

Ettenberg A., Le Moal M., Koob G. F., and Bloom F. E. (1983a) Vasopressin potentiation in the performance of a learned appetitive task: Reversal by a pressor antagonist analog of vasopressin. *Pharmacol. Biochem. Behav.* **18,** 645–647.

Ettenberg A., van der Kooy D., Le Moal M., Koob G. F., and Bloom F. E. (1983b) Can aversive properties of (peripherally injected) vasopressin account for its putative role in memory? *Behav. Brain Res.* **7,** 331–350.

Ettenberg A. (1984) Intracerebroventricular application of a vasopressin antagonist peptide prevents the behavioral actions of vasopressin. *Behav. Brain Res.* **14,** 201–211.

Farese R. V. (1983) Phosphoinositide metabolism and hormone action. *Endocr. Rev.* **4,** 78–95.

Ferrari W., Gessa G. L., and Vargiu L. (1963) Behavioral effects induced by intracisternally injected ACTH and MSH. *Ann. NY Acad. Sci.* **104,** 330–345.

Fitzsimons J. T., Epstein A. N., and Johnson A. K. (1977) The Peptide Specificity of Receptors for Angiotensin-Induced Thirst, in *Central Actions of Angiotensin* (Buckley J. P. and Ferrario C., eds.), Pergamon, New York.

Gessa G. L., Pisano M., Vargiu L., Crabai F., and Ferrari W. (1967) Stretching, and yawning movements after intracerebral injection of ACTH. *Rev. Can. Biol.* **26,** 229–236.

Gibbs J., Young R. C., and Smith G. P. (1973) Cholecystokinin decreases food intake in rats. *J. Comp. Physiol. Psychol.* **84,** 488–495.

Gibbs J., Falasco J. D., and Mc Hugh P. R. (1976) Cholecystokinin-decreased food intake in rhesus monkeys. *Am. J. Physiol.* **230,** 15–18.

Gispen W. H. and Isaacson R. L. (1981) ACTH-induced excessive grooming in the rat. *Pharmacol. Ther.* **12,** 209–246.

Gispen W. H. and Wiegant V. M. (1976) Opiate antagonists suppress ACTH$_{1-24}$-induced excessive grooming in the rat. *Neurosci. Lett.* **2,** 159–164.

Gispen W. H., Wiegant V. M., Greven H. M., and de Wied D. (1975) The induction of excessive grooming in the rat by intraventricular application of peptides derived from ACTH: Structure–activity studies. *Life Sci.* **17,** 645–652.

Gispen W. H., Wiegant V. M., Bradbury A. F., Hulme E. C., Smyth D. G.,

Snell C. R., and de Wied D. (1976) Induction of excessive grooming in the rat by fragments of lipotropin. *Nature* **264**, 794–795.

Gispen W. H., Brakkee J. H., and Isaacson R. L. (1980) Hypophysectomy and novelty-induced grooming in the rat. *Behav. Neural Biol.* **29**, 481–486.

Gmerek D. E. and Cowan A. (1983) Studies on bombesin-induced grooming in rats. *Peptides* **4**, 907–913

Goldman H., Murphy S., Schneider D. R., and Felt B. T. (1979) Cerebral blood flow after treatment with ORG-2766, a potent analog of ACTH4-9. *Pharmacol. Biochem. Behav.* **10**, 883–887.

Green E. J., Isaacson R. L., Dunn A. J., and Lanthorn T. H. (1979) Naloxone and haloperidol reduce grooming occurring as an aftereffect of novelty. *Behav. Neural Biol.* **27**, 546–551.

Greenough W. T. (1976) Enduring Brain Effects of Differential Experience and Training, in *Neural Mechanisms of Learning and Memory* (Rosenzweig M. R. and Bennett E. L., eds.), MIT Press, Cambridge, Massachusetts.

Greven H. M. and de Wied D. (1977) Influence of peptides structurally related to ACTH and MSH on active avoidance behavior in rats. *Front. Horm. Res.* **4**, 140–152.

Guild A. L. and Dunn A. J. (1982) Dopamine involvement in ACTH-induced grooming behavior. *Pharmacol. Biochem. Behav.* **17**, 31–36.

Herz A., Albust K., Matys J., Schubart P., and Taschenachi H. S. (1970) On the central sites for the antinociceptive action of morphine and fetanyl. *Neuropharmacology* **9**, 539–551.

Hoffman W. E. and Phillips M. I. (1976) Regional study of cerebral ventricle sensitive sites to angiotensin II. *Brain Res.* **110**, 313–330.

Hökfelt T., Johansson O., Ljungdahl A., Lundberg J. M., and Schultzberg M. (1980) Peptidergic neurones. *Nature* **284**, 515–521.

Holt J., Antin J., Gibbs J., Young R. C., and Smith G. P. (1974) Cholecystokinin does not produce bait shyness in rats. *Physiol. Behav.* **12**, 497–498.

Hostetter G., Jubb S. L., and Kozlowski G. P. (1977) Vasopressin affects the behavior of rats in a positively-rewarded discrimination task. *Life Sci.* **21**, 1323–1328.

Isaacson R. L. and Green E. J. (1978) The effect of $ACTH_{1-24}$ on locomotion, exploration, rearing and grooming. *Behav. Biol.* **24**, 118–122.

Johnson L. R. and Grossman M. I. (1970) Analysis of inhibition of acid secretion by cholecystokinin in dogs. *Am. J. Physiol.* **218**, 550–554.

Jolles J., Wiegant V. M., and Gispen W. H. (1978) Reduced behavioral effectiveness of $ACTH_{1-24}$ after a second administration: Interaction with opiates. *Neurosci. Lett.* **9**, 261–266.

Jolles J., Rompa-Barendregt J., and Gispen W. H. (1979) Novelty and grooming behavior in the rat. *Behav. Neural Biol.* **25**, 563–572.

Kadar T., Varszegi M., Sudakov S. K., Penke B., and Telegdy G. (1984) Changes in brain monoamine levels of rats during cholecystokinin octapeptide-induced suppression of feeding. *Pharmacol. Biochem. Behav.* **21**, 339–344.

Katz R. J. (1979) Central injection of substance P elicits grooming behavior and motor inhibition in mice. *Neurosci. Lett.* **12**, 133–136.

Katz R. (1980) Grooming elicited by intracerebroventricular bombesin and eledoisin in the mouse. *Neuropharmacology* **19**, 143–146.

Katz R. J. and Roth K. A. (1979) Stress-induced grooming in the rat—an endorphin mediated syndrome. *Neurosci. Lett.* **13**, 209–212.

Kety S. S. (1970) The Biogenic Amines in the Central Nervous System: Their Possible Roles in Arousal, Emotion and Learning, in *The Neurosciences: Second Study Program* (Schmitt F. O., ed.), Rockefeller University Press, New York.

Koob G. F., Le Moal M., Gaffori O., Manning M., Sawyer W. H., Rivier J., and Bloom F. E. (1981) Arginine vasopressin and a vasopressin antagonist peptide: opposite effects on extinction of active avoidance in rats. *Regul. Peptides* **2**, 153–163.

Koob G. F., Dantzer R., Rodriguez F., Bloom F. E., and Le Moal M. (1985) Osmotic stress mimics effects of vasopressin on learned behaviour. *Nature* **316**, 750–752.

Krieger D. T. and Liotta A. S. (1979) Pituitary hormones in brain: Where, how and why? *Science* **205**, 366–372.

Kuffler S. W. and Yoshikami D. (1975) The number of transmitter molecules in a quantum: An estimate from iontophoretic application of acetylcholine at the neuromuscular synapse. *J. Physiol.* **251**, 465–482.

Kuhar M. J. and Yamamura H. I. (1975) Light autoradiographic localization of cholinergic muscarinic receptors in rat brain by specific binding of a potent antagonist. *Nature* **253**, 560–561.

Kulkosky P. J. and Breckenridge C. (1976) Satiety elicited by the C-terminal octapeptide of cholecystokinin pancreozymin in normal and VMH-lesioned rats. *Behav. Biol.* **18**, 227–234.

LaHoste G. J., Olson G. A., Kastin A. J., and Olson R. D. (1980) Behavioral effects of melanocyte stimulation hormone. *Neurosci. Biobehav. Rev.* **4**, 9–16.

Landas S., Phillips M. I., Stamler J. F., and Raizada M. K. (1980) Visualization of specific angiotensin II binding sites in the brain by fluorescent microscopy. *Science* **210**, 791–793.

Lebrun C., Le Moal M., Koob G. F., and Bloom F. E. (1985) Vasopressin pressor antagonist injected centrally reverses behavioral effects of peripheral injection of vasopressin, but only at doses that reverse increase in blood pressure. *Regul. Peptides* **11,** 173–181.

Le Moal M., Koob G. F., Koda L. Y., Bloom F. E., Manning M., Sawyer W. H., and Rivier J. (1981) Vasopressor receptor antagonist prevents behavioural effects of vasopressin. *Nature* **291,** 491–493

Lewis R. E. and Phillips M. I. (1984) Localization of the central pressor action of bradykinin to the cerebral third ventricle. *Am. J. Physiol.* **247,** R63–R68.

Liotta A. S., Li C. H., Schussler G. C., and Krieger D. T. (1978) Comparative metabolic clearance rate, volume of distribution and plasma half-life of human β-lipotropin and ACTH. *Life Sci.* **23,** 2323–2330.

Lorenz D. N. and Goldman S. A. (1982) Vagal mediation of the cholecystokinin satiety effect in rats. *Physiol. Behav.* **29,** 599–604.

Maclagan N. F. (1937) The role of appetite in the control of body weight. *J. Physiol.* **90,** 385–394.

Maddison S. (1977) Intraperitoneal and intracranial cholecystokinin depress operant responding for food. *Physiol. Behav.* **19,** 819–824.

McLaughlin C. L., Baile C. A., and Buonomo F. C. (1985) Effect of CCK antibodies on food intake and weight gain in Zucker rats. *Physiol. Behav.* **34,** 277–282.

Meisenberg G. and Simmons W. H. (1983) Centrally mediated effects of neurohypophyseal hormones. *Neurosci. Biobehav. Rev.* **7,** 263–280.

Morley J. E. (1982) The ascent of cholecystokinin (CCK)—from gut to brain. *Life Sci.* **30,** 479–493.

Morley J. and Levine A. S. (1982) Corticotropin releasing factor, grooming and ingestive behavior. *Life Sci.* **31,** 1459–1464.

Morley J. E., Levine A. S., Kneip J., and Grace M. (1982) The effect of vagotomy on the satiety effects of neuropeptides and naloxone. *Life Sci.* **30,** 1943–1947.

Myers R. D. and McCaleb M. L. (1981) Peripheral and intrahypothalamic cholecystokinin act on the noradrenergic "feeding circuit" in the rat's diencephalon. *Neuroscience* **6,** 645–655.

Nemeroff C. B., Osbahr A. J., Bissette G., Jahnke G., Lipton M. A., and Prange A. J. (1978) Cholecystokinin inhibits tail pinch-induced eating in rats. *Science* **200,** 793–794.

Parrott R. F. and Baldwin B. A. (1981) Operant feeding and drinking in pigs following intracerebroventricular injection of synthetic cholecystokinin octapeptide. *Physiol. Behav.* **26,** 419–422.

Passaro E., Debas H., Oldendorf W., and Yamada T. (1982) Rapid appearance of intraventricularly administered neuropeptides in the peripheral circulation. *Brain Res.* **241,** 335–340.

Pedersen C. A. and Prange A. J. (1979) Induction of maternal behavior in virgin rats after intracerebroventricular administration of oxytocin. *Proc. Natl. Acad. Sci. USA* **76**, 6661–6665.

Phillips M. I. (1984) Angiotensin and Drinking: A Model for the Study of Peptide Action in the Brain, in *Peptides, Hormones, and Behavior* (Nemeroff C. B. and Dunn A. J., eds.), Spectrum, New York.

Prusiner S., Doak C. W., and Kirk G. (1976) A novel mechanism for group translocation: Substrate–product reutilization by γ-glutamyl transpeptidase in peptide and amino acid transport. *J. Cell. Physiol.* **89**, 853–863.

Rapoport S. I., Klee W. A., Pettigrew K. D., and Ohno K. (1980) Entry of opioid peptides into the central nervous system. *Science* **207**, 84–86.

Rees H. D., Dunn A. J., and Iuvone P. M. (1976) Behavioral and biochemical responses of mice to the intraventricular administration of ACTH analogs and lysine vasopressin. *Life Sci.* **18**, 1333–1340.

Rioux F., Park W. K., and Regoli D. (1973) Pharmacology of angiotensin antagonists. *Can J. Physiol. Pharmacol.* **51**, 108–113.

Rivier J., Rivier C., and Vale W. (1984) Synthetic competitive antagonists of corticotropin-releasing factor: Effects on ACTH secretion in the rat. *Science* **224**, 889–891.

Rodriguez E. M. (1984) Design and Perspectives of Peptide Secreting Neurons, in *Peptides, Hormones, and Behavior* (Nemeroff C. B. and Dunn A. J., eds.), Spectrum, New York.

Ryan J. P. and Isaacson R. L. (1983) Intra-accumbens injections of ACTH induce excessive grooming in rats. *Physiol. Psychol.* **11**, 54–58.

Sandman C. A. and Kastin A. J. (1981) The influence of fragments of the LPH chains on learning, memory and attention in animals and man. *Pharmacol. Ther.* **13**, 39–60.

Sawyer W. H. and Manning M. (1985) The use of antagonists of vasopressin in studies of its physiological functions. *Fed. Proc.* **44**, 78–80.

Schally A. V., Redding T. W., Lucien H. W., and Meyer J. (1967) Enterogastrone inhibits eating by fasted mice. *Science* **157**, 210–211.

Scheibel M. E. and Scheibel A. B. (1974) Does the nucleus raphe pontis have chemosensor or neuroendocrine functions? *Neurosci. Abst.* Abstracts Society for Neuroscience, 4th Annual Meeting, 409.

Scott D. E., Krobisch-Dudley G., Paull W. K., and Kozlowski G. P. (1977) The ventricular system in neuroendocrine mechanisms. III. Supraependymal neuronal networks in the primate brain. *Cell Tiss. Res.* **179**, 235–254.

Shashoua V. E. (1973) Seasonal changes in the learning and activity patterns of goldfish. *Science* **181**, 572–574.

Smith G. P., Jerome C., Cushin B. J., Eterno R., and Simansky K. J. (1981) Abdominal vagotomy blocks the satiety effect of cholecystokinin in the rat. *Science* **213**, 1036–1037.

Smith G. P., Jerome C., and Norgren R. (1983) Vagal afferent axons mediate the satiety effect of CCK-8. *Soc. Neurosci. Abst.* **9**, 902.

Sokoloff L. (1981) The relationship between function and energy metabolism: Its use in the localization of functional activity in the nervous system. *Neurosci. Res. Progr. Bull.* **19**, 159–210.

Strupp B. J. and Levitsky D. A. (1985) A mnemonic role for vasopressin: The evidence for and against. *Neurosci. Biobehav. Rev.* **9**, 399–411.

Sutcliffe J. G., Milner R. J., Shinnick T. M., and Bloom F. E. (1983) Identifying the protein products of brain-specific genes with antibodies to chemically synthesized peptides. *Cell* **33**, 671–682.

Tanaka K., Nicholson W. E., and Orth D. N. (1978) Diurnal rhythm and disappearance half-time of endogenous plasma immunoreactive β-MSH (LPH) and ACTH in man. *J. Clin. Endocrinol. Metab.* **46**, 883–890.

Ungar G. (1973) Evidence for Molecular Coding of Neural Information, in *Memory and Transfer of Information* Plenum, New York.

van Nispen J. W., Tesser G. I., Barthe P. L., Maier R., and Schenkel-Hulliger L. (1977) Biological activities of ACTH-analogues varied in the active site. *Acta Endocrinol.* **84**, 470–484.

van Wimersma Greidanus T. B. and de Wied D. (1971) Effects of systemic and intracerebral administration of two opposite acting ACTH-related peptides on extinction of conditioned avoidance behavior. *Neuroendocrinology* **7**, 291–301.

van Wimersma Greidanus T. B., Dogterom J., and de Wied D. (1975) Intraventricular administration of anti-vasopressin serum inhibits memory consolidation in rats. *Life Sci.* **16**, 637–643.

van Wimersma Greidanus T. B., van Dijk A. M. A., de Rotte A. A., Goedermans J. H. J., Croiset G., and Thody A. J. (1978) Involvement of ACTH and MSH in active and passive avoidance behavior. *Brain Res. Bull.* **3**, 227–230.

van Wimersma Greidanus T. B., Donker D. K., Van Zinnicq Bergmann F. F. M., Bekenkamp R., Maigret C., and Spruijt B. (1985) Comparison between excessive grooming induced by bombesin or by ACTH: The differential elements of grooming and development of tolerance. *Peptides* **6**, 369–372.

Versteeg D. H. G. (1980) Interaction of peptides related to ACTH, MSH and β-LPH with neurotransmitters in the brain. *Pharmacol. Ther.* **11**, 535–557.

Weindl A. (1973) Neuroendocrine Aspects of Circumventricular Organs, in *Frontiers in Neuroendocrinology* (Ganong W. F. and Martini L., eds.), Oxford University, New York.

Wiegant V. M. and Gispen W. H. (1977) ACTH-induced excessive grooming in the rat: Latent activity of $ACTH_{4-10}$. *Behav. Biol.* **19**, 554–558.

Wiegant V. M., Gispen W. H., Terenius L., and de Wied D. (1977a) ACTH-like peptides and morphine: Interaction at the level of the CNS. *Psychoneuroendocrinology* **2**, 63–70.

Wiegant V. M., Cools A. R., and Gispen W. H. (1977b) ACTH-induced excessive grooming involves brain dopamine. *Eur. J. Pharmacol.* **41**, 343–345.

Wiegant V. M., Jolles J., Colbern D. L., Zimmerman E., and Gispen W. H. (1979) Intracerebroventricular ACTH activates the pituitary-adrenal system: Dissociation from a behavioral response. *Life Sci.* **25**, 1791–1796.

Williams N. S. and Scampoli D. L. (1984) Handling, ACTH, $ACTH_{1-24}$, and naloxone effects on preening behavior in domestic chickens. *Pharmacol. Biochem. Behav.* **20**, 681–682.

Witter A., Greven H. M., and de Wied D. (1975) Correlation between structure, behavioral activity and rate of biotransformation of some ACTH $_{4-9}$ analogs. *J. Pharmacol. Exp. Ther.* **193**, 853–860.

Young R. C., Gibbs J., Antin J., Holt J., and Smith G. P. (1974) Absence of satiety during sham feeding in the rat. *J. Comp. Physiol. Psych.* **87**, 795–800.

Zwiers H., Aloyo V. J., and Gispen W. H. (1981) Behavioral and neurochemical effects of the new opioid peptide dynorphin-(1-13): Comparison with other neuropeptides. *Life Sci.* **28**, 2545–2551.

Integrative Physiological Studies of Peptides in the Central Nervous System

A. M. Naylor, W. D. Ruwe, and W. L. Veale

1. Introduction

Numerous peptides have been localized with immunocyto-chemical techniques in neuronal cell bodies, axons, and nerve endings of the central nervous system (CNS). The widespread distribution of peptides within the brain and spinal cord implies that these substances may function under physiological conditions as neurotransmitters or neuromodulators. Consequently, a wide variety of techniques for administration to, and recovery from, the CNS for analysis has been utilized to investigate their putative physiological roles.

The use of conscious and behaving animals for such studies enables the determination of peptide actions while physiological control systems are functioning and interacting. Furthermore, experiments concerning the physiological release of peptides may be correlated with certain behaviors or alterations in homeostasis. In addition, an intact blood supply provides the required nutrients for cellular activity and a means of transporting byproducts of metabolism away for further processing. The conscious animal, therefore, serves as a model for investigating (1) how the brain controls and integrates physiological responses and (2) the putative roles of peptides in CNS control mechanisms. The use of isolated neural tissue for investigating peptide actions in the CNS is covered by other chapters in this volume. In this chapter, the aim is to assess critically the methodologies that are used currently to examine the central actions of peptides in conscious animals and to describe the kind of information that can be obtained from studies that utilize these techniques.

2. Critical Analysis of Some Techniques Used for an Integrative Physiological Approach in Conscious Animals

The majority of current techniques used in unanesthetized animals for investigating the physiological roles of centrally administered and released peptides requires a surgical procedure involving the stereotaxic implantation of chronic guide cannulae. These guide cannulae may be above either a cerebral ventricle, a tissue locus, or both. Prior surgical implantation means that during experimentation the animals are free of anesthetic effects—a feature that is of importance when evaluating the physiological and behavioral effects of putative neurotransmitters or neuromodulators.

2.1. Introduction of Peptides Into the Central Nervous System

Demonstration within the CNS of peptides that have biologic activity and are found in the gut or pituitary gland has initiated investigations into their putative physiological roles within central neuronal networks. Methods that have been used commonly to administer chemical substances into the brain include infusion into the ventricular system and direct microinjection into brain tissue. In addition, the push–pull perfusion technique has been utilized since it offers some decided advantages over bolus microinjection methods; also, the returned perfusate can be analyzed for endogenously released neurochemicals. The following sections will assess critically the methodologies currently used to investigate the physiological roles of some peptides in the conscious and behaving animal. Specifically, the relative advantages and disadvantages of the various techniques will be discussed in relation to the information that can be derived from studies in which these methodologies have been employed.

2.1.1. Infusion of Peptides Into the Ventricular System

Injection or infusion of putative neurotransmitters into the fluid-filled spaces of the brain affords a means whereby a first approximation of the effect of a substance can be determined.

When a peptide is introduced into the ventricular system of the brain, it is not possible to localize neuroanatomically a specific site of action. Indeed, it will not be apparent whether the observed effects are caused by an action of the peptide on one specific locus or on several different brain loci. The reason for this is that intracerebroventricularly administered peptides reach many, if not all, neuroanatomical structures lining the ventricles, as well as those on the outside surface of the brain that are bathed by cerebrospinal fluid (CSF). The ventricular route of administration is used frequently to examine the effects of yet uninvestigated peptides on thermoregulatory responses, among others, since the changes produced are relatively easy to measure. Indeed, recent reviews (Ruwe et al., 1983; Clark and Lipton, 1983) cite numerous studies on the thermoregulatory effects of putative neuropeptides.

Although ventricular injections may be useful for initial approximations of central actions of peptides, data should be interpreted carefully. The absence of effect of a peptide should not always be interpreted as such since actions on different brain structures may represent a summation of opposing responses. In addition, neuroanatomical loci not accessible from the CSF may not be influenced even by high concentrations of peptides within the ventricular fluid since it has been observed that some chemical substances may diffuse no more than a millimeter from the ventricular space into the brain tissue (Fuxe and Ungerstedt, 1968). However, evidence has been provided recently supporting the presence of a "paravascular" fluid circulation in the CNS (Rennels et al., 1985). Thus, solutes in the CSF may have access to extracellular spaces within the brain remote from the ventricles via fluid-filled pathways closely associated with the cerebral vasculature.

In addition to the problem of accessibility of substances in the CSF to brain tissue, the concentration of peptides within the extracellular spaces may be influenced by other factors, including degradative enzymes (Meisenberg and Simmons, 1984a) and leakage into the periphery. Indeed, recent observations indicate that hormonal peptides that are introduced into the CSF may reach the bloodstream in sufficient amounts to produce biological effects (Robinson, 1983). For example, after intracerebroventicular injection, measurable amounts of arginine vasopressin (AVP)-like material have been detected in the periphery in concentrations sufficient to evoke an antidiuretic response (Clark et al., 1983). Direct measurements in plasma have also shown elevated levels of α-MSH (Rotte et al., 1980), LHRH (Ben-Jonathan et al., 1974), and AVP (Pittman et al., 1985) in the periphery within minutes after intracerebroventricular or intrathecal administration. Such observations serve to illustrate the importance of controlling for peripheral effects (by iv injections of similar quantities of peptide) when examining the putative physiological role of centrally (into CSF) administered peptides.

Despite these pitfalls, the acute ventricular administration of peptides has provided much preliminary information regarding their physiological roles in central control mechanisms. A variation of the acute method involves the chronic infusion of chemicals into the CSF over a period of days or weeks, thus enabling the long-term rather than acute effects to be examined (Myers, 1977).

2.1.2. Microinjection of Peptides Directly into Brain Tissue

Direct microinjection of peptides into discrete brain loci is used frequently to localize site(s) of action within the CNS. In this regard, it is often possible to find a specific neuroanatomical locus of action for peptide effects previously observed after intracerebroventricular infusion (Naylor et al., 1985b) or even to discover actions that had not been apparent after ventricular administration (Ferris et al., 1984; Naylor et al., 1985a).

The neuroanatomical locus of injection is usually not chosen in a random fashion. Prior identification of a peptide in a specific area and/or the demonstration of specific receptor binding sites are appropriate reasons for microinjecting a peptide into a particular brain area. In addition, prior knowledge of neuroanatomical sites from which physiological processes are controlled primarily (e.g., the hypothalamus and thermoregulation) may provide a brain area from which to initiate studies utilizing microinjection. In localizing sites of action, "misses" are inevitable and important. A lack of response after microinjection of a neuropeptide into one brain area, but not another, indicates that the peptide-induced effects are specific to a certain brain site. In addition to the application of peptide-containing solutions to brain tissue, appropriate controls with carrier vehicle alone or with inactive analogs should be carried out in order to ascertain whether the observed effects are caused by the peptide or by the injection process *per se.*

A concern when injecting peptides into specific brain sites is the extent of the area of tissue affected. In order to arrive at a neuroanatomical localization of peptide action, small volumes of less than 1.0 μL should be used, thereby keeping the diffusion distance as low as possible. Myers (1971), using injections of dye or radiolabeled neurochemicals, has demonstrated that the spread of injectate is closely related to volume. With injection volumes of 0.5–1.0 µL, the area of stained or labeled tissue will encompass approximately 1.0 mm in diameter. The spread of injection can also be limited by employing a slow infusion rate and by leaving the injector needle in place for up to a minute after completion of the microinjection. This procedure reduces the likelihood of the microliter droplet flowing upward around the needle. Another concern is the amount of tissue damage at the tip of the injector needle, since tissue destruction may alter the biological response of a peptide. This can be minimized by using small injection volumes and may be more relevant to those experiments involving continuous perfusion techniques. The extent of tissue damage may also be observed after histological sectioning of the brain tissue. This verification is also essential in order to determine the precise locus of the microinjection since slight variations in injection sites may produce widely differing responses.

A variation of the microinjection technique involves the continuous microinfusion of substances into brain sites over a longer time course (hour-weeks) than is possible using the bolus microinjection (minutes) approach. Thus, rather than administering peptides in a single injection, a similar amount of peptide may be delivered to the brain tissue over a longer period of time.

2.1.3. Perfusion of Peptides Within the Ventricular System and Brain Tissue

Perfusion methods have been used instead of bolus microinjection to administer peptides into the CNS. These involve injecting and withdrawing an identical amount of perfusate from either the ventricles or the brain tissue. The use of both ventricular and brain tissue perfusion techniques (push–pull perfusion; Gaddum, 1961; Myers, 1972; also *see* section 2.2.2) is advantageous in that both techniques allow a constant amount of peptide to be delivered to the tissue over a relatively long period of time, a feature that cannot be mimicked by a bolus microinjection (Table 1). In addition to this, the dynamic characteristics of perfusion may lead to a more rapid and efficient interaction of peptide with the brain tissue. This may explain, in part, the observation that during push–pull perfusion a much lower dose of peptide is required for effect than during continuous microinfusion (W. D. Ruwe, unpublished observation). Determination of the actual amount of peptide reaching the

Table 1
Advantages of the Push–Pull Perfusion Technique

Correlation of local neurotransmitter release with changes in physiological function or behavior.

Ongoing and repetitive analysis of endogenous neurotransmitter release
over long time periods (hours) in the same animal.

Repetitive use of the conscious, behaving animal.

Neuroanatomical localization of peptide actions in all brain areas.

Long-term exposure of neurochemicals to brain tissue.

Concurrent use with other techniques (extracellular/single-unit recording).

---

tissue during push–pull perfusion suggests that at perfusion rates
in the range of 16–50 μL/min, there is an interchange of approximately 10% (Veale, 1971) or less (Naylor, Ruwe, and Bauce, unpublished observations) of the perfusing solution with the brain
tissue. This problem of exchange makes it difficult to draw conclusions on the actual amount of peptide administered over any
given time period. The relative merits of tissue as opposed to
ventricular perfusions are similar to those already described for
microinjections. However, the direct perfusion of peptides within
the brain tissue may reduce the problem of leakage of intracerebroventricularly applied neuropeptide into the periphery.

Numerous variations of the concentric cannulae have been
designed and used in physiological studies (*see* Myers, 1972; Pittman et al., 1985 for review), along with a number of different pump
systems. According to Myers, and also our own experience, the
infusion/withdrawal pump is superior to gravity for withdrawing
the perfusate since the use of gravity alone does not provide the
constancy in vacuum in the pull side required for withdrawing the
perfusate (Myers, 1972). In view of the potential for severe tissue
damage resulting from pull cannula dysfunction, careful monitoring of the "pull side" is essential. Any sign of bubbles (indicative of
occlusion) or blood discoloration (indicative of tissue damage)
should result in termination of the experiment.

The potential of perfusion techniques for investigating central
nervous functions is great, permitting a longer time course of
administration than the more frequently used microinjection techniques. However, the investigation of short-term effects may be

better served using bolus microinjection methods. In addition, in situations in which chronic administration causes desensitization, acute injection may be more relevant. These and other points will be discussed further in section 3, using examples cited in the literature.

2.2. Analysis of Endogenous Levels of Peptides in the Central Nervous System

One of the most important criteria to establish the validity of a neurotransmitter within the CNS is the demonstration of its release into either CSF or push–pull perfusates of brain tissue during an appropriate physiological event. In addition to release studies, techniques that measure changes in tissue concentrations of peptides have frequently been used to obtain evidence for their physiological roles within the brain. In the following paragraphs, these techniques will be discussed with particular reference to the types of physiological information that can be derived from their use and how each is interpreted.

2.2.1. Cerebrospinal Fluid Samples

Collection of CSF, often easier in larger laboratory animals such as cats, monkeys, and sheep, may provide a gross estimate of transmitter release from a number of different neural systems that are likely within close proximity to the ventricular walls. Direct sampling of CSF from different fluid-filled spaces in the brain (lateral, III and IV ventricle, cisterna magna, and subarachnoid space) is possible in either the conscious or anesthetized animal. The relative stability of peptides within CSF means that it is possible to measure the amount of neuropeptide during various physiological manipulations (Robinson and Jones, 1982; Kasting et al., 1983; Stark et al., 1985). Another approach for determining the presence of peptides within the CSF is to perfuse an artificial physiological solution and analyze the effluent for peptide content. A potential drawback with this kind of approach is that the low amount of peptide available for measurement is diluted further with the perfusion medium. However, in spite of this, perfusion techniques have been successfully utilized to measure peptide release into the CSF (Pittman et al., 1984).

The relative ease of ventricular sampling and the contact of the perfusate with large areas of the brain are features that have resulted in the use of this technique. Thus, neurochemical analysis of ventricular and cisternal perfusates is not dependent on knowledge of the precise locus of release. This point, however, is the main drawback regarding the use of CSF samples. Cerebrospinal fluid sampling may, because of the collection from a greater area and because of the possible involvement of one neurotransmitter in a number of neural circuits, reveal a neurochemical profile that is different from that at the site of its release, i.e., ventricular perfusions may smooth out the troughs and peaks of neurotransmitter release (Levine and Ramirez, 1980; Pittman et al., 1985). In spite of this, CSF sampling remains a valuable tool for studying whole animal physiology, for instance, those involving humans (Dogterom et al., 1978) and circadian rhythms (Jenkins et al., 1980; Reppert et al., 1981). In the case of humans, because of the availability of CSF, it is among the most useful of approaches that can be used to study brain neurochemistry.

2.2.2. Push–Pull Perfusates

The method of brain push–pull perfusion (see Tables 1 and 2) allows for the sampling of chemical substances within a circumscribed region of the brain of unanesthetized/unrestrained animals. The advantage of localized perfusion in intact animals is that the animal does not have to be killed before a neurochemical analysis can be carried out. In addition, analysis of endogenous neurotransmitter can be correlated: (1) to an anatomical site and (2) to changes in physiological processes.

A major concern with respect to the use of push–pull perfusion techniques is the potential for tissue destruction at the site of neurotransmitter release (Izquierdo and Izquierdo, 1971). This concern is valid, but should be put in perspective. Any cannulation or microinjection procedure results in tissue damage. Perfusion techniques may produce more damage since occlusion at either the push or pull side of the system results in imbalances in flow and pressure at the cannulae tip. For example, if the outflow (pull) is no longer functioning, then the fluid normally escaping via this route will collect in the brain tissue and cause damage. Problems of this nature can be minimized or abolished if careful attention is used to maintain a constancy of flow. Thus, improper use of the push–pull perfusion technique can result in severe tissue damage at the cannulae tip. However, as long as appropriate care is taken, the push–pull perfusion technique can be a valuable tool for investigating CNS neurochemistry (see Myers, 1972). Measurements of the extent of tissue necrosis indicate that, when isotonic solutions are

Table 2
Disadvantages of the Push–Pull Perfusion Technique

Efficiency of exchange of neurochemicals between the perfusate and tissue is low.

Alterations in the local characteristics of the perfusion site, i.e., proliferation of non-neuronal cells.

Formation of lesion cavities and leukocyte infiltration.

Contamination of perfusion medium with endotoxin.

Compromise or circumvention of the blood–brain barrier.

---

perfused, tissue damage does not occur much beyond the cannula tip (Yaksh and Yamamura, 1974).

Another approach to determining the extent of tissue damage is to measure the concentrations of intracellular components such as lactate dehydrogenase in the returned perfusate. In this regard, after an initial trauma as a consequence of cannula insertion, the levels of such intracellular enzymes remain low (Honchar et al., 1979).

Push–pull cannulae, as well as providing a means of sampling the local concentration of neurotransmitters, also can be used to verify that the release is from a neuronal source. In this regard, altering the ionic composition of the inward perfusion medium by raising potassium should stimulate release, and lowering calcium should reduce this stimulated release. This procedure is necessary to demonstrate also that the release is indeed local and that the presence of neurochemicals in the perfusate is not caused by diffusion from the circulation.

In addition to concern over tissue damage caused by cannulae insertion and pull cannula obstruction, there is also concern as to how extensive the area of perfusion is, since push–pull technology is often used to localize sites of peptide release within the brain. Systematic evaluations of the performance of push–pull cannulae over a variety of conditions have revealed that a tip extension of 0.75–1.0 mm and a perfusion rate of 20–100 μL/min bathes a sphere of brain tissue approximately 1.2–1.5 mm in diameter. Using these specifications, the greatest recovery of neurochemicals into the perfusate has been observed (Myers, 1972; Yaksh and Yamamura, 1974). Even under optimum conditions, the recovery of neurohumors at the cannula tip is normally approximately 10–15% of the amount available (Morris et al., 1984). This fact, coupled with the very small amounts of neurotransmitter released and accessible to the push–pull perfusate, make the measurement of endogenously released peptides difficult. However, push–pull perfusion has been used successfully to demonstrate the release of peptides from the brains of behaving animals (Cooper et al., 1979; Levine and Ramirez, 1980; Ruwe et al., 1985a).

2.2.3. Tissue Content Analysis

The measurement of putative peptide neurotransmitters within specific brain nuclei can be carried out using the tissue punch technique described by Palkovits (1973). This methodology has been used frequently to provide evidence for physiological roles of neurotransmitters (see section 3.2.2.) within the brain. It involves removing and slicing the brain, punching out a small piece of tissue, solubilizing it, and finally measuring the amount of peptide present by a specific radioimmunoassay. A potential drawback of this technique is that the peptides are analyzed from dead tissue. Thus, by its very nature, such an analysis is limited to a single point in time, and this may make it difficult to correlate the measured levels of the peptide to any specific physiological or behavioral event. However, it does not appear to seriously limit the validity of the analysis since it has been demonstrated, for example, that levels of vasopressin immunoreactivity in the human brain are not affected by the delay usually preceding autopsy (Rosser et al., 1981).

Complications may arise in tissue content analysis concerning the interpretation of data. Since actual release is not measured, but rather the presynaptic storage, changes in peptide levels during various physiological challenges may be accounted for by a number of reasons. For instance, increases in neuropeptide levels may be attributed to a decrease in release, an increase in synthesis, or an increase in transport from the cell body to the terminal (or all three combined). Similarly, decreases in peptide levels may be caused by reasons that are the opposite of the above. Moreover, a lack of change within a given brain region may reflect a summated response such as increased release and increased axonal transport from the cell body. Thus, before firm conclusions are formulated using data obtained from tissue content analysis, the reasons for such alterations in peptide levels should be elucidated—perhaps utilizing axonal transport inhibitors. However, with careful interpretation, tissue content analysis will continue to provide information regarding the physiological role of peptides in the brain (*see* section 3.2.2).

2.2.4. Immunocytochemistry

The theory underlying the immunocytochemical localization of neuropeptides within the brain relies on the capacity of immunoglobulins to bind with high selectivity and affinity to comparatively small molecular weight peptides in brain tissue. Once bound, visualization can be achieved directly with a fluorophore or via a secondary antibody and enzyme markers such as horseradish peroxidase (*see* vandeSande, 1979, for review). Potential sources of error using this technique arise primarily from problems with immunoglobulin specificity and also nonspecific staining. For a more comprehensive analysis of immunocytochemistry as an investigative tool in neurobiology, the reader is directed to a recent review (Landis, 1985).

Immunocytochemistry is a powerful research tool that is used commonly to localize and identify peptide-containing cell bodies, axons, and nerve terminals within the CNS. The demonstration of peptides within specific brain areas by immunocytochemistry is used frequently as evidence to support physiological roles for exogenously administered peptides. In addition to localizing peptides neuroanatomically, immunocytochemistry also can be used to illustrate altered central peptide distribution during certain physiological manipulations (*see* section 3.2.2). Direct quantification of actual concentrations of peptides is difficult using this technique, but its high sensitivity means that the central distribution of peptide neurotransmitter systems may be readily demonstrated. Using immunocytochemistry, Zeisberger and coworkers have demonstrated the possible role AVP plays in the antipyretic response at term in the pregnant guinea pig (Merker et al., 1980).

2.3. Pharmacological Investigation of Peptide Effects in the Central Nervous System

Previous sections have detailed the various techniques that are used commonly to investigate peptide actions within the CNS. In addition to these techniques, pharmacological investigations can provide valuable information concerning the mechanisms involved in peptide actions. One of the reasons that a considerable amount is known about monoaminergic systems in the brain is the abundance of pharmacological tools available for investigation, such as specific receptor agonists and antagonists, presynaptic neuron-blocking drugs, drugs that deplete presynaptic neuro-transmitter stores, and drugs that inhibit neuronal uptake or enzymatic degradation. Since peptides are relatively recent transmitter candidates, most have no specific analogs from which their physiological role can be investigated thoroughly. An exception to this, among others, is vasopressin. In this regard, relatively specific antagonists of both the vasopressor (Kruszynski et al., 1980) and antidiuretic (Manning et al., 1982) actions of AVP are available.

Application of pharmacological antagonists to brain tissue can be used either to prevent the effect of exogenously administered peptide or to block the action of endogenously released neuropeptide at the level of the receptor. Successful demonstration of the latter is good evidence to support the hypothesis that a given peptide functions physiologically. As well as antagonists, specific antibodies raised against the peptide in question also can be used to investigate the physiological significance of peptide actions. Thus, exogenously administered antibody would sequester endogenously released neuropeptide, thereby preventing its physiological effects. Antagonists and antibodies may therefore produce the same end result, but by completely different mechanisms. For control experiments, antagonists and antibodies directed toward other molecules from the one in question should be used to guard against potential nonspecific actions.

In conjunction with the use of antagonists, the investigation of dose–response relationships provides further evidence of receptor-mediated events. This is important since it may ultimately reveal the difference between a pharmacological action and an interesting physiological phenomenon. Under normal circumstances, the higher the dose the greater the response, until a maximum is reached. Dose–response curves carried out in the presence of antagonists may reveal the nature of antagonism, i.e., parallel shifts to the right are indicative of competitive antagonism; whereas, a reduction in the maximum response may be accounted for by noncompetitive antagonism. Once a receptor-mediated event has been characterized, structure–activity relationships may be carried out to establish the portion of the peptide molecule necessary for the physiological effect. The potential for use of pharmacological tools in the physiology of peptidergic neurotransmission within the CNS is enormous. The use of antagonists and other pharmacological tools will be discussed later in relation to their usefulness in physiological studies using conscious animals.

2.4. Electrical Stimulation

Electrical stimulation, precisely delivered to distinct nuclei within the brain, may be used to investigate further the involvement of neuropeptides in CNS physiology. In this regard, stimulation of a particular area of the brain may mimic the effect observed after exogenous administration of a peptide. Any potential connection between the two events can be investigated. For instance, electrical stimulation may release the peptide in question from presynaptic nerve terminals within another area of the brain—an event that may be verified with push–pull cannulae and a suitable assay. Furthermore, if the physiological response to electrical stimulation is prevented by the prior administration of an antagonist in the same sites in which electrical stimulation released the peptide, a strong piece of evidence has been obtained to support a connection between electrical stimulation and the action of the exogenously administered peptide (e.g., Pittman and Franklin, 1985). Electrical stimulation experiments may be further validated if stimulation of areas outside the nuclei in question does not evoke a response (similar to microinjection "misses"). However, the results obtained using electrical stimulation methods should be carefully interpreted since it is possible that more than one neurotransmitter will be released from nerve terminals.

Electric current or various neurotoxins may be used to lesion nuclei or fiber passages within the brain. Since peptides are synthesized within the cell body, ablation of these structures effectively depletes them from neuronal pathways. This provides another type of manipulation for investigating the central actions of peptides. For example, alterations in behavioral or physiological control mechanisms may be evident following the removal of endogenous supplies of peptides. However, its specificity is not guaranteed since cell body ablation may not be complete. Furthermore, multiple inputs from a number of brain areas may compensate for the absence of one. Consequently, appropriate controls to verify the absence of peptide after lesion should be undertaken. At a time when pharmacological tools for investigation into peptidergic transmission are few, studies utilizing electrical stimulation and lesion of specific nuclei may be useful for investigating the actions of neuropeptides within the CNS.

2.5. Concluding Remarks

In choosing techniques for evaluating physiological roles of peptides within the CNS, it is necessary to use those most appropriate for solving the question being asked. The relative advantages and disadvantages of a number of techniques for both introduction to and analysis from the brain have been discussed. However, for an integrated understanding of peptide actions, it may be pertinent to examine the effects of a number of different modes of introduction and analysis before conclusions are made about a particular physiological role of any peptide within the CNS.

3. Physiological Studies in Conscious Animals

The current interest in neuropeptides and the demonstration of these putative neurotransmitters within the brain by immunocytochemistry have resulted in numerous investigations into the possible physiological roles they may play. Since our laboratory has been engaged in studying peptide effects on the thermoregulatory system, we will illustrate a number of the above described methodologies by reviewing our data on the thermoregulatory effects of vasopressin. The thermoregulatory response is frequently monitored after central administration of peptides since this physiologically controlled mechanism is relatively easy to measure and a considerable amount is known about the neuroanatomical sites from which it is controlled. In addition, the actions of vasopressin on thermoregulatory mechanisms have been examined using a wide variety of techniques, thus enabling an integrated analysis of its effects in the conscious animal.

3.1. Thermoregulation: Heat Production and Heat Loss

Changes in thermoregulation are often determined by measuring peptide-induced alterations in core temperature over a variety of environmental temperatures. These peptide-induced changes can be broadly classified based on whether they cause a shift in the level about which temperature is regulated or whether they depress thermoregulation (Clark and Lipton, 1983). In addition to measurements of core temperature, changes associated with behavior or specific effectors may provide further information on the thermoregulatory effects of peptides. Since body temperature is affected greatly by centrally or peripherally administered pyrogen, it is essential that sterile precautions are taken and the appropriate controls are carried out to guard against potential contamination.

It is not the aim of this review to outline all the known thermoregulatory actions of peptides, but rather to evaluate the kinds of information that can be obtained from the different techniques used to investigate them. An exhaustive study of the thermoregulatory effects of peptides can be found in two recent reviews (Ruwe et al., 1983; Clark and Lipton, 1983).

3.1.1. Arginine Vasopressin

Initial observations by Cushing (1931) demonstrated that injection of a crude pituitary extract (likely containing vasopressin) into the cerebral ventricles of humans induced a decrease in body temperature, which was coincident with marked sweating and vasodilation. Infusion of synthetic AVP into a lateral cerebral ventricle (icv) of rats (Kruk and Brittain, 1972; Kasting et al., 1980; Meisenberg and Simmons, 1984b) or gerbils (Lee and Lomax, 1983) induces a short-latency, short-lived hypothermia. In contrast, icv administration of AVP into rabbits results in mild hyperthermia (Lipton and Glyn, 1980) or no alterations in core temperature (Naylor et al., 1985a). This discrepancy may be explicable in terms of a genuine species difference. However, further investigations using direct microinjection of AVP into specific brain areas have revealed that this peptide can evoke marked hyperthermic responses when introduced into the preoptic area of the anterior hypothalamus of the rat. This occurs over a wide range of ambient temperatures (Lin et al., 1983), but not in nearby areas (Naylor et al., 1986). It does not appear to be caused by contamination with pyrogens since the hyperthermic response is completely abolished in the presence of a vasopressin antagonist [d(CH$_2$)$_5$Tyr(Me)AVP] (Naylor et al., 1986).

The reason for the differences observed after icv and intrahypothalamic AVP in the rat are not immediately apparent. The hypothermia reported after icv AVP may be nonspecific in nature since the behavioral effects of this peptide may briefly render the animal unable to thermoregulate (Kasting et al., 1980). For example, short periods of immobility and staring are commonly observed after icv vasopressin (Kasting et al., 1980). The high surface area : mass ratio of rats would result in a greater loss of body heat, especially when the rats are immobile. Thus, the short latency hypothermia may be accounted for in terms of a physical phenomenon rather than a physiological response to centrally administered AVP. An investigation of the hypothermia at low (3.0°C), neutral (25.0°C), and high (33.0°C) ambient temperatures may help to clarify this point. However, these anomalous thermoregulatory effects serve to illustrate the importance of the route and site of administration in evaluating peptide actions in the CNS. Alternatively, the hypothermia may be caused by an action of AVP on an as yet unidentified brain site. However, following bilateral microinjection of AVP into a number of brain loci known to contain vasopressin, including the nucleus accumbens, the ventral septal area (VSA; where AVP is antipyretic), the preoptic area, the anterior hypothalamus, and the dorsomedial hypothalamus, hypothermic responses were never observed (Naylor et al., 1986). It is not clear as yet whether AVP functions physiologically in the control of normal body temperature, as it does to modulate the febrile response (see section 3.2.2.). Future studies concerned with the release of endogenous stores of AVP, and with specific antagonists to ensure the observed effects are receptor mediated, may help to clarify the situation.

3.2. Fever and Antipyresis

Fever is a pathophysiological elevation in body temperature that is regulated actively and defended by the host. This elevated temperature is one of the most prominent features of infection since it is the most commonly observed clinical sign of disease. Fever can be initiated by a wide number of agents, including endotoxins, viruses, Gram-positive bacteria, hypersensitivity, and immune responses. A common process is believed to be responsible for the fever induced by these activators, that being the production by various phagocytic cells of a pyrogenic material named endogenous pyrogen or interleukin-I (see Dinarello, 1984 for review).

Antipyresis is the reduction of febrile temperatures toward normal or prefebrile levels. Under afebrile conditions, antipyretics do not lower resting body temperature. That is, they are capable of reducing fever in doses that do not alter normal body temperature. A number of peptides, including AVP (Kasting et al., 1979a), ACTH, and α-MSH (Glyn and Lipton, 1981), possess antipyretic activity against pyrogen-induced fevers. In this regard, AVP has been proposed to be an endogenous antipyretic acting within the CNS to modulate fever (Veale et al., 1981; Kasting et al., 1979a; Cooper et al., 1979; Ruwe et al., 1985a). This section will deal primarily with the evidence relating to an involvement of AVP in the thermoregulatory mechanisms controlling fever suppression. This is an aspect of thermoregulatory physiology that has been extensively investigated. Consequently, many of the different approaches and techniques described previously have been used to examine the problem.

3.2.1. Arginine Vasopressin and Endogenous Antipyresis

The young of a number of species, including lambs (Pittman et al., 1974) and guinea pigs (Blatteis, 1975), either do not develop a fever, or only a very small one, in response to bacterial or endogenous pyrogen administered intravenously. Similarly, neonatal humans often do not develop fever despite the clinical presence of infection (Epstein et al., 1951; Smith et al., 1956). In addition, fever elicited by either endotoxin or endogenous pyrogen is reduced in the pregnant ewe from about 4 d prepartum to about 5 h after birth (Kasting et al., 1978). It has been suggested that these decreased pyrogenic responses are caused by increased levels of an "endogenous antipyretic substance," since white blood cells retain their ability to produce endogenous pyrogen during this periparturient period (Kasting et al., 1979b).

Based on observations by Alexander and coworkers (Alexander et al., 1974), it was determined that of the many hormonal fluctuations occurring during pregnancy, that of AVP most closely correlated to the period of diminished febrile response. Indeed, anatomical evidence suggests that activation of central vasopressinergic neurons occurs in pregnant guinea pigs at the perinatal period (Merker et al., 1980). That is, the levels of AVP immunoreactivity in the paraventricular nucleus rise, and an increase in the number of vasopressin-containing nerve endings in the septum and amygdala have been observed. This observation is of particular interest since it is in these same regions that exogenously administered AVP suppresses the febrile response in adult guinea pigs (*see* Fig. 1).

The functional significance of AVP in fever control in the adult has been demonstrated in the sheep using push–pull perfusion to apply the vasopressin to the brain tissue. Push–pull perfusion was used since it allows a constant, low amount of peptide to be delivered to the tissue over a long period of time—a feature that

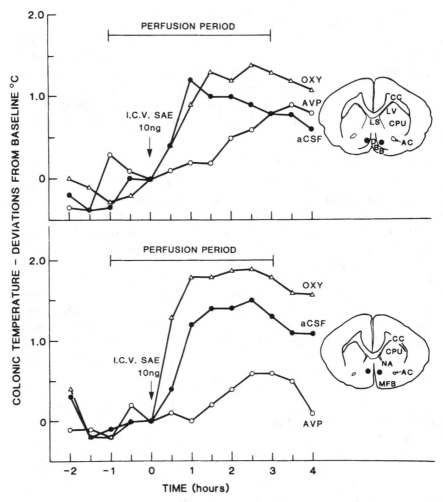

Fig. 1. Colonic temperature in deviation from baseline (°C) in two conscious guinea pigs demonstrating where perfusion of AVP suppressed pyrogen fever (see insets). Arginine-vasopressin, (AVP, open circles), oxytocin (OXY, open triangles) (both at 6.5 μg/mL), or artificial cerebrospinal fluid (aCSF) were perfused (30.0 μL/min) into the ventral septal area for 1.0 h prior to, and 3.0 h following, an intracerebroventricular (icv) injection of *Salmonella abortus equi* (SAE, 10.0 ng/10.0 μL). Abbreviations: AC, anterior commissure; CC, corpus callosum; CPU, caudate putamen; DBB, diagonal band of Broca; LS, lateral septum; LV, lateral ventricle; NA, nucleus accumbens.

makes it particularly useful for physiological investigations into events, such as fever, that occur over a number of hours. Consequently, when vasopressin is perfused within the ventral septal area (VSA) of the brain, the fever typically evoked by a bacterial pyrogen is suppressed in a dose-dependent fashion (Kasting et al., 1979a). The antipyretic effect of AVP is only apparent when perfused within a very circumscribed region of the brain. Thus, perfusion of AVP outside the VSA, into the lateral, anterior, or posterior hypothalamus, the preoptic area, the fornix (Cooper et al., 1979), or the dorsal septum (Ruwe et al., 1985b), is without effect on the normal development of the febrile response (*see* Fig. 2). Similarly, intravenously administered AVP does not result in suppression (Cooper et al., 1979). In accordance with the definition of an antipyretic, perfusion of AVP within the sensitive ventral septal sites, in the afebrile animal, is without effect on the body temperature (Kasting et al., 1979a; Naylor et al., 1985a), i.e., the observed reduction in fever is not caused simply by a nonspecific drop in core temperature.

The antipyretic action of AVP has been examined recently in other species including the rabbit, rat, and cat. Perfusion of vasopressin within the VSA of the rabbit (Naylor et al., 1985a) suppresses both peripherally and centrally induced fevers. Similarly, in the cat, push–pull perfusion of the VSA with AVP suppresses centrally induced fevers in a dose-dependent manner (Ruwe et al., 1985c). It is important to note that the AVP-sensitive antipyretic site is similarly located over a number of species, that is, lateral to the ventral limb of the diagonal band of Broca and immediately ventral to the septum. In the rat, icv administration of low doses of AVP (0.1–100 ng) also can suppress endotoxin fever (Kovacs and deWied, 1983; Kandasamy and Williams, 1983). In the former study, it was observed that it was the hormonally active AVP that was antipyretic and not a behaviorally active fragment DG-AVP (des-9-glycinamide-arginine-vasopressin) or the structurally similar peptide, oxytocin. In other studies employing oxytocin, this peptide was similarly found to be ineffective in attenuating fever in tissue sites where AVP was found to reduce fever in the cat (Ruwe et al., 1985c) and guinea pig (Fig. 1). Other neuropeptides found to be ineffective in reducing fever include somatostatin, substance P, and angiotensin II (Kasting, 1980). However, the very similar AVP analog, arginine vasotocin, partially reduces the febrile response.

Although the exact neurochemical mechanisms responsible for fever development have yet to be delineated clearly, it is be-

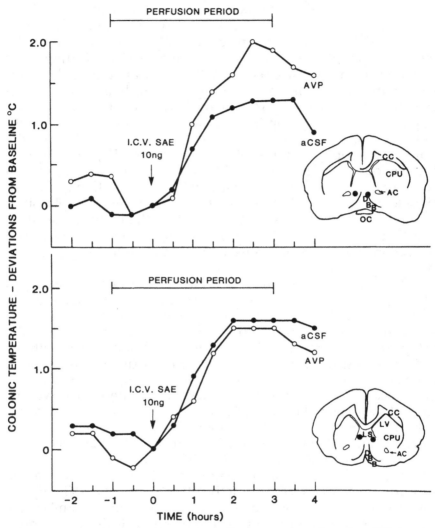

Fig. 2.   Colonic temperature in deviations from baseline (°C) in two
conscious guinea pigs demonstrating where perfusion of AVP did not
suppress pyrogen fever (see insets). Arginine vasopressin (AVP, open
circles; 6.5 μg/mL) or aCSF (closed circles) were perfused (30.0 μL/min)
into sites outside the ventral septal area for 1.0 h prior to, and 3.0 h
following, an icv injection of SAE (10.0 ng/10.0 μL). Abbreviations: AC,
anterior commissure; CC, corpus callosum; CPU, caudate putamen; DBB,
diagonal band of Broca; LS, lateral septum; LV, lateral ventricle; OC, optic
chiasm.

lieved that prostaglandins of the E series (PGEs) are involved (Milton and Wendlandt, 1971). Indeed, these metabolites of arachidonic acid feature in most of the schema proposed to describe the pathogenesis of fever (Milton, 1982). In accordance with the putative role of prostaglandins of the E series as mediators of fever, perfusion of AVP within the VSA of the rat was found to significantly reduce the hyperthermia following an icv infusion of PGE_2 (Ruwe et al., 1985b). AVP is therefore able to prevent the fever caused by two pyrogenic agents, namely bacterial pyrogen and prostaglandins of the E series.

Evidence for the hypothesis that AVP might function in the brain, under physiological conditions, as an endogenous antipyretic agent (Veale et al., 1981) has been demonstrated in the sheep. During fever, the release of AVP is modified within the same brain sites where perfusion of vasopressin suppresses fever (Cooper et al., 1979). Thus, when push–pull perfusates were assayed for AVP content, the amount of peptide present correlated negatively with changes in body temperature. That is, as body temperature rose, smaller amounts of AVP were present in the extracellular fluid; and as temperature fell, AVP secretion rose. In keeping with the hypothesis that AVP is a normal modulator of the febrile response, CSF concentrations of vasopressin measured during fever in the sheep were significantly raised during pyrexia (Kasting et al., 1983). In contrast to the push–pull perfusates, the AVP concentrations in CSF were significantly correlated to increases in body temperature. Although these findings are anomolous, it is likely that peptides collected from push–pull cannulae, proximal to the site of secretion, are a better indication of the dynamics of transmitter release (Pittman et al., 1985). Thus, such observations emphasize the importance of tissue as opposed to ventricular sampling when formulating conclusions about the temporal release of neurotransmitters. Another approach to investigating the physiological roles of peptides is to measure the changes in tissue concentrations in brain areas. In this regard, in response to an injection of endotoxin, concentrations of immunoreactive vasopressin have been shown to decrease in the septum, amygdala, and caudate nucleus of the rat (Kasting and Martin, 1983). In contrast, concentrations of AVP-like material remained unaltered in other areas, supporting the suggestion that AVP may function as a neurotransmitter during fever. Although tissue content analysis may provide little information concerning the dynamics of transmitter release, it may reflect alterations in turnover, in this case those related to the febrile rise in body temperature.

In addition to measuring endogenous levels of neurotransmitters during physiological manipulations, pharmacological agents can be used to alter the normal actions of endogenously released peptides. Thus, perfusion of a specific AVP antiserum, an antagonist analog (desaminodicarba-AVP) or a V_1 receptor antagonist (d(CH$_2$)$_5$Tyr(Me)AVP) in sites where AVP suppresses fever results in a markedly enhanced febrile response (Kasting, 1980; Naylor and Veale, 1986). Therefore, preventing the action of endogenously released AVP from reaching its receptor, either by sequestering it with an antibody or blocking the receptor site, effectively produces the same result. Similarly, augmenting AVP release by hemorrhage results in a reduction in the magnitude of the febrile response (Kasting et al., 1981).

Despite evidence supporting a role for AVP in natural fever suppression in a number of species, a report concerning the antipyretic action of AVP in the rabbit indicates that microinjections of this peptide into the dorsal septum are not antipyretic, but actually enhance the febrile response to leukocyte pyrogen administered intravenously (Bernardini et al., 1983). The sites of maximum sensitivity to the antipyretic effect of AVP in the rabbit are located within the ventral septum, proximal to the diagonal band of Broca (Naylor et al., 1985a). This may, in part, explain why this aforementioned report failed to describe the antipyretic properties of centrally administered AVP, because these injections were located in much more dorsal aspects of the septal area. Indeed, in another species, the rat, the dorsal septum has been shown to have little or no AVP-sensitive antipyretic actions (Ruwe et al., 1985b). In accordance with the hypothesis that AVP acts within a ventral septal neuronal system to modulate fever, dense networks of fibers immunoreactive for vasopressin are present in this area (Buijs et al., 1978; Sofroniew, 1983; Swanson, 1977), along with specific AVP receptor binding sites (Baskin et al., 1983) and potential AVP projections from the paraventricular nucleus, bed nucleus of the stria terminalis, and suprachiasmatic nucleus (Disturnal et al., 1985).

In addition to the dorsal septum, infusion of AVP into a lateral ventricle is ineffective in reducing fever in the rabbit (Bernardini et al., 1983; Naylor et al., 1985a) and only inconsistently attenuates pyrogen fever in the macaque monkey (Lee et al., 1985). That AVP exerts an antipyretic action when perfused directly within the

VSA, but not when it is infused into the ventricles, does not represent a unique situation. Indeed, a similar response has been observed in the golden hamster, in which flank-marking behavior is triggered by a direct microinjection of vasopressin into the medial hypothalamus, but not by ventricularly administered AVP (Ferris et al., 1984). Data of this kind indicate that specific physiological roles for neuropeptides are unlikely to be determined if only the ventricular route of administration is utilized to investigate their actions.

Recently, a potential mechanism of AVP-induced antipyresis has been described (Banet and Wieland, 1985). The conclusions of this study were derived from observations on the effects of intraseptally applied vasopressin on the thermoregulatory responses to preoptic cooling and exposure to different ambient temperatures. Based on this, it was determined that AVP acts to inhibit thermoregulatory heat production, rather than to alter normal set-point temperature. However, the relationship between these experiments and the AVP-induced antipyresis requires further investigation since the septal injections (Banet and Wieland, 1985) were located in regions of the septum (dorsal) not responsive to the antipyretic effect of AVP.

3.2.2. Concluding Remarks

In conclusion, these data indicate the types of information that can be obtained from experiments in the intact animal. They demonstrate how an integrated approach, using a number of different techniques to look at one problem, can be used to obtain information about the physiological roles of peptides within the brain—in this case, vasopressin and endogenous antipyresis.

4. Summary

The antipyretic potency of vasopressin within a distinct region of the CNS, but outside the classical thermoregulatory areas in the anterior hypothalamic/preoptic area, has attracted considerable research interest. Based on physiological and pharmacological studies, there is much evidence to support the hypothesis that AVP may function in the natural suppression of fever, that is, as an endogenous antipyretic. Data have been presented that indicate the antipyretic action of AVP is receptor-mediated, peptide-specific, and site-specific. Furthermore, the release of AVP shows demonstrable changes prior to and during the pathogenesis of fever. The information to support this hypothesis has been obtained using a number of different techniques involving the exogenous administration of, and endogenous analysis of, the peptide. Thus, use of an integrated approach to investigate a peptide action within the conscious animal has resulted in the demonstration of a physiological role for vasopressin in the central mechanisms controlling core temperature.

Acknowledgments

The work reported here is supported by the Medical Research Council of Canada. AMN holds a studentship and WDR a postdoctoral fellowship from the Alberta Heritage Foundation for Medical Research. The vasopressin antagonist [d(CH$_2$)$_5$Tyr(Me)AVP] was kindly supplied by Dr. M. Manning, Toledo, Ohio. Thanks to Grace Olmstead for typing the manuscript.

References

Alexander D. P., Bashore R. A., Britton H. G., and Forsling M. A. (1974) Maternal and fetal arginine vasopressin in the chronically catheterized sheep. *Biol. Neonate* **25**, 242–248.

Banet M. and Wieland U. E. (1985) The effect of intraseptally applied vasopressin on thermoregulation in the rat. *Brain Res. Bull.* **14**, 113–116.

Baskin D. G., Petracca F., and Dorsa D. M. (1983) Autoradiographic localization of specific binding sites for [^{3}H] [Arg^8] vasopressin in the septum of the rat brain with tritium-sensitive film. *Eur. J. Pharmacol.* **90**, 155–157.

Ben-Jonathan N., Mical R. S., and Porter J. C. (1974) Transport of LRF from CSF to hypophysial portal and systemic blood and the release of LH. *Endocrinology* **95**, 18–25.

Bernardini G. L., Lipton J. M., and Clark W. G. (1983) Intracerebroventricular and septal injections of arginine vasopressin are not antipyretic in the rabbit. *Peptides* **4**, 195–198.

Blatteis C. M. (1975) Postnatal development of pyrogen sensitivity in guinea pigs. *J. Appl. Physiol.* **39**, 251–257.

Buijs R. M., Swaab D. F., Dogterom J., and vanLeeuwen F. W. (1978) Intra- and extrahypothalamic vasopressin and oxytocin pathways in the rat. *Cell Tiss. Res.* **186**, 423–433.

Clark W. G. and Lipton J. M. (1983) Brain and pituitary peptides in thermoregulation. *Pharmacol. Ther.* **22**, 249–297.

Clark R. G., Jones P. M., and Robinson I. C. A. F. (1983) Clearance of vasopressin from cerebrospinal fluid to blood in chronically cannulated Brattleboro rats. *Neuroendocrinology* **37**, 242–247.

Cooper K. E., Kasting N. W., Lederis K., and Veale W. L. (1979) Evidence supporting a role for vasopressin in natural suppression of fever in sheep. *J. Physiol.* **195**, 33–45.

Cushing H. (1931) The reaction to posterior pituitary extract (putuitrin) when introduced into the cerebral ventricles. *Proc. Natl. Acad. Sci. USA* **17**, 163–170.

Dinarello C. A. (1984) Interleukin-1. *Rev. Infect. Dis.* **6**, 51–95.

Disturnal J. E., Veale W. L., and Pittman Q. J. (1985) Electrophysiological analysis of potential arginine vasopressin projections to the ventral septal area of the rat. *Brain Res.* **342**, 162–167.

Dogterom J., Van Wimersma Greidanus Tj. B., and de Wied D. (1978) Vasopressin in cerebrospinal fluid and plasma of man, dog and rat. *Am. J. Physiol.* **234**, E463–E467.

Epstein H. C., Hochwald A., and Ashe A. (1951) Salmonella infections of the newborn infant. *J. Pediatr.* **38**, 723–731.

Ferris C. F., Albers H. E., Wesolowski S. M., Goldman B. D., and Luman S. E. (1984) Vasopressin injected into the hypothalamus triggers a stereotypic behavior in Golden Hamsters. *Science* **224**, 521–523.

Fuxe K. and Ungerstedt V. (1968) Histochemical studies on the distribution of catecholamines and 5-hydroxytryptamine after intraventricular injections. *Histochemie* **13**, 16–28.

Gaddum J. H. (1961) Push–pull cannulae. *J. Physiol.* **155**, 1–2P.

Glyn J. R. and Lipton J. M. (1981) Hypothermic and antipyretic effects of centrally administered ACTH (1-24) and α-melanotropin. *Peptides* **2**, 177–187.

Honchar M. P., Hartman B. K., and Sharpe L. G. (1979) Evaluation of *in vivo* brain site perfusion with the push–pull cannula. *Am. J. Physiol.* **236**, R48–56.

Izquierdo I. and Izquierdo J. A. (1971) Effects of drugs on deep brain centers. *Ann. Rev. Pharmacol.* **11**, 189–208.

Jenkins J. S., Mather H. M., and Ang V. (1980) Vasopressin in human cerebrospinal fluid. *J. Clin. Endocrin. Metab.* **50**, 364–367.

Kandasamy S. B. and Williams B. A. (1983) Absence of endotoxin fever but not hyperthermia in Brattleboro rats. *Experientia* **39**, 1343.

Kasting N. W. (1980) An antipyretic system in the brain and the role of vasopressin. Ph.D. thesis, University of Calgary, Alberta, Canada.

Kasting N. W. and Martin J. B. (1983) Changes in immunoreactive vasopressin concentrations in brain regions of the rat in response to endotoxin. *Brain Res.* **258**, 127–132.

Kasting N. W., Veale W. L., and Cooper K. E. (1978) Suppression of fever at term of pregnancy. *Nature* (Lond.) **271**, 245–246.

Kasting N. W., Cooper K. E., and Veale W. L. (1979a) Antipyresis following perfusion of brain sites with vasopressin. *Experientia* **35**, 208–209.

Kasting N. W., Veale W. L., and Cooper K. E. (1979b) Endogenous pyrogen release by fetal sheep and pregnant sheep leukocytes. *Can. J. Physiol. Pharmacol.* **57**, 1453–1456.

Kasting N. W., Veale W. L., and Cooper K. E. (1980) Convulsive and hypothermic effects of vasopressin in the brain of the rat. *Can. J. Physiol. Pharmacol.* **58**, 316–319.

Kasting N. W., Veale W. L., Cooper K. E., and Lederis K. (1981) Effect of hemorrhage on fever: The putative role of vasopressin. *Can. J. Physiol. Pharmacol.* **59**, 324–328.

Kasting N. W., Carr D. B., Martin J. B., Blume H., and Bergland R. (1983) Changes in cerebrospinal fluid and plasma vasopressin in the febrile sheep. *Can. J. Physiol. Pharmacol.* **61**, 427–431.

Kovacs G. L. and deWied D. (1983) Hormonally active vasopressin suppresses endotoxin-induced fever in rats. Lack of effect of oxytocin and a behaviorally active vasopresin fragment. *Neuroendocrinology* **37**, 258–261.

Kruk Z. L. and Brittain R. T. (1972) Changes in body, core and skin temperature following intracerebroventricular injection of substances in the conscious rats: Interpretation of data. *J. Pharm. Pharmacol.* **24**, 835–837.

Kruszynski M., Lammek B., Manning M., Seto J., Haldar J., and Sawyer W. H. (1980) [1-(β-mercapto-β,β-cyclopentamethylenepropionic acid)2-(0-methyl)tyrosine] arginine vasopressin and [1-(β-mercapto-β,β-cyclopentamethylenepropionic acid)] arginine vasopressin, two highly potent antagonists of the vasopressor response to arginine vasopressin. *J. Med. Chem.* **23**, 364–368.

Landis D. M. D. (1985) Promise and pitfalls in immunocytochemistry. *Trends Neurosci.* **8**, 312–317.

Lee R. J. and Lomax P. (1983) Thermoregulatory, behavioral and seizure modulatory effects of AVP in the gerbil. *Peptides* **4**, 801–805.

Lee T. F., Mora F., and Myers R. D. (1985) Effect of intracerebroventricular vasopressin on body temperature and endotoxin fever of macaque monkey. *Am. J. Physiol.* **248**, R674.

Levine J. E. and Ramirez V. D. (1980) *In vivo* release of lutenizing hormone-releasing hormone estimated with push–pull cannulae from the mediobasal hypothalami of ovariectomized, steroid primed rats. *Endocrinology* **107**, 1782–1790.

Lin M. T., Wang T. I., and Chan H. K. (1983) A prostaglandin-adrenergic link occurs in the hypothalamic pathways which mediate the fever induced by vasopressin in the rat. *J. Neural. Trans.* **56**, 21–31.

Lipton J. M. and Glyn J. R. (1980) Central administration of peptides alters thermoregulation in the rabbit. *Peptides* **1**, 15–18.

Manning M., Klis W. A., Olma A., Seto J., and Sawyer W. H. (1982) Design of more potent and selective antagonists of the antidiuretic response to arginine vasopressin devoid of antidiuretic agonism. *J. Med. Chem.* **25**, 414–419.

Meisenberg G. and Simmons W. H. (1984a) Factors involved in the inactivation of vasopressin after intracerebroventricular injection in mice. *Life Sci.* **34**, 1231–1240.

Meisenberg G. and Simmons W. H. (1984b) Hypothermia induced by centrally administered vasopressin in rats. A structure–activity study. *Neuropharmacology* **23**, 1195–1200.

Merker G., Blahser S., and Zeisberger E. (1980) The reactivity pattern of vasopressin containing neurons and its relations to the antipyretic reaction of the guinea pig. *Cell Tiss. Res.* **212**, 47–61.

Milton A. S. (1982) Prostaglandins in Fever and the Mode of Action of Antipyretic Drugs, in *Pyretics and Antipyretics. Handbook of Experimental Pharmacology* (Milton A. S., ed.) Springer, Berlin.

Milton A. S. and Wendlandt S. (1971) Effects on body temperature of prostaglandins of the A, E and F series on injection into the third ventricle of unanaesthetized cats and rabbits. *J. Physiol.* **218**, 325–336.

Morris M., Barnard R. R., and Sain L. E. (1984) Osmotic mechanisms regulating cerebrospinal fluid vasopressin and oxytocin in the conscious rat. *Neuroendocrinology* **39**, 377–383.

Myers R. D. (1971) Methods for Chemical Stimulation of the Brain, in *Methods in Psychobiology* vol. 1 (Myers R. D., ed.) Academic, New York.

Myers R. D. (1972) Methods for Perfusing Different Structures of the Brain, in *Methods in Psychobiology* vol. 2 (Myers R. D., ed.) Academic, New York.

Myers R. D. (1977) Chronic Methods: Intraventricular Infusion, Cerebrospinal Fluid Sampling, and Push–Pull Perfusion, in *Methods in Psychobiology* vol. 3 (Myers R. D., ed.) Academic, New York.

Naylor A. M. and Veale W. L. (1986) Microinjection of a vasopressin antagonist into the ventral septal area: Effects on Interleukin-1 fever. *Proc. West. Pharmacol. Soc.* **29**, 349–352.

Naylor A. M., Ruwe W. D., Kohut A. F., and Veale W. L. (1985a)

Perfusion of vasopressin within the ventral septum of the rabbit suppresses endotoxin fever. *Brain Res. Bull.* **15,** 209–213.

Naylor A. M., Ruwe W. D., Burnard D. M., McNeely P. D., Turner S. L., Pittman Q. J. and Veale W. L. (1985b) Vasopressin-induced motor disturbances: Localization of a sensitive forebrain site in the rat. *Brain Res.* **361,** 242–246.

Naylor A. M., Ruwe W. D., and Veale W. L. (1986) Thermoregulatory actions of centrally administered vasopressin in the rat. *Neuropharmacology* **25,** 787–794.

Palkovits M. (1973) Isolated removal of hypothalamic or other brain nuclei of the rat. *Brain Res.* **59,** 449–450.

Pittman Q. J. and Franklin L. G. (1985) Vasopressin antagonist in nucleus tractus solitarius/vagal area reduces pressor and tachycardia responses to paraventricular nucleus stimulation in rats. *Neurosci. Lett.* **56,** 155–160.

Pittman Q. J., Cooper K. E., Veale W. L., and Van Petten G. R. (1974) Observations on the development of the febrile response to pyrogens in the sheep. *Clin. Sci. Mol. Med.* **46,** 581–592.

Pittman Q. J., Riphagen C. L., and Lederis K. (1984) Release of immunoassayable neurohypophyseal peptides from rat spinal cord, *in vivo. Brain Res.* **300,** 321–326.

Pittman Q. J., Disturnal J. E., Riphagen C. L., Veale W. L., and Bauce L. (1985) Perfusion Techniques for Neural Tissue, in *Neuromethods* vol. 1. *General Techniques* (Boulton A. A. and Baker G. B., eds.) Humana, Clifton, New Jersey.

Rennels M. L., Gregory T. F., Blaumanis O. R., Fujimoto K., and Grady P. A. (1985) Evidence for a "paravascular" fluid circulation in the mammalian central nervous system, provided by the rapid distribution of tracer protein throughout the brain from the subarachnoid space. *Brain Res.* **326,** 47–63.

Reppert S. M., Artman H. G., Swaminathan S., and Fisher D. A. (1981) Vasopressin exerts a rhythmic daily pattern in cerebrospinal fluid but not blood. *Science* **213,** 1256–1257.

Robinson I. C. A. F. (1983) Neurohypophysial Peptides in the Cerebrospinal Fluid, in *The Neurohypophysis: Structure, Function and Control* vol. 60 *Progress in Brain Research* (Cross B. A. and Leng G., eds.) Elsevier, Amsterdam.

Robinson I. C. A. F. and Jones P. M. (1982) Oxytocin and neurophysin in plasma and CSF during suckling in the guinea pig. *Neuroendocrinology* **34,** 59–63.

Rosser M. N., Iversen L. L., Hawthorn J., Ang V. T. Y., and Jenkins J. S. (1981) Extrahypothalamic vasopressin in human brain. *Brain Res.* **214,** 349–355.

Rotte A. A. de, Bouman H. J., and van Wimersma Greidanus Tj. B. (1980) Relationship between α-MSH levels in blood and cerebrospinal fluid. *Brain Res. Bull.* **5**, 375–381.

Ruwe W. D., Veale W. L., and Cooper K. E. (1983) Peptide neurohormones: Their role in thermoregulation and fever. *Can. J. Biochem. Cell. Biol.* **61**, 579–593.

Ruwe W. D., Veale W. L., Pittman Q. J., Kasting N. W., and Lederis K. (1985a) Release of Arginine Vasopressin from the Brain: Correlation with Physiological Events, in *In Vivo Perfusion and Release of Neuroactive Substances* (Drucker-Colin R. and Bayon A., eds.) Academic, New York.

Ruwe W. D., Naylor A. M., and Veale W. L. (1985b) Perfusion of vasopressin within the rat brain suppresses prostaglandin E hyperthermia. *Brain Res.* **338**, 219–224.

Ruwe W. D., Naylor A. M., and Veale W. L. (1985c) Neurohypophysial Peptides Alter the Characteristics of the Febrile Response in the Cat, in *Pharmacology of Thermoregulation. Homeostasis and Thermal Stress: Experimental and Therapeutic Advances* (Cooper K., Lomax P., Schonbaum E., and Veale W., eds.) Karger, Basel.

Smith R. T., Platou E. S., and Good R. A. (1956) Septicemia of the newborn. *Pediatrics* **17**, 549–575.

Sofroniew M. V. (1983) Morphology of Vasopressin and Oxytocin Neurones and Their Central and Vascular Projections, in *The Neurohypophysis: Structure, Function and Control* vol. 60 *Progress in Brain Research* (Cross B. A. and Leng G., eds.) Elsevier, Amsterdam.

Stark R. I., Daniel S. S., Husain M. K., Trapper P. J., and James L. S. (1985) Cerebrospinal fluid and plasma vasopressin in the fetal lamb: Basal concentration and the effect of hypoxia. *Endocrinology* **116**, 65–72.

Swanson L. W. (1977) Immunohistochemical evidence for a neurophysin-containing autonomic pathway arising in the paraventricular nucleus of the hypothalamus. *Brain Res.* **128**, 346–353.

Vandesande F. (1979) A critical review of immunocytochemical methods for light microscopy. *J. Neurosci. Meth.* **1**, 3–23.

Veale W. L. (1971) Behavioral and physiological changes caused by the regional alteration of sodium and calcium ions in the hypothalamus of the unanesthetized cat. Doctoral thesis, Purdue University, West Lafayette, Indiana.

Veale W. L., Kasting N. W., and Cooper K. E. (1981) Arginine vasopressin and endogenous antipyresis: Evidence and significance. *Fed. Proc.* **40**, 2750–2753.

Yaksh T. L. and Yamamura H. I. (1974) Factors affecting performance of the push–pull cannula in brain. *J. Appl. Physiol.* **37**, 428–434.

In Vivo Electrophysiological Techniques in the Study of Peptidergic Neurons and Actions

Alastair V. Ferguson and Leo P. Renaud

1. Introduction

The past decade has witnessed an ever-increasing understanding of the involvement of peptides in diverse physiological functions, including the perception of pain, regulation of the enteric and autonomic nervous systems, body temperature control, and memory retention. Over the same time period, our understanding of the role of peptides in the control of endocrine physiology has been greatly advanced using a combination of anatomical, biochemical, pharmacological, and physiological techniques. There is now considerable evidence to suggest that, in addition to their classical endocrine effects, many peptides may also act as neurotransmitters in transferring neuronal information from a pre- to an adjacent postsynaptic membrane (Kuffler et al., 1984). Recent studies (Tramu et al., 1983; Kiss et al., 1984; Sawchenko et al., 1984) demonstrating that more than one peptide may be contained within a single neuron, and that single neurons may preferentially modify immunological characteristics in favor of one peptide over another in response to relevant physiological stimuli, indicate an intriguing complexity in the coordinated physiological functions of these biologically active substances.

Since the initial visualization using immunocytochemical methods of oxytocin and vasopressin in the neurosecretory cells of the hypothalamic supraoptic nuclei (SON) and paraventricular nuclei (PVN), and the subsequent confirmation that these cells behave as true neurons in terms of their physical characteristics (*see*

Renaud, 1978, for review), many other neuronal cell groups have been shown to contain a variety of different peptides (see Hokfelt et al., 1980, for review).

This demonstration of the widespread occurrence of peptidergic neurons has stimulated efforts to understand their physiological role in neural tissue. In particular, single-cell electrophysiological recording techniques have been utilized as a research tool to investigate both the effects of peptides on neuronal excitability, as well as the mechanisms controlling the excitability of the peptide-producing cells themselves. In vitro studies have contributed greatly to our understanding of the intrinsic properties of these neurons and are described in detail elsewhere in this volume. A major limitation of such in vitro studies, however, is that they provide little information as to either the endogenous afferent and efferent connections of these neurons, or the influences of physiological manipulation on their excitability. It is in these specific areas that valuable information may be gained from in vivo electrophysiological studies. The major emphasis on this chapter will be the review of such techniques, specifically as they pertain to both the study of peptidergic neurons and the investigation of the possible functions of peptides as neurotransmitters.

We will first consider the methods by which it is possible to identify specific peptidergic cell groups under investigation according to their anatomical location, their projection sites as identified by antidromic activation, and their physiological properties. Combined intracellular marking and immunocytochemistry to positively identify single neurons that have been studied by electrophysiological means will also be considered.

In vivo electrophysiological techniques for the study of afferent pathways to such identified neurons will then be described. Studies utilizing these techniques have provided considerable information regarding the nature of synaptic inputs to identified peptidergic cells; some of these will be discussed in detail.

As already mentioned, there is now considerable evidence implicating peptides as neurotransmitters. Much of this evidence has been obtained by studies using microiontophoresis or pressure-ejection techniques to evaluate changes in excitability of single cells following administration of peptides into the immediate vicinity of the cell body. The advantages and limitations of these techniques will be reviewed.

Finally we will outline single-unit recording techniques presently being used in chronic preparations to investigate the physiology of oxytocin- and vasopressin-containing neurons in conscious animals.

2. Cell Identification

A major concern in the utilization of electrophysiological techniques in the in vivo study of peptidergic neurons relates to the ability to identify the specific group of neurons under observation. In select instances, the combination of different techniques for cell identification may permit a definitive classification of specific peptidergic cell groups. The neurohypophysial oxytocin- and vasopressin-synthesizing neurons of the rat hypothalamus will be introduced here as a model peptidergic system.

2.1. Identification by Anatomical Location

An initial step in cell identification entails the accurate definition of the anatomical location of that neuron. Should the area from which electrophysiological recordings are being made consist of highly specialized cell groups, localization of a cell body in such a region provides clues as to the chemical nature of that neuron. For instance, the SON of the hypothalamus contain neurons that synthesize either oxytocin or vasopressin (Vandesande and Dierickx, 1975; Swaab et al., 1975) and send their axons almost exclusively to the posterior pituitary (Sherlock et al., 1975). Accordingly, cells recorded in this region may be assumed to contain and secrete one of these two peptides.

The topography of the recording electrode's position may also provide data relevant to neuronal function. This is particularly true when dealing with cerebellar or cerebral cortical neurons, or basal hypothalamic structures when accessing the brain from the ventral surface. With deeper brain structures, visualization of specific regions may require that the overlying tissue be first removed, usually by suction, in order to place recording electrodes directly in brain regions containing peptidergic neurons.

2.1.2. Reconstruction of Electrode Tracts

A second relatively simple technique for determining the anatomical location of individual cells involves reconstruction of recording electrode tracks through the brain from histological sections (Fig. 1). The accuracy of this procedure is dependent on keeping precise records as to the distance from the brain surface at which each cell is recorded, as well as the relative positions of sequential recording electrode penetrations. This approach to locating recording electrode sites can provide more accurate information if it is combined with a procedure to mark each recording site. Metal recording electrodes permit small lesions to be placed in tissue by passing direct current (cathodal) through the electrode. The intensity (mA) and duration (s) necessary to cause appropriate lesions obviously vary according to the tip exposure. When recording with fluid-filled glass micropipets, it is also possible to pass sufficient current to make identifiable lesions. More often, however, specific dyes (e. g., fast green, pontamine blue, lucifer yellow), are dissolved in the recording electrolytes and then ejected at recording sites by current application. When using this technique, appropriate choices of histological stain and dye should be made to ensure that the former will not mask the dye spot. For example, a combination of pontamine blue as a marker, and subsequent cresyl violet staining of the section, would suffer from this problem.

Although the above techniques permit reasonably accurate determination of the anatomical location of cells, they do not provide definitive information as to their chemical identity, i. e., the specific peptides contained in these neurons. A combination of single-cell marking and immunocytochemistry can provide such information. There are problems in developing such techniques for routine use, however, in part related to the present difficulties associated with in vivo intracellular recordings of adequate quality and duration. Fine-tipped, high-resistance (60–100Ω, with KCL) micropipets are needed to impale cells. When the pipets are filled with solutions that contain intracellular markers, for example lucifer yellow or horseradish peroxidase, recordings may be of poor quality because of the presence of even higher resistances. In the hypothalamus, where a high percentage of neurons are peptidergic, these problems combined with the inherent instability of a highly vascular region suggest that at present this identification technique is more likely to succeed when used in in vitro preparations (Yamashita et al., 1983).

2.2. Identification by Antidromic Invasion

The axonal projections of certain groups of peptidergic neurons in the hypothalamus permit reliable identification by antidromic activation (Renaud, 1978). Since all axons are capable of

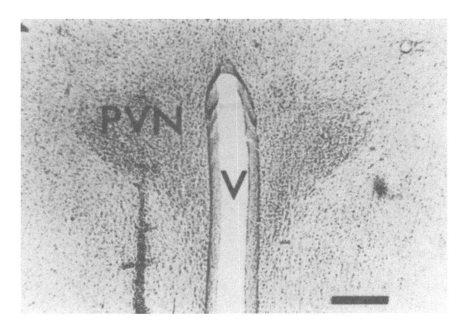

Fig. 1.   Photomicrograph of a 100-μm cresyl violet-stained coronal section through the medial hypothalamus of a rat illustrates the upper portion of third ventricle (V) and the paraventricular nuclei (PVN). Below left the PVN is a blood-filled micropipet trajectory produced from a recording electrode inserted from the ventral hypothalamus. Calibration bar, 100 μm.

propagating action potentials in both orthodromic (from cell body to terminal) and antidromic (from terminal to cell body) directions, the following criteria serve to establish that an action potential recorded from a cell soma in response to a distant electrical stimulus has been antidromically propagated (Fig. 2):

1.   Evidence for all-or-nothing constant latency response at threshold intensities of stimulation. The involvement of any synapse between a cell body and a projection site would result in obvious variability in the response of that cell body to electrical stimulation in the projection site.
2.   Ability to follow high-frequency (up to 200 Hz) stimulation.

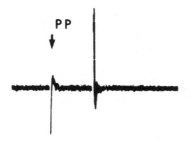

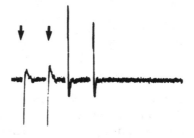

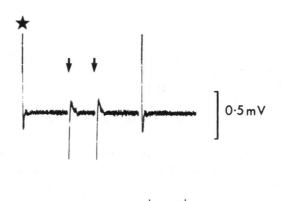

0·5 mV

10 msec

3. The presence of collision cancellation between antidromically evoked spikes and spontaneous orthodromic action potentials within a critical period of one latency interval of the stimulus.

When recording from quiescent cells that display no spontaneous activity, it is obviously not possible to demonstrate collision cancellation; in this event satisfaction of the first two of these criteria may be considered proof of antidromic activation. In contrast, these two criteria become redundant in spontaneously active cells in which the critical period for collision cancellation is clearly demonstrated; by itself this demonstration becomes the most compelling proof of antidromic activation.

Thus the technique of antidromic activation may be utilized to identify a single neuron as having an axon that projects to, or through, the region where the stimulating electrode is located. It should be emphasized, however, that apart from some specialized cases (see below), this does not provide information as to the termination sites of that axon. In experiments using this technique, it is also important to establish the specificity of the stimulation sites. Placement of electrodes in adjacent anatomical regions resulting in lack of antidromic activation provides strong evidence as to such specificity, as well as evidence against excessive current spread from stimulation sites. The arguments for and against the use of specific electrode types for the stimulation of brain tissue have been reviewed elsewhere (Ranck, 1975) and so will not be covered in detail here.

The anatomical projection site of the axons of virtually all oxytocin- and vasopressin-synthesizing magnocellular neurons of the SON and PVN is the posterior pituitary (Sherlock et al., 1975). This is a unique and convenient arrangement in that the posterior

←

Fig. 2. Oscilloscope traces of action potentials recorded from a hypothalamic neurohypophysial cell in the supraoptic nucleus of the rat. Criteria for antidromic activation from a posterior pituitary (PP) stimulation electrode include constant latency all-or-none response at threshold (upper trace), ability to follow a pair of stimuli presented at frequencies greater than 100 Hz (middle trace), and evidence for cancellation of antidromic spikes by spontaneously occurring action potentials (*) within the critical interval (one latency period on each side of the stimulus), as observed for the first stimulus in the lower trace.

pituitary can be readily approached with stimulation electrodes without fear of stimulus spread to adjacent brain regions. This homogeneity of projections permits the positive identification of these specific peptidergic cells by antidromic activation.

A second major group of hypothalamic peptidergic neurons that may be identified according to their antidromic responses are those projecting to the median eminence (Renaud, 1981). Cells that may be antidromically activated from this region, but not from the posterior pituitary (axons of oxytocin and vasopressin neurons pass close to the median eminence), may be identified as possible peptidergic neurosecretory cells controlling the anterior pituitary. Further clues as to the specific peptides such antidromically identified cells contain may be obtained according to the anatomical locations of the cells; for example, TRH- and CRF-immunoreactive cells are located in different regions of the PVN (Swanson and Sawchenko, 1983).

A further use for the technique of antidromic activation is in the identification of multiple projection sites for the axons of single peptidergic neurons. For example, single PVN cells that demonstrated antidromic activation following electrical stimulation in both the posterior pituitary (oxytocin- or vasopressin-containing) and the PVN of the thalamus (Fig. 3B,E) would be viewed as projecting to both of these sites (Ferguson et al., 1984a). In addition, interaction studies and an assessment of the intervals between stimuli that achieve collision cancellation (Fig. 3C,D) indicate the likely site of axonal bifurcation. The latter provides information that cannot be obtained from anatomical double-labeling studies. Thus the cell from which data in Fig. 3 were obtained had a branch point close to the cell body.

2.3. Identification According to Physiological Properties

Cells from which electrophysiological recordings are made may also be classified according to their intrinsic properties and responses to physiological manipulations of an animal. The usefulness of such classifications is perhaps best demonstrated by considering its use in the hypothalamic neurohypophysial system.

The spontaneous activity patterns of antidromically identified oxytocin- and vasopressin-containing neurons fall into one of three major categories: phasic, continuous, or slow irregular (including quiescent cells) (see Poulain and Wakerley, 1982, for review). Physiological experiments have shown that occlusion of the carotid

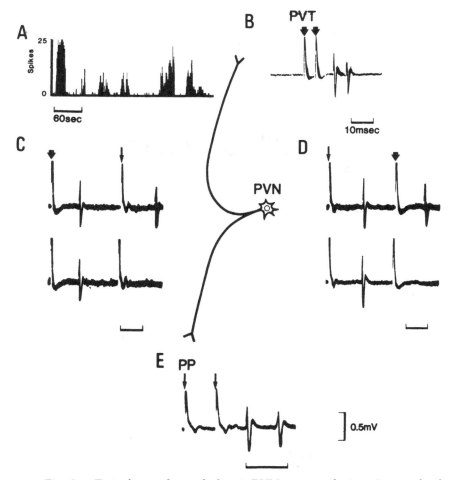

Fig. 3. Data from a hypothalamic PVN neuron that projects to both the posterior pituitary (PP) and the region of the paraventricular nucleus of the thalamus (PVT). The ratemeter record in A illustrates intermittent spontaneous activity that may be termed "phasic." Superimposed oscilloscope sweeps in B and E illustrate constant latency responses to a pair of stimuli presented in the PVT (B) or PP (E); arrowheads depict shock artifacts. In C, oscilloscope traces illustrate the critical interval for cancellation of the PP-evoked antidromic spike following PVT stimulation. In D, similar features are presented, but with the order of stimuli reversed. These data are interpreted to indicate a PVN cell whose axon is branched near the cell soma and that innervates both PVT and PP.

PUTATIVE VP NEURONS :

● **Phasic.**

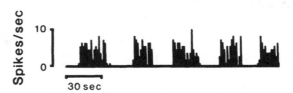

● **Continuous, BP−sensitive.**

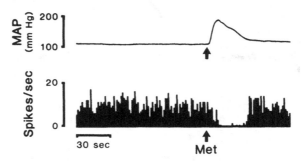

PUTATIVE OXY NEURONS :

● **Continuous, BP−insensitive.**

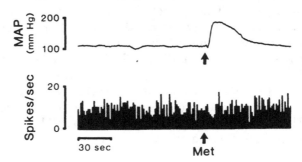

arteries causes independent release of vasopressin (without oxytocin) from the posterior pituitary (Clark and Rocha e Silva, 1967). Electrophysiological experiments demonstrated a specific activation of the phasically active neurohypophysial cells in response to such carotid occlusion (Dreifuss et al., 1976; Harris, 1979). These data suggest that phasically active cells are vasopressinergic. This view was further supported by electrophysiological studies in anesthetized lactating rats with suckling pups. Periodic milk ejection occurs in this preparation and is related to a pulse release of oxytocin from the posterior pituitary. Resulting increases in intramammary pressure may be monitored following cannulation of a mammary gland and, thus, oxytocin release may be measured. These experiments showed a population of antidromically identified neurohypophysial cells that demonstrated a high-frequency discharge 10–15 s prior to the rise in intramammary pressure (*see* Poulain and Wakerley, 1982). These cells were, therefore, classified as oxytocinergic. It was an extremely rare observation where neurons in this group showed a "phasic" pattern of spontaneous activity (Lincoln and Wakerley, 1974; Summerlee, 1982).

These classifications, however, leave a proportion of continuously active or slow irregular cells unclassified as to their peptidergic content. It now appears, however, that these neurons may be classified according to their responses to baroreceptor activation. As already mentioned, occlusion of the carotid arteries specifically activates vasopressin-secreting cells. Conversely, an increase in arterial blood pressure is associated with a reduction in the firing frequency of this same group of neurons (Harris, 1979). Such increases in blood pressure may be induced by iv administration of the alpha-adrenergic agonist metarminol, and, as expected, this stimulus inhibits the activity of phasic cells as well as of a proportion of continuously active or slow irregular cells (Fig. 4). These neurons are, therefore, putatively classified as vasopressinergic.

←——————————————————————————————

Fig. 4.   Ratemeter records and corresponding mean arterial pressure (MAP) measurements illustrating features that are of value during in vivo extracellular recordings of SON or PVN neurohypophysial neurons to distinguish between vasopressin (VP)- and oxytocin (OXY)-synthesizing cells. VP-synthesizing neurons may fire in a phasic or continuous mode and demonstrate a reduction in firing following baroreceptor activation achieved through an intravenous injection of metaraminol (MET). In contrast, OXY-synthesizing neurons fire continuously and are insensitive to baroreceptor manipulations (from Renaud et al., 1985).

As the above account indicates, a combination of identification techniques permits a relatively clear distinction to be made between oxytocin- and vasopressin-secreting neurons of the SON and PVN. Unfortunately, at present such a clear classification is not possible for other groups of peptidergic neurons showing less anatomical as well as functional homogeneity. Identification of such groups of cells awaits the definition of *specific* physiological stimuli to which they respond [for example, TRH-containing neurons of the PVN *may be* specifically activated by cold exposure (Ishikawa et al., 1984)] and/or technical advances allowing for a routine combination of intracellular labeling and immunocytochemistry.

3. Afferent Inputs to Peptidergic Neurons

The ability to determine that recordings arise from specific types of peptidergic neurons on the basis of their anatomical localization, antidromic activation, and firing patterns is a prelude to the use of electrophysiological techniques to investigate their afferent connections. The following examples pertain specifically to the oxytocin- and vasopressin-containing neurons of the SON and PVN. In most cases, the first stage of such electrophysiological analysis of afferent inputs to identified peptidergic cells is dependent on information derived from anatomical tracing studies. For example, anterograde and retrograde tracing studies have described widespread afferent projections to both the SON and PVN (Silverman et al., 1981; Sawchenko and Swanson, 1983), although such studies provide no information as to the functional nature, for example, excitatory or inhibitory, of such synaptic inputs. The advantage of the electrophysiological approach lies in the information it provides as to the nature of these synaptic inputs (excitatory, inhibitory, latency, and duration) and whether such inputs may specifically influence oxytocin or vasopressin neurons, or both.

The basic techniques employed in such experiments are illustrated in Fig. 5. Extracellular single-unit recordings are made from identified peptidergic neurons, and the effects of electrical stimulation in discrete brain regions known to send efferent projections to these neurons are examined. Changes in cells' excitability may be assessed as either an increase, decrease, or no effect, by using either ratemeter records or peristimulus histograms (PSHs). There are two major technical considerations in this experimental protocol: (1) the method of electrical stimulation, and (2) the method of analysis. In consideration of the former, much has been written regarding the respective merits of monopolar as compared to bipolar or concentric bipolar electrodes for electrical stimulation of the brain (for review, *see* Ranck, 1975). It would appear that use of either type of electrode should prove satisfactory as long as the appropriate controls are part of these experimental protocols. That is, it must be shown that the effects being attributed to the area of interest are only observed when the stimulating electrode is in this region, and that they are no longer seen when the electrode is positioned in immediately adjacent anatomical regions. Such rigorous controls will also ensure that current intensities for stimulation are not excessive. At this time it should also be pointed out that intensity of stimulation is related to both the current and the pulse width of the stimulus. Thus, both should always be defined when reporting experimental protocols.

The analysis of data obtained in this type of experiment is probably best served using both approaches previously mentioned, although it should be stressed that peristimulus histograms provide more definitive data for analysis in terms of both latency and duration of observed synaptic effects. The ratemeter analysis in most cases will only be used to see whether a high-frequency or repetitive stimulation has similar effects to those observed in a PSH. Also, such information may be of particular relevance in a cell from which the PSH shows both excitatory and inhibitory effects at different latencies.

Of the various inputs to the hypothalamic SON and PVN, those arising from the limbic system, the subfornical organ, and the noradrenergic cell groups of the medulla have been extensively examined. The influences of these different inputs on the excitability of neurosecretory cells differ in such a way as to demonstrate the variety of afferent inputs through which other brain regions may influence the excitability of peptidergic neurons. The findings of these studies will thus be reviewed in some detail.

3.1. Limbic System

Anatomical tracing studies have demonstrated that the amygdala and septum project to areas immediately surrounding the SON and PVN (Sawchenko and Swanson, 1983), indicating a pathway through which emotional stimuli may influence the release of

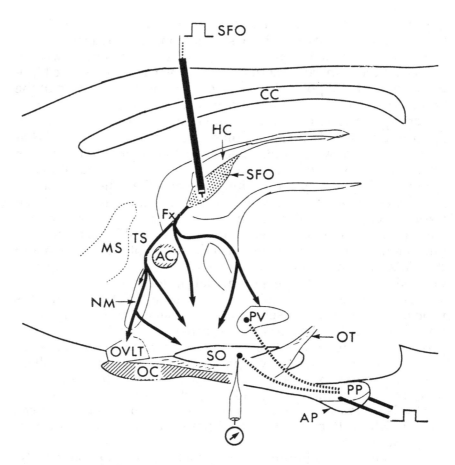

Fig. 5.   A schematic sagittal drawing of the rat diencephalon depicts
the location of the subfornical organ (SFO), lying in the rostral-dorsal
third ventricle under the hippocampal commissure (HC) and fornix (Fx).
Efferent pathways (solid heavy lines) are illustrated coursing rostrally
and caudally to the anterior commissure (AC) to innervate nucleus
medianus (NM), organum vasculosum of the lamina terminalis (OVLT),
hypothalamic supraoptic (SO), and paraventricular (PV) nuclei. In order
to activate SFO efferents, concentric bipolar electrodes may be inserted
from the dorsal surface. A ventral approach to the hypothalamus permits
insertion of a bipolar stimulating electrode in the posterior pituitary (PP)
and a recording micropipet in the SON. Abbreviations: AP, anterior
pituitary; CC, corpus callosum; MS, medial septal nucleus; OC, optic
chiasm; OT, optic tract; TS, triangular septal nucleus.

posterior pituitary hormones (Mirsky et al., 1954; Cross, 1955; Taleisnik and Deis, 1964; Keil and Severs, 1977).

Electrophysiological studies have since examined the effects of electrical stimulation in these limbic structures on the excitability of identified neurosecretory cells. Stimulation in the amygdala has been reported to cause excitation or inhibition among antidromically identified neurohypophysial putative oxytocin- and vasopressin-secreting neurons, although the inhibitory effects appear to predominate (Negoro et al., 1973; Pittman et al., 1981; Thomson, 1982; Hamamura et al., 1982; Ferreyra et al., 1983; Cirino and Renaud, 1985). Attempts to dissect out the apparent functional heterogeneity of these projections by stimulating in different regions of the amygdala indicate a lack of functionally distinct projections (Hamamura et al., 1982; Ferreyra et al., 1983).

Electrical stimulation in the ipsilateral septum has been shown to elicit both inhibitory and excitatory responses from oxytocin and vasopressin cells of the PVN (Koizumi and Yamashita, 1972; Negoro et al., 1973). In contrast, septal inputs to the SON have been reported to be predominantly inhibitory to both oxytocin- and vasopressin-secreting cells (Poulain et al., 1980; Cirino and Renaud, 1985). Thus, in the case of the septum, electrophysiological studies indicate that there *may be* different functional inputs to the SON as compared to the PVN neurohypophysial cells.

3.2. Noradrenergic Afferents

The development of techniques for the visualization of catecholamines in nervous tissue resulted in the demonstration of dense terminal plexi in both the SON and magnocellular region of the PVN (Carlsson et al., 1962; Fuxe, 1965). Later, anatomical studies showed this innervation (1) to be primarily noradrenergic (Palkovits et al., 1974), (2) to be concentrated in SON and PVN regions containing the majority of vasopressin-containing neurons (McNeill and Sladek, 1980), and (3) to be derived primarily from the A1 cell group of the caudal ventrolateral medulla (Sawchenko and Swanson, 1981, 1982). Until recently, however, much contradictory evidence derived from a combination of lesion, microinjection, and iontophoretic techniques (see Renaud et al., 1985) existed as to the functional role of these noradrenergic afferents in the control of hormone release from the posterior pituitary. In vivo electrophysiological techniques have now been utilized to examine the effects of electrical stimulation of the A1 cell group on the excitability of putative oxytocin- and vasopressin-containing neurons. These studies demonstrated that discrete electrical stimulation in the A1 cell group enhanced the activity of SON and PVN neurosecretory cells, supporting the view of a facilitatory role of the noradrenergic afferents in controlling the excitability of these cells (Day and Renaud, 1984; Day et al., 1984). Interestingly, these effects were observed only on those cells putatively identified as vasopressinergic according to the criteria reviewed in section 2 (Fig. 6). Thus, in this case these in vivo electrophysiological techniques have clearly demonstrated a selective facilitatory innervation of vasopressin-secreting neurons by noradrenergic afferents from the A1 cell group. These findings are supported by the anatomical demonstration of the catecholaminergic terminal plexus specifically in the regions of SON and PVN where vasopressin cells predominate (McNeill and Sladek, 1980), as well as by recent iontophoretic studies (Day et al., 1986). It may well be that previous iontophoretic demonstration of inhibitory effects of noradrenaline may have been related to the high concentrations of the monoamine that have now been shown to occur during iontophoretic ejection (Armstrong-James and Fox, 1983). The problems associated with this technique will be discussed in further detail in section 4.

3.3. Subfornical Organ

The subfornical organ (SFO) is a highly vascularized circumventricular structure that lacks a normal blood–brain barrier (Dellman and Simpson, 1979). Among the predominantly hypothalamic efferent projections of the SFO are the SON and PVN (Miselis, 1981; Lind et al., 1982), and at least a portion of these efferents have been shown using ultrastructural techniques to synapse directly on the neurosecretory cells of these regions (Renaud et al., 1983). A study of SFO efferent projections to the SON and PVN is of particular relevance to studies on peptidergic function, since it has been demonstrated that systemic angiotensin II acts on neurons within this structure to cause drinking behavior (Simpson et al., 1978; Eng and Miselis, 1981), an increase in plasma vasopressin concentrations (Knepel et al., 1982), and increases in arterial blood pressure (Mangiapane and Simpson, 1983; Ferguson and Renaud, 1984). Furthermore, it has been demonstrated recently, using anatomical double-labeling studies, that angiotensin II-containing

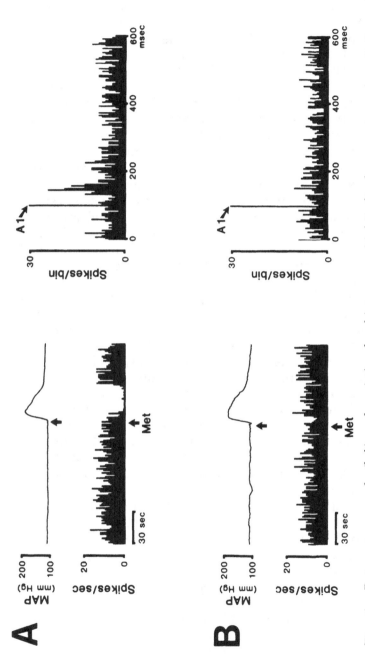

Fig. 6. Ratemeter records (left) and peristimulus histograms (right) display spontaneous activity and response to A1 region stimulation, respectively, from two PVN neurohypophysial neurons in the same experiment. (A) Putative VP-synthesizing neuron displays continuous activity, depressed by a metaraminol-induced rise in blood pressure. Delivery of a cathodal 100-μA pulse to the A1 region at 1 Hz increases the probability of cell firing, as seen in the peristimulus histogram. (B) Putative OXY-synthesizing neuron displays continuous firing and lack of response to metaraminol as well as to A1 stimulation.

neurons in the SFO project directly to the region of the PVN (Lind et al., 1985). A variety of other peptides has also been localized in either nerve fibers or cell bodies within the SFO.

Recent studies of the connections of the SFO with SON and PVN neurohypophysial neurons using electrophysiological techniques (Sgro et al., 1984; Ferguson et al., 1984b) have demonstrated that the predominant response to electrical stimulation in the SFO is an *increase* in the excitability of both oxytocin- and vasopressin-synthesizing neurons (Fig. 7). The fact that different latencies and durations of excitation, as well as a small number of inhibitory responses (Fig. 7), were observed within single experiments suggests that several functionally distinct connections may exist between the SFO and the hypothalamic neurohypophysial system. This clearly demonstrates one major advantage, as well as a problem, with this experimental approach. The differences in latency and duration obviously provide information regarding a functional heterogeneity that cannot be derived from anatomical tracing techniques alone. This heterogeneity, however, also serves to illustrate an important limitation of electrical stimulation techniques in that they result only in the activation of anatomically specific rather than functionally specific groups of neurons. A further interesting finding is the long duration of the predominant excitatory response on the neurosecretory cells. Such an effect is not typical of a simple monosynaptic projection, although other demonstrations of long-duration neurotransmitter action of peptides combined with the observation of angiotensin II immunoreactive projections from SFO to PVN (Lind et al., 1985) raise the possibility that these effects could be caused by the action of this peptide as a neurotransmitter. Evidence either in support of or against such a role for angiotensin II may best be obtained by the iontophoretic or pressure-ejection techniques to be described in more detail below.

In this section, we have examined some of the electrophysiological techniques that may be used to study afferent neural inputs to identified peptidergic neurons. Although the example of the hypothalamic oxytocin- and vasopressin-containing neurons has been used extensively to illustrate the information regarding the nature of inputs that may be obtained using these techniques, it should not be assumed that only these groups of cells are accessible to such investigation. The present limitations, however, are mainly associated with the difficulties surrounding the positive identification of other peptidergic cell groups (for example, tuberoinfundibular cells in the PVN projecting to the median eminence). Fu-

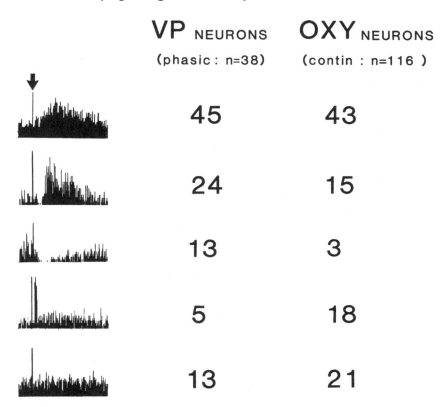

| VP NEURONS | OXY NEURONS |
| (phasic: n=38) | (contin : n=116) |

45 43

24 15

13 3

5 18

13 21

Fig. 7. Representative peristimulus histograms on the left illustrate the variety of patterns of responses recorded from rat VP- and OXY-synthesizing SON and PVN neurosecretory neurons following single-pulse electrical stimulation in the subfornical organ. The arrow illustrates the stimulus artifact occurring 100 ms after the onset of each histogram. The values on the right indicate the percentage of neurons corresponding to each category of response.

ture advances in such identification techniques should permit a more definitive investigation of the role that other CNS structures play in controlling the excitability of these peptide-containing neurons.

4. Assessment of Peptide Actions on Neuronal Function

The potential for peptides to function as neurotransmitters has been the subject of several recent reviews (Kelly, 1982; Renaud,

1983; Kreiger, 1983; Iversen, 1984). The general strategy utilized for in vivo electrophysiological experiments involves obtaining recordings (almost all extracellular) of spike frequency in association with peptide administration. Changes in central neuronal activity as a result of peripheral peptide administration leaves open the question as to how much (if any) of the peptide actually reaches the neuron under observation, given the lack of permeability of the blood–brain barrier to polypeptides and larger molecular species. Although intracerebroventricular administration eliminates the blood–brain barrier issue, one still faces the uncertainties related to peptide diffusion, concentration, and precision as to its locus of action. These problems may be avoided by the utilization of microiontophoretic or pressure-ejection techniques for the direct application of agents in the region of individual neurons. This section examines some of the technical considerations relevant to the use of these experimental approaches.

4.1. Microiontophoresis vs Pressure Ejection

Two techniques routinely used to apply specific agents in the immediate vicinity of individual neurons being recorded from are microiontophoresis and pressure ejection. The former technique relies on the movement of charged molecules within an electric field to eject specific particles from the mouth of one barrel of a micropipet (Kelly, 1975), whereas the latter employs direct pressure to eject the solution filling a similar micropipet (Sakai et al., 1979). The major advantages of the iontophoretic technique are (1) no mechanical pressure is needed for drug ejection, and (2) a retaining current may be applied to individual barrels to minimize the problems of drug leakage, although it should be realized that this will always result in the ejection of an oppositely charged ion (Hicks, 1984). Specific disadvantages relate to (1) an inability to define precisely the doses of applied drugs, and (2) possible current artifacts. Recent work addressing the former problem now suggests that, at least in some cases, the amount of drug ejected may be considerably higher than was originally believed (Armstrong-James and Fox, 1983). Appropriate controls can minimize the possibilities of observed effects being artifacts of current application (for review, see Hicks, 1984).

Obviously the pressure-ejection technique offers certain advantages over iontophoresis. The fact that drugs are ejected as a result of pressure rather than electrical charge means that the concentration of drug leaving the micropipet will always be known. It should be stressed, however, that there are still problems controlling the volume ejected by each pressure pulse. The pressure-ejection technique also offers the advantage of avoiding the problems of current artifacts, although on the negative side the pressure applied to eject drugs may in some cases cause sufficient mechanical displacement of the electrode tip so that it will no longer be close enough to the cell of interest to record its action potentials. This mechanical displacement will be a more significant problem where intracellular recordings are being made.

A further technical consideration of importance to both of these techniques is the type of micropipet to be used. The aim is to combine a recording electrode giving satisfactory signal-to-noise ratios, with drug ejection barrels that remain patent while being advanced through brain tissue. Many different electrode configurations have been used and their relative advantages and disadvantages have been reviewed extensively elsewhere (Hicks, 1984).

A final and probably most important consideration of the techniques of iontophoresis and pressure ejection relates to the area of the cell under study that will be exposed to substances applied. In view of recent studies demonstrating extensive dendritic trees of many different populations of neurons (Keirstead and Rose, 1983; Randle et al., 1985), it should be stressed that drugs applied by either iontophoresis or pressure ejection will only gain access, even in ideal situations, to the cell body and proximal dendrites. Although a discussion of the relative importance of synaptic inputs on distal as opposed to proximal dendrites is beyond the scope of this presentation, it is quite clear that the greater the significance of the former, the more serious is this limitation.

4.2. In Vivo Evidence for Peptidergic Neurotransmission

Iontophoretic and pressure-ejection techniques have provided considerable in vitro evidence implicating peptides as neurotransmitters. Few in vivo studies, however, have provided information that could not be obtained utilizing the in vitro approach. Although in vivo experiments have demonstrated that a large variety of peptides when administered by iontophoresis or pressure ejection cause changes in neuronal excitability, these studies in isolation do little to promote the role of these substances as neurotransmitters. According to the definition of Kandel and Schwartz (1984), in order to be accepted as a neurotransmitter at a particular synapse a substance must:

1. be synthesized in the presynaptic neuron.
2. be present in the presynaptic terminal and be released from these terminals in sufficient quantities to have its supposed action on the postsynaptic membrane.
3. when applied exogenously in reasonable concentrations, mimic the actions of endogenously released neurotransmitter.
4. have a specific mechanism for removing it from its site of action.

In vivo electrophysiological studies utilizing iontophoresis or pressure ejection may provide evidence toward meeting the second and third criteria.

The major advantage offered by the in vivo approach is that experiments may be carried out to investigate neurons in specific anatomical regions known to be the termination site of identified peptidergic neural projections. Such pathways may also be activated by electrical stimulation in order to establish if endogenous release of the peptide may have similar effects to exogenous application. Finally, it is necessary to demonstrate that pharmacological antagonists administered in the region of the neuron under study will abolish the effects of endogenous release of the peptide.

One peptide that has been proposed to play a neurotransmitter role in the central nervous system is angiotensin II (AII). This substance has been shown to be present in the cell bodies of a number of different neuronal populations, although other regions of the brain such as the PVN, SFO, and organum vasculosum of the lamina terminalis (OVLT) have been reported to contain specific AII receptors (Mendlesohn et al., 1984). Recent double-labeling studies have reported that AII immunoreactive neuronal cell bodies in the SFO send axonal projections to the PVN (Lind et al., 1985). These studies, therefore, suggest that such projections from the SFO to PVN may utilize AII as a neurotransmitter.

In vivo electrophysiological studies in which AII has been iontophoretically applied to PVN neurosecretory cells lend some support to such a view. Such studies have reported that approximately 50% of these cells are activated by iontophoretic application of AII, an effect that may be inhibited by pretreatment with the AII antagonist saralasin, while the remaining cells are unaffected

(Akaishi et al., 1981). These studies indicate a postsynaptic effect of AII in a region containing AII receptors and receiving neural inputs from AII immunoreactive cell bodies. Further studies are, however, necessary to show that the effects of endogenously released AII (release by electrical activation of AII-containing afferents) are blocked by iontophoretic application of saralasin.

The preceding section suggests the potential value of techniques of iontophoresis or pressure ejection in studying the possible role of specific peptides as neurotransmitters in vivo. Possibly the two most important considerations in the design of these experiments are (1) that the neuron under study receive relevant peptidergic afferents, and (2) that specific antagonists of the peptide under study be available.

5. Single-Cell Recordings From Conscious Animals

Our understanding of the role of peptidergic neurons in neuroendocrine reflexes has until recently remained limited. The main reason for such a lack of information is that the majority of these reflexes are inhibited or abolished by the anesthesia necessary to permit stable extracellular single-unit recordings from these neurosecretory cells. Techniques have recently been developed, however, that permit single-cell recordings from identified oxytocin- and vasopressin-secreting cells in unanesthetized rats (Summerlee and Lincoln, 1981), thereby eliminating the problems associated with the use of anesthesia in the study of neuroendocrine reflexes.

In order to obtain such recordings, an array of platinum microwire recording electrodes is chronically implanted such that the tips of these electrodes are located in the region of the PVN. A stimulating electrode is also implanted with its tip in the posterior pituitary for the antidromic activation of PVN oxytocin- and vasopressin-secreting cells that project to this region. Following a recovery period, it is then possible to record extracellular action potentials in antidromically identified cells from a small number of these chronically implanted electrodes. Using these techniques, it has been possible to record single-unit activity from putative oxytocin and vasopressin neurons in conscious rats and rabbits, and to confirm that putative oxytocin neurons demonstrate a brief high-frequency discharge prior to milk ejection (Summerlee and Lincoln, 1981; Summerlee, 1983; Paisley and Summerlee, 1984). Oxytocin neurons have also been studied during the expulsive phase of birth (Summerlee, 1981). In this situation these cells showed a 2–5-fold increase in their level of spontaneous activity 15 min before abdominal contractions began; this was maintained for 45 min after delivery of the last placenta. High-frequency discharges were also recorded from these neurons following forceful abdominal contractions, and in some cases prior to delivery of fetuses or placentae.

Although these studies demonstrate clearly that extracellular recordings may be obtained from oxytocin and vasopressin neurons in conscious, freely moving animals, a major drawback associated with the routine use of these chronic recording methods relates to the limited number (<2%) of implanted electrodes yielding suitable single-unit recordings. Further development and technical refinement should enable future electrophysiological studies to examine the cellular mechanisms underlying a large variety of different neuroendocrine reflexes.

6. Commentary

In this chapter we have described a number of electrophysiological techniques that may be used in order to study peptidergic neurons and actions in vivo. At present, a major limiting factor in our ability to study peptidergic neurons in vivo relates to the difficulties associated with the positive identification of the neuron under study. Certain peptidergic cells may be identified according to anatomical location, axonal projection, and physiological properties. Considerable benefit could be derived from the development of reliable means for routine combined intracellular marking and immunocytochemistry. Our present knowledge of the connections of oxytocin- and vasopressin-secreting cells stems from the ability to use electrophysiological techniques to define their functional nature. Other inputs to any identifiable group of peptidergic cells may be similarly studied. The anatomical description of specific peptidergic neuronal pathways has suggested a neurotransmitter role for these substances within the central nervous system. The ability, using iontophoretic or pressure-ejection techniques, to study the effects of activation of such peptidergic pathways on postsynaptic neurons should permit a definitive analysis of the possible functions of peptides as interneuronal messengers. It is anticipated that the variety of in vivo electrophysiological techniques now available for the study of brain peptides will yield exciting new data as to the physiological roles of these substances in a variety of homeostatic and behavioral functions.

Acknowledgments

Thanks to Wendy Brown for typing this manuscript. Supported by the MRC of Canada.

References

Akaishi T., Negoro H., and Kobayasi S. (1981) Electrophysiological evidence from multiple sites of actions of angiotensin II for stimulating paraventricular neurosecretory cells in the rat. *Brain Res.* **220,** 386–390.

Armstrong-James M. and Fox K. (1983) Effects of iontophoresed noradrenalin on the spontaneous activity of neurones in a rat primary sematosensory cortex. *J. Physiol.* (Lond.) **335,** 427–447.

Carlsson A., Falck B., and Hillarp N. A. (1962) Cellular localization of brain monoamines. *Acta. Physiol. Scand.* **56** (suppl. 196), 1–27.

Cirino M. and Renaud L. P. (1985) Influence of lateral septum and amygdala stimulation on the excitability of hypothalamic supraoptic neurons. An electrophysiological study in the rat. *Brain Res.* **326,** 357–361.

Clark B. J. and Rocha e Silva M. (1967) An afferent pathway for the selection release of vasopressin in response to carotid occulsion and haemorrhage in the cat. *J. Physiol.* **191,** 529–542.

Cross B. A. (1955) Neurohormonal mechanisms in emotional inhibition of milk ejection. *J. Endocrinol.* **12,** 29–37.

Day T. A. and Renaud L. P. (1984) Electrophysiological evidence that noradrenergic afferents selectively facilitate the activity of superoptic vasopressin neurons. *Brain Res.* **303,** 233–240.

Day T. A., Ferguson A. V., and Renaud L. P. (1984) Facilitatory influence of noradrenergic afferents on the excitability of rat paraventricular nucleus neurosecretory cells. *J. Physiol.* **355,** 237–249.

Day T. A., Randle J. C. R., and Renaud L. P. (1986) Opposing alpha- and beta-adrenergic mechanisms mediate dose-dependent actions of noradrenaline on supraoptic vasopressin neurons in vivo. *Brain Res.* (in press).

Dellman H. D. and Simpson J. B. (1979) The subfornical organ. *Int. Rev. Cytol.* **58,** 333–421.

Dreifuss J. J., Harris M. C., and Tribollet E. (1976) Excitation of phasically firing hypothalamic supraoptic neurones by carotid occlusion in rats. *J. Physiol.* (Lond.) **257**, 337–354.

Eng R. and Miselis R. R. (1981) Polydypsia and abolition of angiotensin II-induced drinking after transections of subfornical organ efferent projections in the rat. *Brain Res.* **225**, 200–206.

Ferguson A. V. and Renaud L. P. (1984) Hypothalamic paraventricular nucleus lesions decrease pressure responses to subfornical organs stimulation. *Brain Res.* **305**, 361–364.

Ferguson A. V., Day T. A., and Renaud L. P. (1984a) Connections of hypothalamic paraventricular neurons with the dorsal medial thalamus and neurohypophysis: An electophysiological study in the rat. *Brain Res.* **299**, 376–379.

Ferguson A. V., Day T. A., and Renaud L. P. (1984b) Subfornical organ afferents influence the excitability of neurohypophysial and tuberoinfundibular paraventricular nucleus neurons in the rat. *Neuroendocrinology* **39**, 423–428.

Ferreyra H., Kannan H., and Koizumi K. (1983) Influences of the limbic system on hypothalamo-neurohypophysial system. *Brain Res.* **264**, 31–45.

Fuxe K. (1965) Evidence for the existence of monoamine neurons in the central nervous system. IV. The distribution of monoamine nerve terminals in the central nervous system. *Acta. Physiol. Scand.* **64** (suppl. 247), 37–85.

Hamamura M., Shibuki K., and Yagi K. (1982) Amygdalar inputs to ADH-secreting supraoptic neurones in rats. *Exp. Brain Res.* **48**, 420–428.

Harris M. C. (1979) The effect of chemoreceptor and baroreceptor stimulation on the discharge of hypothalamic supraoptic neurones in rat. *J. Endocrinol.* **82**, 115–125.

Hicks T. P. (1984) The history and development of microiontophoresis in experimental neurobiology. *Prog. Neurobiol.* **22**, 185–240.

Hokfelt T., Johansson O., Ljungdahl A., Lundberg J. M., and Schultzberg M. (1980) Peptidergic neurones. *Nature* **284**, 515–521.

Ishikawa K., Kakegawa T., and Suzuki M. (1984) Role of the hypothalamic paraventricular nucleus in the secretion of thyrotropin under adrenergic and cold stimulated conditions in the rat. *Endocrinology* **114**, 352–358.

Iversen L. L. (1984) Amino acids and peptides: Fast and slow chemical signals in the nervous system? *Proc. Roy. Soc. Lond.* B **221**, 245–260.

Kandel E. R. and Schwartz J. H. (1981) Principles of Neuroscience. Elsevier-North Holland, New York.

Keil L. C. and Severs W. B. (1977) Reduction in plasma vasopressin levels of dehydrated rats following acute stress. *Endocrinology* **100**, 30–38.

Keirstead S. A. and Rose P. K. (1983) Dendritic tree structure of motoneurons in different spinal segments innervating the same muscle: A study of splenius motoneurons intracellularly stained with horseradish peroxidase in the cat. *J. Comp. Neurol.* **219**, 273–284.

Kelly J. S. (1975) Microiontophoretic Application of Drugs onto Single Neurons, in *Handbook of Psychopharmacology* vol. 2 (Iversen, L. L., Iversen S. D., and Snyder S. H., eds.), Plenum, New York.

Kelly J. S. (1982) Electrophysiology of peptides in the central nervous system. *Br. Med. Bull.* **38**, 283–290.

Kiss J. Z., Mezey E., and Skirboll L. (1984) Cortical control in releasing factor in immunoreactive neurons of the paraventricular nucleus become vasopressin positive after adrenaline activity. *Proc. Natl. Acad. Sci. USA* **81**, 1854–1858.

Knepel W., Nutto D., and Meyer D. K. (1982) Effect of transection of subfornical organ efferent projections on varopressin release induced by angiotensin or isoprenaline in the rat. *Brain Res.* **248**, 180–184.

Koizumi J. and Yamashita H. (1972) Studies of antidromically identified neurosecretory cells of the hypothalamus by intracellular and extracellular recordings. *J. Physiol.* (Lond.) **221**, 683–705.

Kreiger D. T. (1983) Brain peptides: What, where and why? *Science* **222**, 975–985.

Kuffler S. W., Nicholls J. G., and Martin A. R. (1984) *From Neuron to Brain* Sinauer, Sunderland, Massachusetts.

Lincoln D. W. and Wakerley J. B. (1974) Electrophysiological evidence for the activation of supraoptic neurones during the release of oxytocin. *J. Physiol.* **242**, 533–554.

Lind R. W., Van Hoesen G. W., and Johnson A. K. (1982) An HRP study of the connections of the subfornical organ of the rat. *J. Comp. Neurol.* **210**, 265–277.

Lind R. W., Swanson L. W., and Ganton D. (1985) Angiotensin II immunoreactivity in a neural afferents and efferents of the subfornical organ of the rat. *Brain Res.* (in press).

Mangiapane M. L. and Simpson J. B. (1983) Drinking and pressor responses after acetylcholine injection into subfornical organ. *Am. J. Physiol.* **244**, R508–R513.

McNeill T. H. and Sladek J. R., Jr. (1980) Simultaneous monoamine histofluorescence and neuropeptide immunocytochemistry. II. Correlative distribution of catecholamine varicosities and magnocellular neurosecretory neurons in the rat supraoptic and paraventricular nuclei. *J. Comp. Neurol.* **193**, 1023–1033.

Mendelsohn F. A. O., Quirion R., Saavedra J. M., Aguilera G., and Catt K. J. (1984) Autoradiographic localization of angiotensin II receptors in rat brain. *Proc. Natl. Acad. Sci. USA* **81,** 1575–1579.

Mirsky I. A., Stein N., and Paulisch G. (1954) The secretion of an antidiuretic substance into the circulation of rats exposed to noxious stimuli. *Endocrinology* **54,** 491–505.

Miselis R. (1981) The efferent projections of the subfornical organ of the rat: A circumventricular organ with a neural network subserving water balance. *Brain Res.* **230,** 1–23.

Negoro H., Visessuwan S., and Holland R. C. (1973) Inhibition and excitation of units in paraventricular nucleus after stimulation of the septum, amygdala and neurohypophysis. *Brain Res.* **57,** 479–483.

Paisley A. C. and Summerlee A. J. S. (1984) Activity of putative oxytocin neurons during reflex milk ejection in conscious rabbits. *J. Physiol.* **347,** 465–478.

Palkovits M., Brownstein M., Saavedra J. M., and Axelrod J. (1974) Norepinephrine and dopamine content of hypothalamic nuclei of the rat. *Brain Res.* **77,** 137–149.

Pittman Q. J., Blume H. W., and Renaud L. P. (1981) Connections of the hypothalamic paraventricular nucleus with the neurohypophysis, median eminence, amygdala, lateral septum and midbrain periaqueductal gray: An electrophysiological study in the rat. *Brain Res.* **215,** 15–28.

Poulain D. A. and Wakerley J. B. (1982) Electrophysiology of hypothalamic magnocellular neurones secreting oxytocin and vasopressin. *Neuroscience* **7,** 773–808.

Poulain D. A., Ellendorff F., and Vincent J. D. (1980) Septal connections with identified oxytocin and vasopressin neurones in the supraoptic nucleus of the rat. An electrophysiological investigation. *Neuroscience* **5,** 379–387.

Ranck J. B. (1975) Which elements are excited in electrical stimulation of mammalian central nervous system: A review. *Brain Res.* **98,** 417–440.

Randle J. C. R., Bourque C. W., and Renaud L. P. (1985) Serial reconstruction of Lucifer Yellow-labelled supraoptic nucleus neurons in the perfused rat hypothalamic explants. *Neuroscience* **17,** 453–467.

Renaud L. P. (1978) Neurophysiological Organization of the Endocrine Hypothalamus, in *The Hypothalamus.* (Reichlin S., Baldessarini R. J., and Martin J. B., eds.), Raven, New York.

Renaud L. P. (1981) A neurophysiological approach to the identification, connections and pharmacology of the hypothalamic tuberoinfundibular system. *Neuroendocrinology* **33,** 186–191.

Renaud L. P. (1983) Role of Neuropeptides in the Regulation of Neural

Excitability, in *Basic Mechanisms of Neuronal Hyperexcitability.* (Jasper H. H. and Van Gelder N., eds.), Alan Liss, New York.

Renaud L. P., Rogers J., and Sgro S. (1983) Terminal degeneration in supraoptic nucleus following subfornical organ lesions: Ultrastructural observations in the rat. *Brain Res.* **275,** 365–368.

Renaud L. P., Bourque C. W., Day T. A., Ferguson A. V., and Randle J. C. R. (1985) Electrophysiology of Mammalian Hypothalamic Supraoptic and Paraventricular Neurosecretory Cells, in *The Electrophysiology of the Secretory Cell* (Poisner A. N. and Trifaro J., eds.), Elsevier, Amsterdam.

Renaud L. P., Day T. A., Randle J. C. R., and Bourque C. W. (1985) In Vivo and In Vitro Electrophysiological Evidence That Central Noradrenergic Pathways Enhance the Activity of Hypothalamic Vasopressinergic Neurosecretory Cells, in *Vasopressin* (Schrier R. W., ed.) Raven, New York.

Sakai M., Swartz B. E., and Woody C. D. (1979) Controlled micro release of pharmacological agents: Measurements of volume ejected in vitro through fine-tipped glass microelectrodes by pressure. *Neuropharmacology* **18,** 209–213.

Sawchenko P. E. and Swanson L. W. (1983) The organization of forebrain afferents to the paraventricular and supraoptic nuclei of the rat. *J. Comp. Neurol.* **218,** 121–144.

Sawchenko P. E. and Swanson L. W. (1981) Central noradrenergic pathways for the integration of hypothalamic neuroendocrine and autonomic responses. *Science* **214,** 685–687.

Sawchenko P. E. and Swanson L. W. (1982) The organization of noradrenergic pathways from the brainstem to the paraventricular and supraoptic nuclei in the rat. *Brain Res. Rev.* **4,** 275–325.

Sawchenko P. E., Swanson L. W., and Vale W. W. (1984) Co-expression of corticotropin releasing factor and vasopressin immunoreactivity in parvocellular neurosecretory neurons of the adrenalectomised rat. *Proc. Natl. Acad. Sci. USA* **81,** 1883–1887.

Sgro S., Ferguson A. V., and Renaud L. P. (1984) Subfornical organ–supraoptic nucleus connections: An electrophysiological study in the rat. *Brain Res.* **303,** 7–13.

Sherlock D. A., Field P. A., and Raisman P. G. (1975) Retrograde transport of horseradish peroxidase in the magnocellular neurosecretory system of the rat. *Brain Res.* **88,** 403–414.

Silverman A.-J., Hoffman D. L., and Zimmerman E. A. (1981) The descending afferent connections of the paraventricular nucleus of the hypothalamus (PVN). *Brain Res. Bull.* **6,** 47–61.

Simpson J. B., Epstein A. N., and Komado J. S. (1978) Localization of dipsogenic receptors for angiotensin II in the subfornical organ. *J. Comp. Physiol. Psychol.* **92,** 581–608.

Summerlee A. J. S. (1981) Extracellular recordings from oxytocin neurons during the expulsive phase of birth in unanesthetized rats. *J. Physiol.* **321,** 1–9.

Summerlee A. J. S. (1982) Phasic patterns of discharge from putative oxytocin neurones in unanaesthetized rats. *J. Physiol.* **327,** 43P.

Summerlee A. J. S. (1983) Hypothalamic neuron activity, hormone release and behaviour in freely moving rats. *Quarterly J. Exp. Physiol.* **68,** 505–515.

Summerlee A. J. S. and Lincoln D. W. (1981) Electrophysiological recordings from oxytocinergic neurons during suckling in the unanesthetized lactating rat. *J. Endocrinol.* **90,** 255–265.

Swaab D. F., Pool C. W., and Nijveldt F. (1975) Immunofluorescence of vasopressin and oxytocin in the rat hypothalamo-neurohypophyseal system. *J. Neural Transm.* **36,** 195–215.

Swanson L. W. and Sawchenko P. E. (1983) Hypothalamic integration: Organization of the paraventricular and supraoptic nuclei. *Ann. Rev. Neurosci.* **6,** 269–324.

Taleisnik S. and Deis R. P. (1964) Influence of cerebral cortex inhibition on oxytocin release induced by stressful stimuli. *Am. J. Physiol.* **207,** 1394–1398.

Thomson A. M. (1982) Responses of supraoptic neurones to electrical stimulation of the medial amygdaloid nucleus. *Neuroscience* **7,** 2197–2205.

Tramu G., Croix C., and Pillez A. (1983) Ability of the CRF immunoreactive neurons of the paraventricular nucleus to produce a vasopressin-like material. *Neuroendocrinology* **37,** 467–469.

Vandesande F. and Dierickx K. (1975) Identification of the vasopressin producing and of the oxytocin producing neurons in the hypothalamic neurosecretory system of the rat. *Cell Tiss. Res.* **164,** 153–162.

Yamashita H., Inenaga K., Kawata M., and Sano Y. (1983) Phasically firing neurons in the supraoptic nucleus of the rat hypothalamus: Immunocytochemical and electrophysiological studies. *Neurosci. Lett.* **37,** 87–92.

In Vitro Preparations for Electrophysiological Study of Peptide Neurons and Actions

Q. J. Pittman, B. A. MacVicar, and W. F. Colmers

1. Introduction

The value of in vivo electrophysiological studies for revealing information about peptidergic neurons and peptide actions has been discussed by Ferguson and Renaud in this volume. Nevertheless, carrying out pharmacological studies in vivo raises a number of problems with respect to both interpretation of data and the ease with which data can be obtained. In particular, the following considerations pertain to working in vivo: (1) the use of anesthetics is often required and their presence may alter the neuronal responses to peptide application; (2) it can be very difficult to manipulate the environment of the cell in order to differentiate between, for example, pre- and postsynaptic effects or to determine the ionic fluxes underlying a peptide action; (3) it is difficult to perform intracellular experiments because pulsation caused by blood pressure and respiration make it difficult to keep the electrode in the cell. Because of these problems, some investigators have turned to simpler in vitro animal models, such as the *Aplysia*, a marine mollusc, the utility of which is described by Murphy and Lukowiak elsewhere in this volume. The evident success achieved by investigators using simple invertebrate preparations has prompted a general movement over recent years to simpler in vitro neuronal preparations for the study of peptide actions in vertebrate tissue. In vitro preparations of vertebrate nervous system can be established that can incorporate some of the features of the intact nervous system, along with many of the advantages of a simple, invertebrate model system. Alternately, they can be manipulated following their removal from the animal to permit defined growth, connections, or physiology such that they can be utilized to test specific hypotheses concerning peptide actions. The aim of this chapter, therefore, is to provide evidence for the utility of various in vitro electrophysiological preparations to reveal information about peptidergic neurons and peptide actions in neuronal tissue. Although it is not our intention to provide a detailed methodological "recipe" for obtaining various in vitro preparations, we will identify some of those preparations that have provided valuable information about peptides, indicate some of the advantages or disadvantages associated with each type, and finally, illustrate some of the data that have been obtained using these preparations. For more detailed accounts of specific methodologies prepared by acknowledged experts in their respective fields, appropriate references will be provided for the interested reader. An introduction to an understanding of the basic electrophysiological techniques illustrated in this chapter can be found in a recent review (Pittman, 1986).

2. General Principles

For in vitro neuronal preparations to exhibit physiological properties akin to those they possess *in situ*, certain features in the preparation and maintenance of neuronal tissue in vitro have proven to be particularly critical. Furthermore, in order to obtain data relevant to our understanding of neuronal communication in the intact animal, certain methodological considerations become important. Three of the issues that will be discussed in this section are choice of incubation media, methods of peptide application, and identification of the cell under study.

2.1. Media

In order to achieve meaningful and relevant data concerning peptide actions and to define the membrane properties of peptidergic cells, it is imperative that cells maintained in vitro be healthy. Ideally, one would hope to maintain the cells in the physiological conditions approximating that *in situ*. In order to accomplish this, a perfusion medium must be chosen that will (1) maintain the appropriate ionic milieu of the cell, (2) provide nutrients necessary for metabolism, and (3) provide an adequate source of oxygen. A wide variety of perfusion media have been chosen for maintaining neuronal tissue in vitro. Depending upon the type of tissue maintained and the length of time it must be supported outside of the nervous system, different considerations apply. For long-term maintenance of nervous tissue in culture, a variety of growth factors and hormones are required to maintain viability, support axonal outgrowth, initiate synaptogenesis, determine transmitter phenotype, and so on. These issues will be addressed in a later section. Acute studies on fully differentiated tissue can be maintained for a few hours in vitro in an apparently healthy state with a simple, buffered salt solution. For peripheral nervous tissue, a solution resembling the extracellular fluid of the body, such as Krebs medium, works well. Central nervous system tissue, on the other hand, has an extracellular environment somewhat different from that of the rest of the body. In particular, protein content is low and the quantity of unbound calcium is lower than that found in the circulation. As calcium stabilizes membranes and thus reduces excitability, slight variations in the quantity of calcium in the perfusate can have profound effects upon neuronal excitability and spontaneous activity (Oliver et al., 1977; Pittman et al., 1981; Fig. 1). There are also differences in the concentration of extracellular potassium between various body tissues, with that of the central nervous system being approximately 3 mM and that of the extracellular environment in nonnervous tissue approximately 5 mM. As neuronal membrane potential varies directly with extracellular potassium concentration, slight changes in potassium concentration can have profound effects on neuronal excitability. Elevated potassium concentrations of the order of 5–6 mM will result in enhanced spontaneous activity, reduced membrane potential, and slower axonal conduction (Pittman, 1983; Fig. 2). Despite these cautionary notes concerning ion constituents of the solution, one must also bear in mind that the composition of the perfusate can be modified to make the preparation more amenable to in vitro analysis. For example, in the hippocampal slice, the extracellular concentration of calcium is often artificially raised in order to reduce excitability in this preparation, since many of the inhibitory inputs are absent in vitro and the neurons have a propensity to generate epileptiform discharges when calcium concentrations are reduced (Yamamoto, 1972). Furthermore, when attempting to record intracellularly from tiny neurons, it is often helpful to raise extracellular calcium, since this appears to facilitate membrane repair and promote healthy recording conditions.

Whereas extracellular levels of glucose in the brain are thought to be 3–5 mM, that of most perfusion media is usually 10 mM. This appears to be a case of "erring on the side of safety," since it is thought advisable to maintain high glucose levels to compensate for the absence of other cellular energy sources (e.g., amino acids,

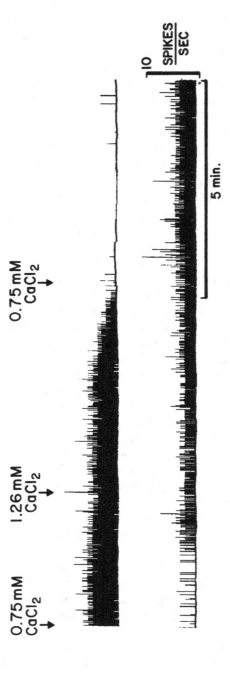

Fig. 1. Rate meter record of a spontaneously active neuron recorded with an extracellular electrode in the paraventricular nucleus of a coronal, hypothalamic slice. At points indicated by the arrows, artificial CSF that differed only in the $CaCl_2$ concentration was introduced into the bath. The spontaneous activity during the initial perfusion with medium containing 0.75 mM $CaCl_2$ declined when the $CaCl_2$ concentration was increased; reduction in $CaCl_2$ concentration was then associated with a reappearance of spontaneous activity (from Pittman et al., 1981, with permission).

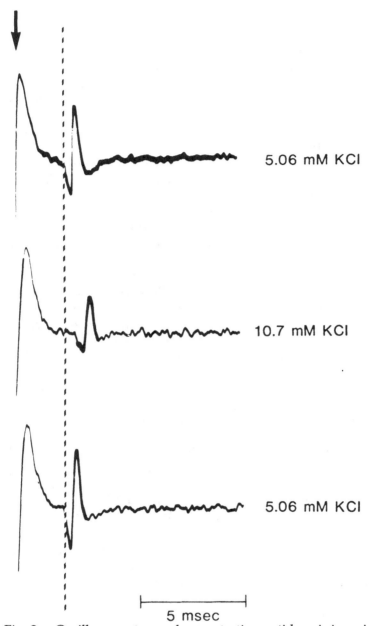

5.06 mM KCl

10.7 mM KCl

5.06 mM KCl

5 msec

Fig. 2.   Oscilloscope traces demonstrating antidromic invasion of a paraventricular neuron, recorded extracellularly, in hypothalamic slice preparation following electrical stimulation (arrow) of the neurohypophysial tract at 1 Hz. Increasing the concentration of KCl caused a reversible increase (middle) in the antrodromic latency, as well as a reduction in the magnitude of the extracellularly recorded potential (from Pittman, 1983, with permission).

fatty acids, and so on) that may be available to neurons in vivo. In tissues in which there are considerable diffusion distances (e.g., chunks or slices), the gradient caused by elevated glucose concentration will facilitate maintenance of sufficient quantities in the interior of the tissue.

The maintenance of pH and delivery of oxygen is often accomplished by bubbling the perfusion media with, or maintaining an atmosphere of, 95% oxygen:5% CO_2 over the tissue. With the use of $NaHCO_3$ in the medium, pH can be balanced to 7.35 (brain pH). At different altitudes and atmospheric pressures, it may be necessary to alter the concentration of bicarbonate by up to 5 mM to maintain this pH. This can and should be tested empirically, since tissue pH is an important variable in maintaining neuronal health.

The delivery of oxygen to the tissue has been a subject of considerable controversy, and discussion and centers around two major considerations: (1) the distance that oxygen can diffuse from a saturated solution into the tissue and (2) the efficacy of the oxygen delivery to the tissue. With respect to brain slices, it has been shown empirically that olfactory cortex slices do not develop anoxia until they have exceeded a thickness of 430 μm (Fujii et al., 1982). In our laboratory, we have successfully maintained hippocampal slices 600 μm thick in vitro for several hours in an electrophysiologically healthy state, although our usual working slice thickness is 400 μm.

The delivery of oxygen to the tissue can be either by direct diffusion from the media in submerged or perfused tissue, or by diffusion into the tissue from the atmosphere. Many investigators maintain their tissue slices at the gas–liquid interface; oxygen thus diffuses directly into the tissue from the atmosphere. In this case, it is necessary to maintain the 95% oxygen:5% CO_2 saturation of the atmosphere above the slice and also to maintain a high humidity in this gas to prevent dehydration of the tissue. The latter is often accomplished by positioning the slice chamber in water bath that both maintains the temperature and moisturizes the gas that is bubbled through the water. A potential disadvantage of an "interface" type chamber lies in the fact that access to the tissue may be somewhat restricted by the necessity of enclosing the chamber (to maintain the humidity and oxygen saturation). In contrast, submerged slices can be maintained in a simple chamber that provides easy access to the slice (Nicoll and Alger, 1981; Fig. 3).

Fig. 3.   Photograph of a "submerged-type" slice chamber that permits easy access to the slice as well as rapid exchange of bath media.

2.2 Peptide Application

An important consideration relevant to studies on peptide action on neurons concerns the method of peptide application. This can take several forms: (1) application of microdrops directly to the surface of the tissue; (2) micropressure application of a peptide solution directly in the vicinity of the neuron under study; (3) microiontophoretic application of peptide solution to the neuron; and (4) delivery of the peptide to the neuron by perfusion. Each of these methods has certain advantages and disadvantages, and the choice of technique for peptide application will depend in part upon the questions to be asked. If one wishes to localize the receptive zones on a neuron, it would appear that iontophoresis or micropressure application would be the method of choice. Alternatively, if one wishes to carry out precise dose–response studies in order to compare the efficacy of various peptide analogs or to study agonist/antagonist interactions, the perfusion method of application may be most appropriate. It is our opinion that the microdrop method of peptide application is to be avoided since considerable artifact formation may result from the flooding of the tissue with a peptide-containing solution of unknown concentration, pH, and ionic constituents. The pitfalls and potential artifacts that may be associated with iontophoresis have been well described elsewhere (Hicks, 1984; MacDonald, 1985), and, similarly, the utility of the micropressure approach has also been well described (Palmer et al., 1980). With respect to iontophoresis, the greatest source of artifact is probably that caused by the current used to eject the peptide, and appropriate controls and safeguards must be incorporated in any pharmacological study to differentiate between peptide- and current-induced responses. With respect to micropressure application, we can not overemphasize the importance of carrying out control applications of the vehicle. Neurons are extremely susceptible to both flow and pressure artifacts that can be mistaken as peptide effects (Gruol et al., 1980; MacDonald, 1985; Pittman and Watson, 1987). Wherever possible, use of inactive analogs or of antagonists should be utilized in conjunction with peptide agonists to verify that any responses are indeed mediated by receptors. It is also important to remember that micropressure or microdrop application of peptides introduces the actual fluid contents of the micropipettes into the vicinity of the cell. Since the extracellular constituents in brain tissue may be buffered by protein binding, and so on, the free ionic constituents of the pipet solution may differ considerably from that of the immediate extracellular environment of the cell. Furthermore, if the peptide is dissolved in the normal bicarbonate-buffered perfusion medium, the pipet contents may be more alkaline than the tissue pH, since it is unlikely that the pipet contents have been bubbled with carbogen (95% CO_2:5% O_2) during the interval between the time such solutions are made and the time that they are applied to the cell. In our laboratory, we have found it necessary to utilize a balanced salt solution whose buffering is not dependent upon exposure to carbogen gas to maintain appropriate pH.

2.3. Identification of Cells

In order to arrive at a conclusion concerning the action of a peptide in a tissue, it is helpful to know the identity of the cell under study. There are a number of different strategies that have been utilized to identify cells in vitro. In most isolated tissues, it has proven possible to record reproducibly from an identified population of cells on the basis of their location. For example, in transilluminated slices, nuclear areas are highlighted, and electrodes can be directed visually to the specific area of interest. Within a nucleus or area of neuronal tissue, neurons can be further identified on the basis of their connectivity. This can be established in at least two ways. If the excised tissue maintained in vitro is large enough and oriented appropriately, antidromic identification (cf. Pittman, 1986) can establish the axonal projection of the cell under study. Orthodromic potentials recorded in neurons can be used to characterize some of the afferent connections of a cell. In some neuronal tissue, however, the removal from the intact nervous system makes identification on the basis of connectivity impossible. To circumvent this problem, it has recently been shown feasible to inject axonally transported dyes into terminal areas, prior to sacrifice of the animal. The labeled cell bodies can then be identified in the isolated tissue following its removal from the animal with the use of fluorescence microscopy (Katz et al., 1984). Similar *post hoc* identification of cells on the basis of their morphology can be made possible by injecting into a recorded neuron a dye such as lucifer yellow or enzymes such as horseradish peroxidase, which, following appropriate processing, reveal the entire cell structure under microscopy (Fig. 4). As an elegant adjunct to the dye injection studies, it has also proven possible to identify such neurons as peptidergic in nature by subsequent immunohistochemistry (Kayser et al., 1982; Reaves et al., 1982; Yamashita et al., 1983; Smithson et al., 1984). In cell cultures of mammalian nervous system, photographic localization of a cell under electrophysiological study can be used along with subsequent immunohistochemistry to unambiguously define a cell type (Fig. 5; cf, MacVicar, 1984).

3. Preparations

A wide variety of in vitro preparations are available for the study of peptide actions. The choice of preparation will depend, in part, on the specific peptide or cell type that one is interested in and whether or not there is evidence for peptidergic neurons or receptors on cells in these tissues. We will attempt to identify some of the major in vitro preparations and illustrate how they have been useful for revealing certain peptide actions.

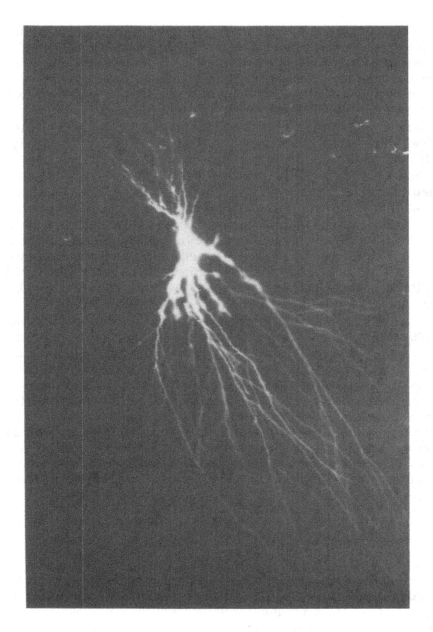

Fig. 4. Pyramidal neuron in area CA_4 of the hippocampal slice injected with the fluorescent dye lucifer yellow.

3.1. Autonomic and Sensory Ganglia

The autonomic nervous system consists of pre- and postganglionic neurons. In the sympathetic component of the autonomic nervous system, short preganglionic fibers synapse with postganglionic cells in encapsulated collections of cell bodies called sympathetic ganglia. In addition to the paravertebral ganglia situated near the spinal cord and a number of other small sympathetic ganglia, a large cervical ganglion situated in the neck area and a large mesenteric ganglion in the abdomen have been particularly well studied. Because these ganglia are easily removed intact from the animal, they have provided useful preparations for electrophysiological recording in vitro. The cells are large and can be easily impaled for intracellular studies. Thus, their electrical properties have been well characterized (e.g., Nishi and Koketsu, 1960). Furthermore, in sympathetic ganglia of frogs, the cell bodies lack dendrites, thereby making the soma amenable for voltage clamp studies. For in vitro studies, sympathetic ganglia along with their pre- and postganglionic fiber bundles are removed intact from animals under anesthesia and placed in a recording chamber. The ganglia are generally surrounded by a tough fibrous sheath, and, for intracellular recordings, this must be either digested with an enzyme such as collagenase or carefully dissected away, after which the cell soma can be visualized under a microscope.

The sympathetic ganglia also provide the advantage that afferents can be stimulated to evoke synaptic responses and the efferent axons can also be electrically activated to provide antidromic invasion for cell identification. In frog and mammalian sympathetic ganglia, complex synaptic responses arise from stimulation of orthodromic fibers. Although some of these appear to be mediated by classical neurotransmitters (e.g., acetylcholine), others have proved resistant to all of the known neurotransmitter antagonists (Weight, 1983). With the immunohistochemical demonstration that the sympathetic ganglia were rich in a variety of peptidergic fibers, interest arose in the possibility that peptides may be the neurotransmitters active at some of the synapses. In guinea pig inferior mesenteric ganglion, a synaptic response called the noncholinergic, slow EPSP is now thought to be mediated by release of the peptide substance P. Otsuka and Konishi (1983) and colleagues have been able to show that pressure application of substance P mimics the synaptic response to nerve stimulation, and that a substance P antagonist reduces in parallel both the synaptically and peptidergically generated response.

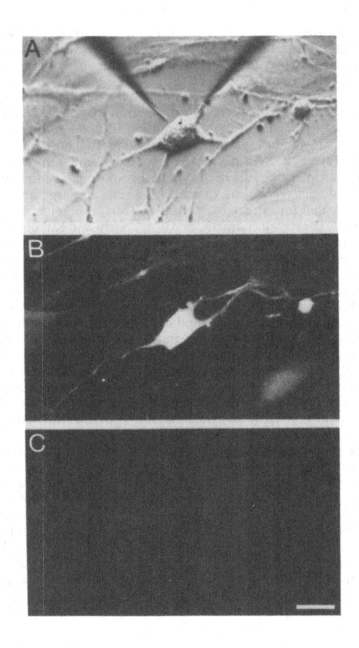

The fact that defined afferents can be stimulated while intracellular recording is in progress in postsynaptic cells makes it possible also to use sympathetic ganglia to reveal presynaptic peptide actions. Konishi et al. (1981) have thus been able to show that met-enkephalin acts presynaptically via naloxone-sensitive receptors to inhibit the release of acetylcholine in inferior mesenteric ganglia of guinea pigs.

An elegant series of experiments has now succeeded in identifying an LHRH-like peptide as a probable transmitter mediating the late, slow EPSP in frog sympathetic ganglia (reviewed in Jan and Jan, 1983). This peptide can be detected immunohistochemically in fibers in the ganglion, and can be released upon electrical stimulation of preganglionic fibers. Furthermore, bath application of LHRH mimics the synaptically evoked late, slow EPSP, and an LHRH antagonist completely blocks both responses. These studies identifying this LHRH peptide as a probable neurotransmitter at this synapse illustrate the use of these isolated in vitro preparations extremely well.

In contrast to the sympathetic ganglia that exist in well-defined groups of cells, parasympathetic ganglia are found directly in the tissue of innervation along with a rich collection of local interneurons. The neuronal tissue investing the guinea pig ileum is called the myenteric plexus and has provided a useful preparation for defining the electrophysiological properties of myenteric neurons (North, 1982) and examining their responses to application of opiates. By removing the muscle and adherent plexus from the ileum and pinning it into a recording chamber where it can be perfused with physiological solution, stable intracellular recordings can be obtained. North and Williams (1983) and colleagues have exploited this preparation to study the actions of enkephalins

---

Fig. 5. Glial cells from which recordings are made in culture contain glial fibrillary acidic protein (GFAP). (A) Micrograph of a glial cell during recording. This cell was impaled with two microelectrodes, one for current injection and the other for voltage recording. (B) The cultures were immunohistochemically stained for GFAP, thus demonstrating that the cell from which the recording was made in (A) was indeed a glial cell. In this micrograph, the monoclonal antibody to GFAP was visualized with avidin-conjugated rhodamine. (C) Similar cultures in which incubation with the antibody to GFAP was omitted demonstrated no staining of the intracellular fibrils in the glial cells. Scale bar, 13 μm (from MacVicar, 1984, with permission).

on these cells and have determined that these opioid peptides directly inhibit cell firing by opening potassium channels and hyperpolarizing the membrane; a similar mechanism is thought to underlie the reduction and release of excitatory transmitter from these neurons that follows application of enkephalin.

Dorsal root ganglion cells also provide useful preparations for intracellular electrophysiological analysis. These cells have been called pseudo-unipolar since they have one axon that then bifurcates to innervate both the periphery and the spinal cord. Although the soma of these neurons have receptors to a variety of neurotransmitter substances, there are no synaptic contacts on the cell soma. It is possible that these receptors respond to circulating substances or to those that diffuse from a distance. Thus, these cells may be particularly appropriate for perfusion application of peptides. Curiously, however, few peptide studies have been reported on acutely isolated dorsal root ganglion cells in vitro (in contrast to numerous studies on cultured dorsal root ganglion cells). Nonetheless, the actions of opiates have been examined in dorsal root ganglion cells acutely isolated and perfused in vitro (Williams and Zieglgansberger, 1981).

3.2. Chunks and Tissues

3.2.1. Pituitary

The pituitary consists of three separate lobes: the anterior pituitary, consisting of trophic hormone (e.g., ACTH, FSH, and others) cells, the intermediate pituitary containing a homogenous population of proopiomelanocortin-containing cells, and the posterior pituitary consisting of nerve endings of hypothalamic neurons and associated glial cells (pituicytes). The pituitary gland can be carefully dissected free from the overlying brain tissue and, along with its stalk, maintained in vitro for several hours. Under appropriate illumination, the three lobes of the pituitary can be easily differentiated (Fig. 6) and a microelectrode directed to the particular area of interest. Anterior pituitary cells on the periphery of the gland appear to receive sufficient nutrients and oxygen by diffusion to permit maintenance of healthy resting potentials and overshooting action potentials. The intermediate pituitary also appears particularly well suited for in vitro investigations. This tissue lacks a vascular supply in vivo and it has a well-defined system of canals and intercellular spaces to facilitate diffusion of nutrients to the interior of the gland. We have successfully re-

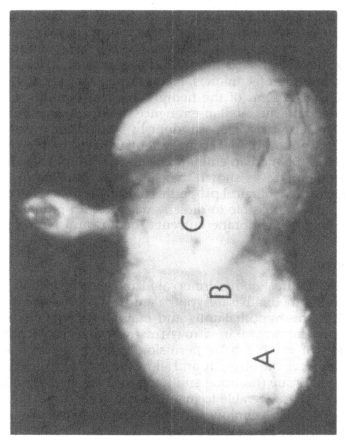

Fig. 6. Photomicrograph of a rat pituitary demonstrating the three lobes: (A) anterior pituitary; (B) intermediate pituitary; (C) posterior pituitary.

corded from these cells within the intermediate pituitary and have been able to carry out pharmacological studies on these glandular cells (MacVicar and Pittman, 1986).

Although the posterior pituitary itself has been studied in vitro with a variety of neurochemical techniques (Pittman et al., 1983), the fine, unmyelinated nerve endings of neurohypophysial neurons have proven difficult for electrophysiological recording in vitro. Nonetheless, the possibility of presynaptic receptors to certain peptides or possibly receptors on the pituicytes in the neural lobe raises the possibility that peptide effects could be investigated electrophysiologically in this tissue. Because the terminals are too small to facilitate standard intracellular recording, it has been necessary to use other techniques. For example, Zingg et al. (1979) have utilized an explant of the neural lobe of the pituitary, the pituitary stalk, and the median eminence in vitro to examine excitability of the terminals. This was done by stimulating the neural lobe and eliciting a compound action potential that is then recorded in the region of the median eminence. The peak amplitude of the compound potential was shown to be altered by substances that alter the electrophysiological properties of the terminals. Recently it has been shown possible to directly patch clamp pituitary terminals and record membrane currents (Lemos and Nordmann, 1986).

3.2.2. Chunks

Since the neural lobe is innervated by neurons whose cell bodies lie in the medial basal hypothalamus, it is also possible to excise part of the hypothalamus and maintain a hypothalamo-neurohypophysial explant in vitro (Armstrong and Sladek, 1982). The addition of intravascular perfusion to such a preparation appears to improve the integrity and allow the use of a much larger piece of neural tissue (Bourque and Renaud, 1983; Fig. 7); in fact, Llinas et al. (1981) were able to maintain the entire brain stem in vitro utilizing similar techniques. With the maintenance of a larger piece of tissue, it is possible to preserve more of the afferent and efferent connections of peptidergic neurons and thereby perform detailed electrophysiological and pharmacological analyses of these connections.

A structure that is rich in peptidergic innervation is the spinal cord, and, fortunately, this structure can also be maintained in vitro for the purpose of electrophysiological analysis of peptide actions. Both neonatal (Otsuka and Konshi, 1974) and juvenile (Bagust and Kerkut, 1981) rodent and frog (Kudo, 1978) spinal cord

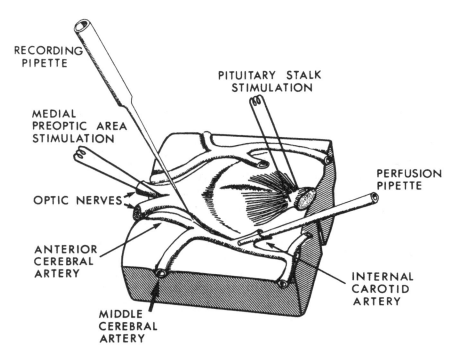

RECORDING
PIPETTE

PITUITARY STALK
STIMULATION

MEDIAL
PREOPTIC AREA
STIMULATION

PERFUSION
PIPETTE

OPTIC NERVES

ANTERIOR
CEREBRAL
ARTERY

INTERNAL
CAROTID
ARTERY

MIDDLE
CEREBRAL
ARTERY

Fig. 7. Schematic illustration of the basal hypothalamic chunk (ventral surface uppermost) in which perfusion with artificial media is accomplished through one of the anterior cerebral arteries. Magnocellular supraoptic nucleus neurosecretory neurons may be activated antidromically via stimulation applied to the pituitary stalk and orthodromically via stimulation to the medial preoptic area. The recording pipet is directed to the area of the supraoptic nucleus (from Bourque and Renaud, 1983, with permission).

have been well characterized electrophysiologically. The large motor neurons are amenable to intracellular recording while the excitability of the afferent fibers coursing through the dorsal roots can be monitored via the sucrose gap method of recording. Stimulation of the dorsal root while recording from the ventral root can give an index of the efficacy of synaptic transmission through the circuit. Thus, this preparation provides one with the opportunity not only to assess the effects of peptides on cell somata, but also to examine their influence upon synaptic transmission. Using such techniques, the excitatory action of TRH on spinal motor neurons has been revealed (Nicoll, 1977), and Otsuka and Konishi (1983) have provided convincing evidence that substance P may be a transmitter in primary afferent fibers in the frog and rat spinal cord.

3.2.3. Retina

The retina is a highly organized, layered neural tissue that has a number of distinct advantages for in vitro studies of peptide actions. It has well-characterized structural (Ramon y Cajal, 1972; Famiglietti et al., 1977), electrophysiological (Wu, 1985) and pharmacological (Lam, 1975; Dowling et al., 1983) properties. The retina is a rich source of peptides, with an extensive innervation of a variety of peptides arising from both local and distant neurons (Brecha, 1983; Stell, 1985). Finally, it is a structure whose physiological input (light) can be applied "in vitro" while its output (ganglion cell activity or electroretinogram) can be monitored (cf., Walker and Stell, 1986). The retina is easily isolated intact from the eye, and can be maintained "in vitro" for many hours. Peptides can be applied in the perfusate in the case of submerged retina, or, because of the short diffusion distances, they can be applied in a "mist" from an atomizer to retinas maintained in an interface chamber.

3.3. Brain Slices

One of the most popular preparations for electrophysiological studies of peptide actions in the mammalian brain is the brain slice preparation. These were originally developed for biochemical studies by McIlwain and colleagues, but since the mid-1960s it has become appreciated that brain slices could be utilized for electrophysiological studies. Brain slices are usually 300–600 μm thick and oxygenation and nutrient delivery is by diffusion from the perfusion medium or atmosphere to the entire tissue. Under appropriate illumination, specific nuclear areas are highlighted, and therefore microelectrodes can be directed to specific cell nuclei. By obtaining slices of appropriate orientation, it is also possible to preserve considerable synaptic circuitry. At the present time, virtually all areas of the central nervous system have been successfully maintained as slice preparations in vitro, and several recent monographs and textbooks provide details on the methodology employed to maintain such slices in vitro (Kerkut and Wheal, 1981; Dingledine, 1984; Jahnsen and Laursen, 1983; Lipton, 1985). For studying defined, peptidergic neurons in vitro, the hypothalamic slice preparation has proven useful (Pittman et al., 1980; Dudek et al., 1982). The most popular slice preparation, however, for studying peptide *actions* has been the hippocampal slice. Its utility for pharmacological analysis of peptide actions is such that it could almost be termed the "guinea pig ileum of the brain," in that it is reminiscent of the contribution that the isolated guinea pig ileum has made to pharmacological studies (cf., Brownlee and Harry, 1963; Couture and Regoli, 1982). It should be kept in mind that, just as the multitude of pharmacological studies that have been carried out on the guinea pig ileum in vitro have provided us with considerable information about how various drugs and putative transmitters may interact, they have provided us with little information about how the guinea pig digests its food; similarly, the extensive literature on peptide and other neurotransmitter actions in the hippocampal slice has revealed little about the actual physiological role of these substances or indeed of the hippocampus in the intact animal.

The reasons for the popularity of the hippocampal slice lie both in the ease with which it can be removed from the brain and in the fact that considerable synaptic circuitry remains in a transverse hippocampal slice. Furthermore, the cell body layer of the pyramidal neurons is particularly well defined in the transilluminated slice. These morphological features facilitate introduction of a recording electrode into the cell body layer where large amplitude orthodromic field potentials or intracellular excitatory postsynaptic potentials can be recorded (Fig. 8).

Under the appropriate perfusion conditions, it is possible to maintain slice preparations with spontaneous neuronal activity that resembles that seen in vivo. Consequently, it is possible to carry out extracellular recording while defined quantities of peptides and/or their antagonists are applied through the medium. Thus, it was possible to identify opioid-mediated inhibitory actions within the paraventricular nucleus of the hypothalamus (Fig. 9). With the stability afforded one in the in vitro slice preparation, it is also possible to carry out intracellular studies and thereby reveal actions of peptides on ionic conductances and synaptic transmission. Both postsynaptic (Fig. 10) and presynaptic (Fig. 11) actions of peptides have thus been revealed.

3.4. Cell Cultures

Cell cultures of nervous systems were used for many years for biochemical and developmental studies. Approximately 20 years ago, it became evident that tissue cultures also provided favorable

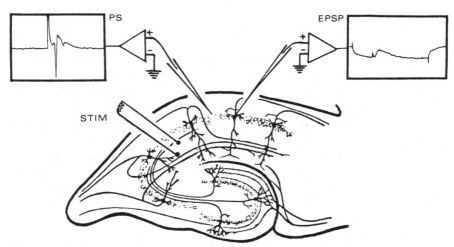

Fig. 8. Schematic illustration of a coronal hippocampal slice with stimulating electrode placed in the stratum radiatum to orthodromically activate pyramidal cells in area CA$_1$. An extracellular recording electrode is utilized to record (upper left) the evoked compound field potential and population spike (PS) arising from the stimulation. On the right is shown the simultaneous intracellular recording from an impaled pyramidal cell in which the excitatory postsynaptic potential (EPSP) can be seen following stratum radiatum stimulation. The EPSP is augmented in this case by superimposing a hyperpolarizing pulse via a bridge circuit.

preparations for examining certain electrophysiological events in cells. The literature on cell culture techniques is voluminous and space does not permit us to adequately address any details of culture methods (*but see* Hertz et al., 1985).

One type of cell culture that has been of considerable utility in examining electrophysiological properties of cells and peptide actions is the clonal cell line (Schubert et al., 1974; Bulloch et al., 1976). These are transformed cell lines of neural tissue that have the advantage that, unlike neurons, the cells divide and under appropriate conditions will reproduce themselves throughout many generations. Thus, cells in these cultures are all identical and maintain similar properties from generation to generation. The disadvantage of the such cultures is that, as they are transformed, the properties expressed may not necessarily resemble those of intact neurons. More commonly, primary cultures (Federoff and Hertz, 1977) are established of normal, usually neonatal nervous tissue and maintained in vitro for a period of days to weeks. Cultures of a variety of brain areas can be established and main-

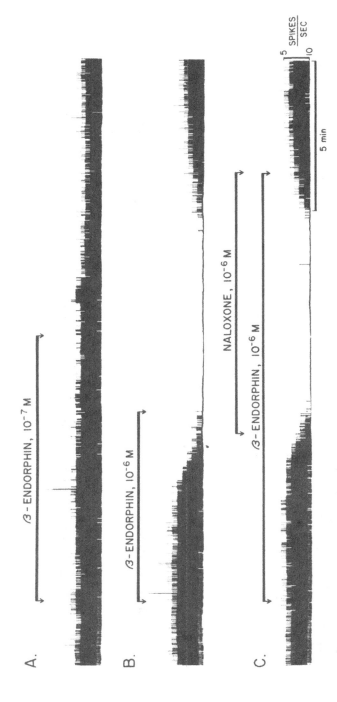

Fig. 9. Rate meter records of a spontaneously active paraventricular nucleus neuron in the hypothalamic slice demonstrating dose-dependent inhibitory actions of β-endorphin (A and B) that could be antagonized with concurrent application of naloxone (C). Cellular responses to the peptide applications are delayed approximately 3 min because of delay in actual admittance of the perfused substance to the slice chamber.

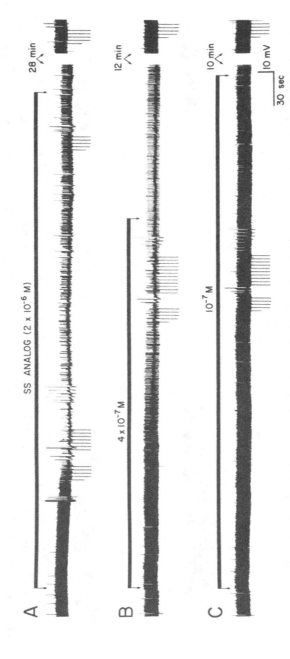

Fig. 10. Intracellular voltage records of a CA_1 hippocampal cell demonstrating dose-dependent hyperpolarizations in response to perfusion application of an active analog of somatostatin at various concentrations.

430

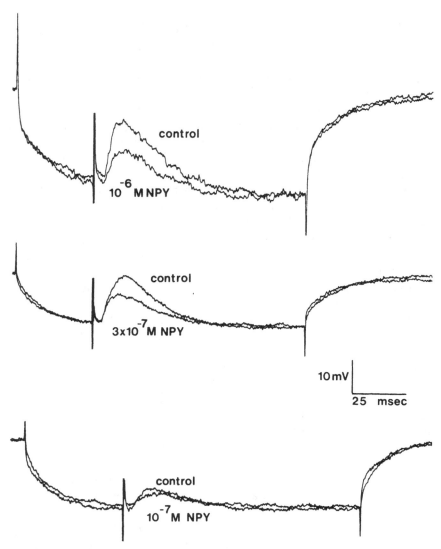

Fig. 11. Oscilloscope traces of intracellular voltage records from three CA_1 pyramidal neurons demonstrating hyperpolarizations of the membrane in response to the passage of hyperpolarizing current through the electrode. Stratum radiatum stimulation (vertical transient) 35 ms after the start of the hyperpolarizing pulse elicited a monosynaptic EPSP in each of these cells, which was reduced when the control media was replaced by that containing the peptide neuropeptide Y (NPY). The action of NPY would appear to be presynaptic in reducing the size of the EPSP, since there was no evidence for a change in the electrophysiological properties of the postsynaptic cell.

tained; in certain lower vertebrate and invertebrate cultures, it is possible to maintain identified neurons in culture and to establish defined connections that are identical to those seen in vivo (Bulloch, 1985). With the use of appropriate culture conditions, it is also possible to establish cell populations enriched in either specific neuronal types or in glial cells. It can be seen, therefore, that culture techniques can be utilized to either establish an in vitro tissue that resembles as closely as possible that seen in vivo or in which the tissue is manipulated to establish defined conditions not attainable in vivo.

As might be expected, excision of nervous tissue and its establishment in culture necessitates certain compromizes. The morphology of the cells can be altered, although it would appear that certain morphological features are maintained in vitro that assist in identifying cell types by their appearance and establishing neuronal or glial identity. This can be accomplished *post hoc* with the use of appropriate strains that can identify cells as to glial or neuronal phenotype (Fig. 12). It must also be remembered that normal synaptic development has been disrupted in culture, and that the culture medium is only an approximation of a true extracellular environment. Possibly because of this, there is indeed evidence that cultured cells demonstrate pharmacological responses in vitro that are not seen in the same cells in vivo. For example, Mudge et al. (1979) were able to demonstrate an action of enkephalin on dorsal root ganglion cells in vitro to reduce the calcium component of the action potential. On these same cells, in vivo, Williams and Zieglgansberger (1981) were unable to demonstrate opiate receptor-mediated actions on cell somata and suggested that the cultured cells of Mudge et al. (1979) possessed receptors on the somata that are situated only on the electrically distant terminals in vivo.

Nonetheless, the use of cell cultures has permitted elegant electrophysiological analyses of membrane actions of peptides. For example, since intracellular records can be made from both pre- and postsynaptic cells in culture, it has been possible to carry out quantal analysis and thereby examine presynaptic actions of opiates (MacDonald and Nelson, 1978). A particular advantage of cultured neurons, and one in particular that is not found in the other in vitro preparations described in this monograph, is that cultured cells are not invested with glial cells. Thus, these clean membranes permit patch clamp experiments (Hamill et al., 1981) in which peptide actions on individual ion channels can be determined.

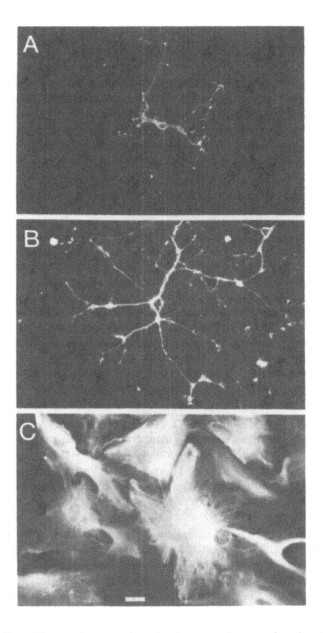

Fig. 12. Photomicrographs of primary cultures of rat brain stain to reveal: (A) oligodendrocytes with the use of an antibody raised against galactocerebroside; (B) a neuron, visualized with tetanus toxin; and (C) astrocytes, visualized with an antibody raised against glial fibrillary acidic protein. Scale bar, 13 μm (Milner and Pittman, unpublished observations).

4. Summary

The use of in vitro electrophysiological techniques to study peptide actions has provided a wealth of useful data about peptide actions in cells resembling those found in vivo (i.e., slices, chunks, or ganglia) or in experimentally altered nervous tissue (i.e., cell cultures). In terms of elegant electrophysiological analysis and the ease with which it can be carried out, in vitro electrophysiological studies have many advantages over similar studies attempted in vivo. Thus, this approach to the study of peptide actions provides a useful means of obtaining information about the roles and actions of peptides in brain.

Acknowledgments

Supported by the MRC of Canada. Q. J. Pittman is an MRC Scientist and AHFMR Scholar, B. A. MacVicar is an AHFMR Scholar and a Sloan Fellow, and W. F. Colmers is an AHFMR Fellow.

References

Armstrong W. E. and Sladek C. D. (1982) Spontaneous 'phasic-firing' in supraoptic neurons recorded from hypothalamo-neurohypophysial explants in vitro. *Neuroendocrinology* **34,** 405–409

Bagust J. and Kerkut G. A. (1981) An *In Vitro* Preparation of the Spinal Cord of the Mouse, in *Electrophysiology Of Isolated Mammalian CNS Preparations* (Kerkut G. A. and Wheal H. V., eds.) Academic, London.

Bourque C. W. and Renaud L. P. (1983) A perfused in vitro preparation of hypothalamus for electrophysiological studies on neurosecretory neurons. *J. Neurosci. Meth.* **7,** 203–214.

Brecha N. (1983) Retinal Neurotransmitters: Histochemical and Biochemical Studies, in *Chemical Neuroanatomy* (Emson P. C., ed.) Raven, New York.

Brownlee G. and Harry J. (1963) Some pharmacological properties of the circular and longitudinal muscle strips from the guinea pig ileum. *Br. J. Pharmacol.* **21,** 544–554.

Bulloch A. G. M. (1985) Development and Plasticity of the Molluscan Nervous System, in *The Mollusca* vol. 8 *Neurobiology And Behavior* part 1 (Willows A. O. D., ed.) Academic, New York.

Bulloch K., Stallcup W. B., and Cohn M. (1976) The derivation and characterization of neuronal cell lines from rat and mouse brain. *Brain Res.* **135**, 25–36

Couture R. and Regoli D. (1982) Minireview: Smooth muscle pharmacology of substance P. *Pharmacology* **24**, 1–25.

Dingledine R., ed. (1984) *Brain Slices* Plenum, New York.

Dowling J. E., Lasater E. M., van Buskirk R., and Watling K. J. (1983) Pharmacological properties of isolated fish horizontal cells. *Vision Res.* **23**, 421–432.

Dudek F. E., Andrew R. D., MacVicar B. A., and Halton G. I. (1983) Intracellular electrophysiology of mammalian peptidergic neurons in rat hypothalamic slices. *Fed. Proc.* **41**, 2953–2958.

Famiglietti E. V., Jr., Kaneko A., and Tachibana M. (1977) Neuronal architecture of on and off pathways to ganglion cells in carp retina. *Science* **198**, 1267–1269

Federoff S. and Hertz L., eds. (1977) *Cell, Tissue and Organ Cultures in Neurobiology* Academic, New York.

Fujji T., Baumgartl H., and Lubberg D. W. (1982) Limiting section thickness of guinea pig olfactory cortical slices studied from tissue pO_2 values and electrical activities. *Pflugers Arch.* **393**, 83–87.

Groul D. L., Barker J. L., Huang M. L., MacDonald J. F., and Smith T. G. (1980) Hydrogen ions have multiple effects on the excitability of cultured mammalian neurons. *Brain Res.* **183**, 247–252.

Hamill O. P., Marty A., Neher E., Sakmann B., and Sigworth F. J. (1981) Improved patch-clamp techniques for high-resolution current recording from cells and cell-free membrane patches. *Pflugers Arch.* **391**, 85–100.

Hertz L., Juurlink B. H. J., Szuchet S., and Walz W. (1985) Cell and Tissue Cultures, in *Neuromethods* vol. 1, General Neurochemical Techniques (Boulton A. A. and Baker G. B., eds.) Humana, Clifton, New Jersey.

Hicks T. P. (1984) The history and development of microiontophoresis in experimental neurobiology. *Prog. Neurobiol.* **22**, 185–240.

Jahnsen H. and Laursen A. M. (1983) Brain Slices, in *Current Methods In Cellular Neurobiology* vol. III, *Electrophysiological and Optical Recording Techniques* (Barker J. L. and McKelvy J. F., eds.) John Wiley, New York.

Jan Y. N. and Jan L. Y. (1983) An LHRH-like peptidergic neurotransmitter capable of "action at a distance" in autonomic ganglia. *Trends Neurosci.* **6**, 320–325.

Katz L. C., Burkhalter A., and Dreyer W. J. (1984) Fluorescent latex microspheres as a retrograde neuronal marker for in vivo and in vitro studies of visual cortex. *Nature* **310**, 498–500.

Kayser B. E. J., Muhlethaler M., and Dreifuss J. J. (1982) Paraventricular neurones in the rat hypothalamic slice: Lucifer yellow injection and immunocytochemical identification. *Experientia* **38**, 391–393.

Kerkut G. A. and Wheal H. V., eds. (1981) *Electrophysiology of Isolated Mammalian CNS Preparations.* Academic, London.

Konishi S., Tsunoo A., and Otsuka M. (1981) Enkephalin as a transmitter for presynaptic inhibition in sympathetic ganglia. *Nature* **294**, 80–82.

Kudo Y. (1978) The pharmacology of the amphibian spinal cord. *Prog. Neurobiol.* **11**, 1–76.

Lam D. M. (1975) Synaptic chemistry of identified cells in the vertebrate retina. *Cold Spring Harbor Symp. Quant. Biol.* **40**, 571–579.

Lemos J. R. and Nordmann J. J. (1986) Ionic channels and hormone release from peptidergic nerve terminals. *J. Exp. Biol.* **124**, 53–72.

Lipton P. (1985) Brain Slices: Uses And Abuses, in *Neuromethods* vol. 1, *General Neurochemical Techniques* (Boulton A. A and Baker G. B., eds.) Humana, Clifton, New Jersey.

Llinas R. Yarom Y., and Sugimori M. (1981) Isolated mammalian brain in vitro: A new technique for analysis of electrical circuit function. *Fed. Proc.* **40**, 2240–2245.

MacDonald J. F. (1985) Identification of Central Transmitters. Microiontophoresis and Micropressure Techniques, in *Neuromethods* vol. 1, *General Neurochemical Techniques* (Boulton A. and Baker G. B., eds.) Humana, Clifton, New Jersey.

MacDonald R. L. and Nelson P. G. (1978) Specific opiate-induced depression of transmitter release from dorsal root ganglion cells in culture. *Science* **199**, 1449–1451.

MacVicar B. A. (1984) Voltage-dependent calcium channels in glial cells. *Science* **226**, 1345–1347.

MacVicar B. A. and Pittman Q. J. (1986) Novel synaptic responses mediated by dopamine and γ-aminobutyric acid in neuroendocrine cells of the intermediate pituitary. *Neurosci. Lett.* **64**, 35–40.

Mudge A. W., Leeman S. E., and Fischbach G. D. (1979) Enkephalin inhibits release of substance P from sensory neurons in culture and decreases action potential duration. *Proc. Natl. Acad. Sci. USA* **76**, 526–530.

Nicoll R. A. and Alger B. E. (1981) A simple chamber for recording from submerged brain slices. *J. Neurosci. Meth.* **4**, 153–156.

Nicoll R. A. (1977) Excitatory action of TRH on spinal motoneurones. *Nature* **265**, 242–243.

Nishi S. and Koketsu K. (1960) Electrical properties and activities of single sympathetic neurons in frogs. *J. Cell. Comp. Physiol* **55**, 15–30.

North R. A. (1982) Electrophysiology of the enteric nervous system. *Neuroscience* **7**, 315–325.

North R. A. and Williams J. T. (1983) How do opiates inhibit neurotransmitter release? *Trends Neurosci.* **6,** 337–339.

Oliver A. P., Hoffer B. J., and Wyatt R. J. (1977) The hippocampal slice: A system for studying the pharmacology of seizures and for screening anticonvulsant drugs. *Epilepsia* **18,** 543–548.

Otsuka M. and Konishi S. (1983) Substance P—The first peptide neurotransmitter? *Trends Neurosci.* **6,** 317–320.

Otsuka K. and Konishi S. (1974) Electrophysiology of mammalian spinal cord *in vitro. Nature* **252,** 733–734.

Palmer M. R., Wuerthelee S. M., and Hoffer B. J. (1980) Physical and physiological characteristics of micropressure ejection of drugs from multibarreled pipettes. *Neuropharmacology* **19,** 931–938.

Pittman Q. J. (1983) Increases in antidromic latency of neurohypophysial neurons during sustained activation. *Neurosci. Lett.* **37,** 239–243.

Pittman Q. J. (1986) How do we listen to neurons? *Discus. Neurosci.* **3,** 1–60.

Pittman Q. J. and Watson T. (1987) Somatostatin Actions in Hippocampus, in *Inactivation Of Hypersensitive Neurons* (Chalazonitis, N., ed.) Alan Liss, New York (in press).

Pittman Q. J., Hatton J. D., and Bloom F. E. (1981) Spontaneous activity in perfused hypothalamic slices: Dependence on calcium content of perfusate. *Exp. Brain Res.* **42,** 49–52.

Pittman Q. J., Hatton J. D., and Bloom F. E. (1980) Morphine and opioid peptides reduce paraventricular neuronal activity: Studies on the rat hypothalamic slice preparation. *Proc. Natl. Acad. Sci. USA* **77,** 5527–5531.

Pittman Q. J., Lawrence D., and Lederis K. (1983) Presynaptic Interactions in the Neurohypophysis: Endogenous Modulators of Release, in *The Neurohypophysis: Structure, Function and Control. Progress in Brain Research.* vol. 60 (Cross B. A. and Leng G., eds.) Elsevier, New York.

Ramon Y. Cajal S. (1972) *The Structure of the Retina* Thomas, Springfield, Illinois.

Reaves T. A., Jr., Cummings R., Libber M. T., and Hayward J. N. (1982) A technique combining intracellular dye-marking, immunocytochemical identification and ultrastructural analysis of physiologically identified single neurons. *Neurosci. Lett.* **29,** 195–199.

Schubert D., Heinemann S., Carlisle W., Tarikas H., Kines B., Patrick J., Steinbach J. H., Culp W., and Branct B. L. (1974) Clonal cell lines from the rat central nervous system. *Nature* **249,** 224–227.

Smithson K. G., Cobbett P., MacVicar B. A., and Hatton G. I. (1984) A reliable method for immunocytochemical identification of lucifer yellow injected, peptide-containing mammalian central neurons. *J. Neurosci. Meth.* **10,** 59–69.

Stell W. K. (1985) Putative Peptide Transmitters, Amacrine Diversity and Function in the Inner Plexiform Layer, in *Neurocircuitry Of The Retina, A Cajal Memorial* (Galbego A. and Govras P., eds.) Elsevier, New York.

Walker S. E. and Stell W. K. (1986) Gonadotropin-releasing hormone (GnRF), molluscan cardioexcitatory peptides (FMRFamide), enkephalin and related neuropeptides affect goldfish retinal ganglion cell activity. *Brain Res.* **384,** 262–273.

Weight F. F. (1983) Synaptic mechanisms in amphibian sympathetic ganglia, in *Autonomic Ganglia* (Elfvin L. G., ed.) John Wiley, New York.

Williams J. and Zieglgansberger W. (1981) Mature spinal ganglion cells are not sensitive to opiate receptor mediated actions. *Neurosci. Lett.* **21,** 211–216.

Wu S. M. (1985) Synaptic transmission from rods to bipolar cells in the tiger salamander retina. *Proc. Natl. Acad. Sci. USA* **82,** 3944–3947.

Yamamoto C. (1972) Activation of hippocampal neurons by mossy fiber stimulation in thin brain sections in vitro. *Exp. Brain Res.* **14,** 423–435.

Yamashita H., Inenaga K., Kawata M., and Sano Y. (1983) Phasically firing neurons in the supraoptic nucleus of the rat hypothalamus: Immunocytochemical and electrophysiological studies. *Neurosci. Lett.* **37,** 87–92.

Zingg H. H., Baertschi A. J., and Dreifuss J. J. (1979) Action of γ-aminobutyric acid on hypothalamo-neurohypophysial axons. *Brain Res.* **171,** 453–459.

Molluscan Model Systems for the Study of Neuropeptides

Ken Lukowiak and A. Don Murphy

1. Introduction

The aim of this chapter is to describe the usefulness of relatively simple model systems to study the roles played by neuropeptides in the mediation of adaptive and homeostatic behaviors. In particular, we will discuss some of the experimental approaches to exploit the unique characteristics of gastropod molluscan nervous systems. The recent discoveries and initial characterizations of neuroactive peptides, peptidergic neurons, and systems of such neurons have fundamentally altered our classical views regarding the transmission of neural information. The view that transmission of chemical signals within the brain largely involves only a very few neural transmitters with one transmitter per neuron and transmitter release only at anatomically stereotyped synapses no longer is tenable.

The explosion of research on neuroactive peptidergic systems is inarguably one of the major developments in neurobiology over the last several years. Application of fractionation and separation techniques to vertebrate neural tissue, in concert with bioassays, has revealed an apparent plethora of neuroactive peptides (e.g., Iversen et al., 1983). Still, we have a meager understanding of (1) how particular peptidergic systems are organized to regulate or influence discrete physiological processes or behaviors (e.g., feeding) and (2) the roles peptidergic neurons may play in physiological and behavioral plasticity (e.g., nonassociative and associative learning).

The relative immensity, complexity, and inaccessability of the vertebrate brain provide formidable obstacles to addressing these problems (*see* Pittman et al., this volume). It is in addressing similar kinds of problems in nonpeptidergic neuronal systems that molluscan and other invertebrate model systems have proven their greatest utility (e.g., Kandel, 1976; Koester and Byrne, 1980; Willows, 1985).

The same characteristics (see below) of molluscan, especially gastropod, nervous systems that have proven advantageous for studies of membrane biophysics, synaptic transmission, and the neuronal organization and plasticity of behavior make these organisms promising for analyses of peptidergic systems. There is no single technique that can be applied to invertebrates as opposed to vertebrates. Rather, the major advantage offered by certain invertebrate and, in particular, gastropod molluscs, is the facility with which behavioral, physiological, anatomical, and biochemical techniques can be applied in unison to uniquely identifiable peptidergic neurons and neuronal systems.

Our goal is not a comprehensive review of peptides and peptidergic systems studied in molluscs. There have been a few well-written reviews concerned with peptides in the invertebrate nervous system. Of particular note are the comprehensive studies on the roles of families of peptides involved in integrating a series of discrete behaviors into the complex behavioral sequence associated with egg laying in the marine opisthobranch *Aplysia* and in the fresh water pulmonate *Lymnaea* (e.g., Scheller et al., 1984; Joose and Geraerts, 1983; Kaldany et al., 1985).

Here, we will describe briefly the general characteristics of gastropod nervous systems that make them good models. We will then discuss several techniques and preparations that we have begun to utilize in approaching the general questions referred to above. We will concentrate on the approaches we are taking to characterize the roles of peptidergic neurons in the feeding behavior of the pond snail *Helisoma*. We will then discuss the evidence that peptides are involved in neurophysiological and behavioral plasticity in the context of the gill withdrawal response of *Aplysia*. Finally, we will briefly discuss the still largely untapped potential offered by molluscs for the study of cellular and molecular mechanisms of actions of peptides.

We hope that the readers of this volume will come to appreciate the utility of the molluscan nervous system for studies of peptide actions in the nervous system. We further hope that some of our excitement with recent findings on the roles played by these substances in the mediation of important homeostatic and adaptive behaviors will be infectious. We fully realize that these model systems cannot answer all questions in which neurobiologists and neurophysiologists are interested, but they do offer very substantial benefits. We feel that an integrative approach utilizing a variety of invertebrate and vertebrate systems ranging up to intact preparations will lead to an understanding of how neuropeptides exert modulatory or direct effects on neurons that ultimately control behavior.

2. Molluscan Neural Systems as Models

The model system will be defined as part of a nervous system that is directly amenable to neurophysiological, anatomical, or biochemical analyses, and that mediates simple yet nontrivial behaviors or discrete physiological processes. Our basic assumption in using these molluscan model systems is that what we find will be applicable to higher, more complicated organisms (e.g., humans). Nature appears to be conservative, and what works in simpler systems is often utilized in the higher organisms. May it not, in fact, be that invertebrates "invented" neuropeptides for use in mediating or modifying nervous system activity and the vertebrates just made slight adjustments to the basic model? If this is so, then finding out how peptides function in a simpler system will enable us to unravel their functions and mechanisms of action in more complex organisms like mammals.

The most obvious technical advantages offered by molluscan nervous systems arise from the extremely large size of many neurons and neural processes. The squid giant axon, for example, is a classical preparation used to define the mechanisms of action potential generation and propagation (Hodgkin and Huxley, 1952). Similarly, the squid giant synapse has provided a major model for studies of excitation–secretion coupling and mechanisms of synaptic transmission (e.g., Llinas et al., 1981). For a number of reasons, gastropods have proven to be of great utility for neuronal analyses of behavior (*see* Kandel, 1976). Gastropod brains have relatively few, perhaps 10^5 to 10^6, neurons located in several discrete ganglia. The neuronal somata are found in an outer rind of cortex of the ganglia, with neuronal processes extending inward to form a neuropile containing most of the synapses within the ganglia. In many species (e.g., *Helisoma*), the ganglia contain red pigment against which the white, yellow, and orange hues of the somata are readily contrasted, facilitating individual identification and accurate penetration with microelectrodes.

Many of these somata are comparatively large (e.g., greater than 100-μm diameter in *Helisoma*). In some species, such as *Aplysia*, individual neurons may be greater than 1 mm in diameter. Many gastropod neurons have a unique identity. That is, when numerous individuals of the species are examined, each will have individual neurons distinguishable from all the other neurons in the animal on the basis of their morphological, biochemical, biophysical, and pharmacological characteristics, as well as their synaptic relationships with other neurons and effectors. Several such neurons are identifiable in ganglia, simply by visual inspection, based upon their relative size, position, and as mentioned above, pigmentation. In addition, gastropods are hardy individuals and discrete behaviors (e.g., feeding) can be evoked in partially dissected individuals while intracellular microelectrode recordings are made from identifiable neurons. *Thus, behavioral, electrophysiological, pharmacological, morphological (i.e., intracellular dye injections), and immunocytochemical analyses can be readily applied in concert to individually identifiable neurons.* In recent years, it has come to light that many such invertebrate neurons are peptidergic.

3. Peptides and Feeding in *Helisoma*

To demonstrate how molluscan model systems can facilitate our understanding of the peptidergic regulation of discrete, physiological processes and behaviors, we will discuss the roles of peptides, particularly small cardioactive peptide B (SCP$_B$), and peptidergic neurons with respect to feeding in the snail *Helisoma*. Our basic protocol is to combine electrophysiological recordings with intracellular dye injections and subsequent immunocytochemical staining.

Helisoma typically feeds by rhythmically scraping the substratum with a file-like radula (Kater, 1974). Movements of the radula are mediated by groups of muscles that are activated in more or less stereotyped sequences. These muscles collectively form the buccal mass, and they are driven by motor neurons in the buccal ganglia. Patterned activity in these motor neurons, as well as in the glandular effector neurons, is driven by synaptic inputs from interneurons, which collectively comprise the central pattern generator (CPG) (Kater, 1974; Kaneko et al., 1978).

We routinely record such activity intracellularly simultaneously from up to four identified or identifiable neurons (Fig. 1). The CPG in this system is analgous to pattern generators mediating a wide variety of rhythmic behaviors (e.g., respiratory, locomotory, and feeding movements, as well as the activity of neurogenic hearts) in both vertebrates and invertebrates. Peptides and other neuroactive substances can modulate these rhythmic behaviors by influencing the CPG itself, as well as by directly and selectively affecting individual motor neurons, muscles, and gland cells (e.g., Fig. 2).

We have developed a multidisciplinary approach to determine which endogenous neuroactive substances influence patterned motor activity (PMA) of the buccal ganglia, to localize these to candidate regulatory neurons, and then to characterize these neurons physiologically and morphologically. The sequence of studies involved (1) exogenous application of suspected endogenous peptides and monitoring for effects on buccal ganglion PMA; (2) localization of candidate regulatory peptidergic neurons by immunoreactive staining with antisera to the effective peptide; and (3) unequivical identification of regulatory neurons by (a) intracellular electrophysiological analysis of neurons in the locations in which candidate regulatory neurons were previously stained, (b) intracellular injection of Lucifer Yellow into the electrophysiologically analyzed neuron, and (c) subsequent double staining with primary antisera to the neural peptides and a rhodamine-conjugated second antibody. Neuroactive peptides can be introduced into the perfusion chamber either by a constant flow system or by direct injection with a syringe, while maintaining constant chamber volume by suction from the surface. When the latter method is used, multiple washes of the chamber with normal physiological saline are performed immediately prior to application of the test solution to control for possible nonspecific effects (e.g., from unstirred layers). All solutions are maintained at room temperature to prevent complications from temperature differences. We are aware that data obtained by bath application of neuroactive peptides could be misinterpreted. That is, the effects of general access of substances to somata and neuropile could differ from those evoked by potentially more localized physiological release. However, this is not a serious problem for our studies. For instance, the effects of bath application of 5-hydroxytryptamine (5-HT, serotonin) on PMA are essentially the same as those evoked by stimulation of the serotonergic neurons (C1) (Granzow

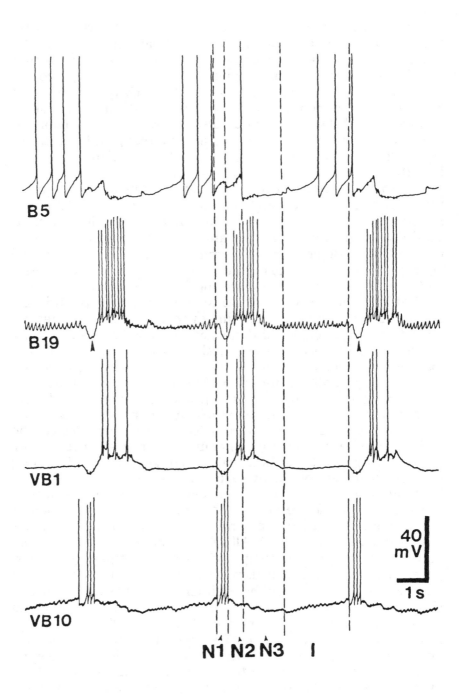

B5

B19

VB1

VB10

40
mV

1 s

N1 N2 N3   I

and Kater, 1977). More importantly, the exogenous application of neuropeptides serves mainly as a key for identification of physiologically relevant neuroregulatory neurons. The main goal of this research is the characterization of these physiologically important peptidergic neurons and their interactions with the other neurons and effectors involved in feeding and other oral behaviors. In addition, these accessible systems will allow characterization of peptide receptors with dose–response curves and displacement of these curves by agonists and antagonists.

We use standard intracellular and extracellular recording, stimulation, and display techniques (e.g., Edstrom and Lukowiak, 1985; Murphy and Kater, 1980). A key aspect of our studies is the multiple staining of physiologically characterized neurons, with both intracellular injection of Lucifer Yellow and indirect immunofluorescent staining with antisera to various endogenous neuropeptides. Routinely, both of these techniques are used successfully to stain neurons in ganglion whole mounts (Murphy et al., 1985a,b) and thick sections, and the techniques have been successfully combined to double stain individual identified neurons following electrophysiological recordings (Fig. 3). These staining procedures have been described in detail (e.g., Murphy and Kater, 1980; Murphy et al., 1983, 1985a,b; Mackie et al., 1985). Optimization of results of the double-staining protocol involves principally a shortening of incubation times with the primary antisera to 8–12 h and with the secondary antisera to 4 h from 2 d and overnight, respectively. By switching between the filter combinations for Lucifer Yellow and rhodamine on the fluorescence microscope, it can be determined unequivocally whether the electrophysiologically analyzed neurons were immunoreactive for the

---

Fig. 1. Simultaneous intracellular recordings from four identified neurons during patterned motor activity (PMA). Each cycle of electrical activity underlying a single bite or rasp in *Helisoma* and most other gastropods can be divided into three major phases of synaptic activity (N1, N2, and N3) separated by a relatively inactive period (I). Particular identified neurons can receive either excitatory, inhibitory, or no synaptic inputs during a given phase of the cycle. These synaptic inputs cause different identified neurons to fire bursts of action potentials during different phases of the cycle to orchestrate the overall pattern of motor activity. PMA [shown here as evoked by serotonin (5-HT)] can be initiated, terminated, or modified by neuropeptides (from Murphy and Lukowiak, unpublished).

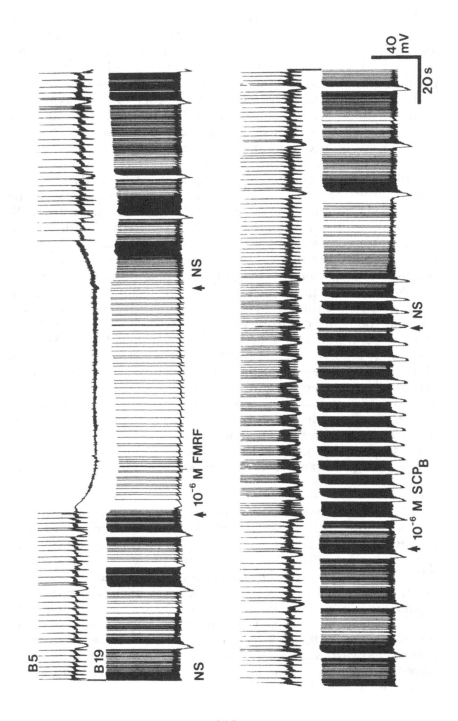

NS

B5

B19

NS

↑ 10⁻⁶ M FMRF

↑ NS

↑ 10⁻⁶ M SCP_B

↑ NS

40 mV

20 s

446

Fig. 2. The effects of two endogenous neuropeptides upon PMA in *Helisoma*. Simultaneous intracellular recordings from neurons B5 and B19. Upper and lower pairs of traces are continuous. Initially in normal physiological saline (NS), PMA occurred at a low rate. Note three N2 phase IPSPs occurring prior to application of 10^{-6} *M* FMRFamide at the first arrow. The FMRFamide stopped PMA and also caused a pronounced (>20 mV) hyperpolarization of neuron B5. Washing with NS (second arrow) restored PMA to its "spontaneous" rate. Application of SCP$_B$ at a concentration of 10^{-6} *M* (third arrow) increased the rate of PMA, and its effects were also completely reversible upon washout with NS (fourth arrow). Action potentials of neuron B5 are clipped (Murphy, et al., 1985b).

447

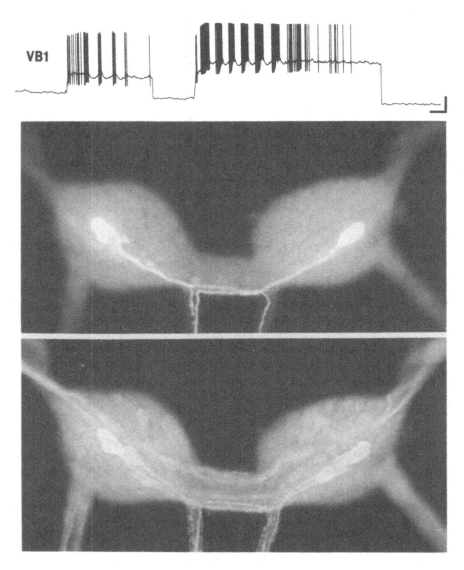

Fig. 3. Identification by "double staining" of electrophysiologically characterized SCP$_B$-immunoreactive neurons. (Top panel) Intracellular recordings from neuron VB1. Neuron VB1 displayed no spontaneous activity. Constant dc depolarization evoked burst of action potentials that quickly accommodated. When the current injected was increased, a greater number of bursts with an increased number and frequency of action potentials per burst was generated. These also accommodated.

particular neuropeptide. Background fluorescence (especially with the rhodamine-conjugated secondary antisera) is reduced by pre-incubation of the secondary antisera with tissue that has not been exposed to a primary antibody prior to use on the experimental tissue. Specificity controls routinely involve liquid preabsorption of primary antisera with the corresponding antigen. We are aware of the inherent limitations of immunohistochemical staining techniques with regard to identification of particular antigens (Landis, 1985). This technique will not, of course, give a biochemical identification of the neuroactive agent contained in the stained neuron. Some neurons that contain antigens whose effects, if any, are dissimilar from those of the neuroactive agent to which the antiserum was made may be localized by this technique. Nonetheless our protocol has already proven quite effective for the identification of neurons modulating PMA. These techniques used in different sequences can reveal both (1) whether candidate peptidergic neurons exert physiological effects that mimic, in total or in part, the effects of exogenously applied peptides and (2) whether identified neurons that have been previously characterized physiologically and morphologically are likely to contain a particular neuropeptide.

We found that PMA of buccal ganglia is controlled by a highly distributed regulatory system. At least nine different neuroactive agents are capable of modulating PMA. These include the peptides FMRFamide, SCP_B, and arginine vasotocin (AVT), the monoamines 5-HT, dopamine, and octopamine, plus acetylcholine, glutamate, and γ-aminobutyric acid. Of these, FMRFamide is clearly inhibitory, suppressing PMA (Murphy et al., 1985b), whereas glutamate (A. G. M. Bulloch and P. Jones, personal communication) and AVT (Richmond et al., 1985) have complex effects. The others are excitatory or inhibiting, or increase the rate

---

Calibration: 20 mV; 5s. (Middle panel) Intracellular Lucifer Yellow stains of the neuron VB1 from which the recordings in the top panel were taken (right side of photograph) and of its contralateral homolog. (Lower panel) SCP_B-immunoreactive staining of the same preparation examining the top and middle panels. Neuron VB1 clearly shows rhodamine irSCP$_B$ as do neurons VB2, VB3, and VB4. Different filter combinations were used on the fluorescent microscope to show Lucifer Yellow fluorescence in the middle panel and rhodamine fluorescence in the lower panel (from Murphy and Lukowiak, unpublished).

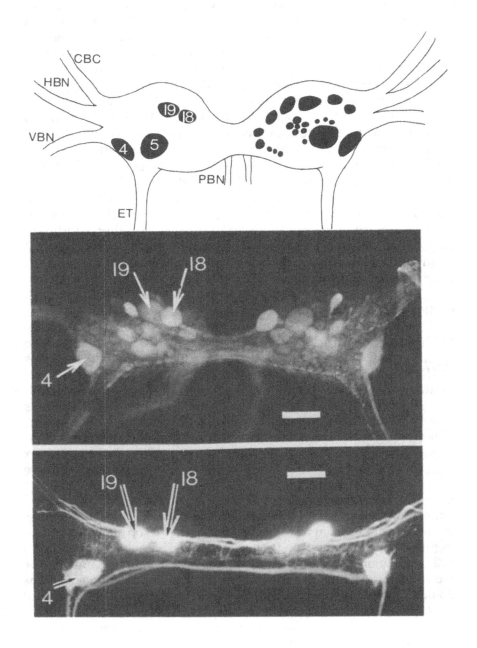

of PMA (e.g., Gramsow and Kater, 1977; Trimble and Barker, 1984; Murphy et al., 1985b; Richmond et al., 1986).

Here we will discuss our findings with respect to some of the SCP$_B$ immunoreactive neurons of the buccal ganglia and their roles in feeding. Though these data are in some ways quite preliminary, they are, nevertheless, very interesting to us and serve to illustrate the potential of a concerted application of physiological, morphological, and immunohistochemical techniques to peptidergic systems composed of individually identifiable neurons.

A schematic "map" showing the relative sizes and positions of several identified neurons involved in feeding is depicted in Fig. 4 (*see* Kater and Rowell, 1973; Kater, 1974; for the numerical "identities" of the smaller neurons and for their physiological roles during feeding). The four largest pairs of neurons, B4, B5, B18, and B19, are readily identifiable by visual inspection of the living ganglia. Intracellular recordings are readily made from these neurons (Fig. 5) and their morphology can be determined by the injection of dyes such as LY (Fig. 4). Indirect immunocytochemical staining of the buccal ganglion with an antibody to SCP$_B$ suggested the presence of immunoreactivity in neurons B4, B18, and B19 (Fig. 4). This was subsequently confirmed with the double-staining technique combining Lucifer Yellow injections with rhodamine indirect immunofluorescence (data not shown, but see Fig. 3). Neurons B4, B18, and B19 are primary effector neurons involved in feeding. Neuron B4 innervates and controls the activity of electrogenic

---

←

Fig. 4. Identification of putative peptidergic neurons. (Upper panel) Schematic map showing the relative size and positions of somata of several buccal neurons involved in feeding (*see* Kater and Rowel, 1973; Kater, 1974). These neurons have unique identities based upon their membrane characteristics, synaptic interactions, morphology, and transmitter content. (Middle panel) Dorsal view of SCP$_B$-like immunoreactivity and a buccal ganglia whole mount preparation. Three of the four largest pairs of neurons in the ganglia (B4, B18, and B19) have SCP$_B$-like immunoreactivity (arrows). (Lower panel) Morphologies of both left and right neurons B4, B18, and B19 as revealed by Lucifer Yellow dye injections. Calibrations: 100 μM, middle panel and 75 μM, lower panel. Abbreviations: CBC, cerebral buccal connective; ET, esophageal nerve trunk; HBN, heterobuccal nerve; PBN, posterior buccal nerve; VBN, ventral buccal nerve (from Murphy and Lukowiak, unpublished observations).

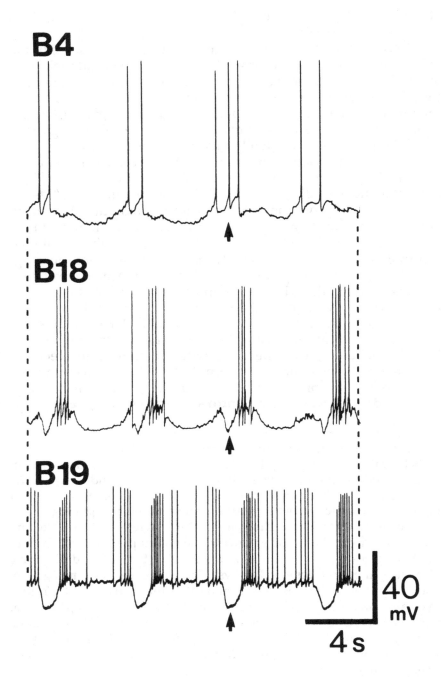

salivary gland secretory cells (Kater et al., 1978; Bahls et al., 1980; Coates and Bulloch, 1985). Neurons B18 and B19 are motor neurons that innervate muscles of the buccal mass (Kater, 1974). These three neurons have little, if any, effect upon PMA in the buccal ganglia and appear to utilize SCP$_B$ (and/or related peptides such as SCP$_A$; Lloyd et al., 1985) for self-modulation of their own peripheral efferent connections (Coates and Bulloch, 1985; Lloyd et al., 1984). We thus began a search for previously unidentified SCP$_B$ immunoreactive neurons whose effects might mimic the effects of bath application of SCP$_B$ on PMA. The most interesting neurons identified to date belong to a cluster of neurons (the VB1 cluster) in the medial part of the ventral surface of the buccal ganglia. In whole mounts of the ganglia, we routinely stained the four largest pairs of these neurons (VB1–VB4) immunocytochemically (Fig. 6, top panel). Often, additional smaller neurons were stained and staining of paraffin sections has revealed at least nine immunoreactive neurons in the cluster. Neurons VB1 to VB4, and possibly the rest of the cluster, have major axons in each of the paired posteriobuccal nerves that project to the muscular buccal mass (e.g., Fig. 6, lower panel). It is yet to be determined whether these neurons function peripherally as motor neurons. In any event, they have very interesting effects upon patterned activity within the buccal ganglion and have given us insights into the organization of neural circuitry underlying PMA associated with feeding and possibly other oral activities.

An intracellular recording from neuron VB1 in a quiescent (i.e., not producing PMA) buccal ganglion is shown in Fig. 3. Depolarizing dc current injection caused neuron VB1 to generate cyclical bursts of action potentials reminiscent of those seen in

---

Fig. 5.   Simultaneous intracellular recordings from three SCP$_B$ immunoreactive neurons during a spontaneous PMA. Arrows point out the coincident compound IPSPs in neurons B18 and B19 and corresponding EPSPs in neurons B4. These are typical N2 phase synaptic potentials. This recording illustrates variability in phase relationships. Note that the onset of EPSPs characteristic of N3 inputs in neurons B18 and B19 precedes the N2 phase compound IPSPs so that the N3 phase encompasses the N2 phase (*see* Fig. 1). This results in bursts of action potentials in neuron B19 both before and after the N2 phase IPSPs. Neuron B4 was held slightly hyperpolarized to eliminate occurrence of spontaneous action potentials during interburst intervals and to enhance the cyclical EPSPs (from Murphy and Lukowiak, unpublished).

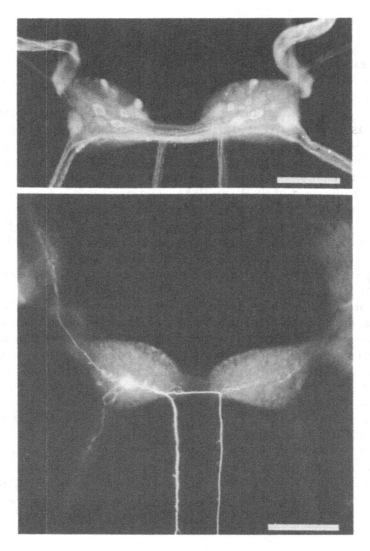

Fig. 6. SCP_B-immunoreactive neurons on the ventral surface of the buccal ganglia. (Upper panel) Ventral view of a whole-mount preparation of the buccal ganglia showing SCP_B-like immunoreactivity. A cluster of four medium-size neurons in the central portion of the ganglia was consistently stained. These neurons designated VB1 to VB4 all have major axons in the posteriobuccal nerves. (Lower panel) Lucifer Yellow stain of neuron VB1, the largest and most lateral of the VB1 cluster (different preparation from upper panel). A dendritic arborization extends into the ipsilateral esophageal nerve trunk, and a small process extends up the ipsilateral cerebral buccal connective to the cerebral ganglion. Calibration bars: 200 μM (from Murphy and Lukowiak, unpublished).

certain neurons during feeding PMA. Both number of bursts and frequency of action potentials per burst were dependent upon the level of current injected. In each case, accommodation to the current injection occurred relatively rapidly, resulting in a cessation of action potential generation. After recording from neuron VB1, Lucifer Yellow dye was injected into it and into its contralateral homolog (Fig. 3, middle panel). Subsequently, the same preparation was processed for immunoreactive staining using a monoclonal antibody to SCP_B (*see* Murphy et al., 1985b) and a rhodamine-conjugated secondary antibody. This procedure confirmed that the neuron analyzed was indeed the largest neuron of the VB1 cluster. To determine whether bursts of action potentials generated in neuron VB1 could evoke synaptic activity similar to feeding PMA in previously identified motor neurons, the commissures of buccal ganglia were twisted 180° so that simultaneous recordings could be made from neurons VB1 in one buccal ganglion and from identified neurons on the dorsal surface of the contralateral buccal ganglion. In some cases, stimulation of neuron VB1 evoked synaptic activity in follower neurons resembling the complete PMA seen during feeding (*see* Fig. 1). More often, however, only the subset of synaptic activity that normally occurs during the N3 phase of PMA (compare Figs. 1 and 7) was evoked by bursts of action potentials generated in neuron VB1. Similarly, hyperpolarization of neuron VB1 during ongoing PMA eliminates the N3 phase of synaptic activity in primary effector neurons involved in feeding (Fig. 8). Thus these identified SCP_B immunoreactive neurons only rarely mimic the total effects of bath application of SCP_B, but they play a prominent role in the production of PMA. The VB1 cluster appears to be either a part of the N3 group of interneurons that provides synaptic input to motor neurons during the N3 phase of PMA, or to exert intimate control over the N3 interneurons. These studies have already taught us a great deal about the neural organization underlying feeding and other oral activities in *Helisoma*. The central pattern generator driving patterned motor output consists of three loosely coupled subsystems of interneurons (driving the N1, N2, and N3 phases of synaptic activity; *see* Fig. 1), and these subsystems can be independently activated or suppressed. This can result in qualitatively different patterns of activity in follower feeding neurons.

Recent preliminary evidence suggests that two additional SCP_B immunoreactive neurons may evoke PMA similar to that seen during feeding. Trains of action potentials in a neuron in the location of the bright immunoreactive neuron just lateral to neuron

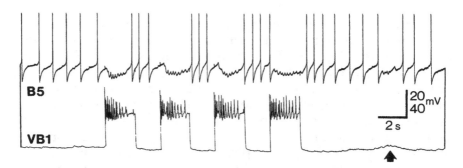

Fig. 7.   Action potentials in neuron VB1 activate the N3 subunit of
the central pattern generator in the buccal ganglion of *Helisoma*. (Top)
Simultaneous intracellular recordings from neuron B5 and VB1. Burst of
action potentials evoked in neuron VB1 by depolarizing pulses elicits
polysynaptic IPSPs in neuron B5 similar to those occurring during the N3
phase of PMA. The N3 interneuron network that projects EPSPs back
onto neuron VB1 is apparently entrained by the pulses and generates a
single cycle of synaptic output (arrow) at the termination of the pulses.
(Bottom) Simultaneous intracellular recordings from neurons B19 and
VB1. Burst of action potentials evoked by depolarizing pulses in neuron
VB1 activates the N3 interneuron network to elicit corresponding burst of
action potentials in neuron B19. The first burst of a train also elicited a
large compound IPSP in neuron B19 similar to N2 phase synaptic activity.
The train of pulses in neuron VB1 entrain the N3 network to generate five
additional cycles of synaptic activity with increasing intervals (arrows)
(from Murphy and Lukowiak, unpublished).

B19 (*see* Fig. 4), as well as in a neuron in the location of another
SCP$_B$ immunoreactive neuron in the cerebral ganglion, evoked
PMA similar to that seen during feeding. The double-staining
protocol described above for neuron VB1 will be applied to determine if these are SCP$_B$-like regulatory neurons for feeding PMA.
    The results obtained using the *Helisoma* preparation demon-
strate very well the advantages offered by an invertebrate system
for determining the functional roles of peptidergic neurons in
mediating discrete behaviors. In the next section, we examine
another invertebrate model system, the *Aplysia* siphon, mantle,
gill, and abdominal ganglion preparation (Fig. 9), and show how
this system can be used to study the role played by peptides in the
plasticity of adaptive behaviors.

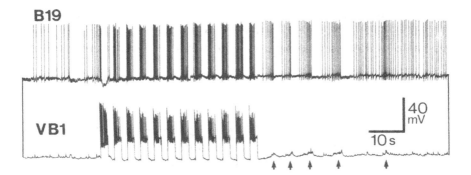

Fig. 8. Hyperpolarization of neuron VB1 during PMA eliminates the N3 phase of synaptic activity. Simultaneous intracellular recordings from neurons B5, B19, VB1, and VB10 during PMA elicited by $10^{-6}M$ 5-HT. Hyperpolarization of neuron VB1 (between arrows) inactivates the N3 subunit of the pattern generator, eliminating the N3 EPSPs in neuron B19 and the N3 IPSPs in neuron B5 (*see* Fig. 1). The N2 phase synaptic inputs are not affected. Thus, neuron B19 continues to generate bursts of action potentials caused by "anode break" after the N2 phase IPSPs, but the number and frequency of action potentials per burst is greatly decreased. Neuron B5 fires tonically during the hyperpolarization of neuron VB1. Activity in neuron VB10 is unaffected since it receives EPSPs from the N2 subunit of the pattern generator, but no synaptic inputs from the N3 subunit (from Murphy and Lukowiak, unpublished).

4. The *Aplysia* Model System

The *Aplysia* preparation shown in Fig. 9 has proven to be an extremely successful preparation for gaining an understanding of the neuronal mechanisms that underly adaptive behavior (Mptisos and Lukowiak, 1985). Both nonassociative (habituation and sensitization) and associative (classical conditioning) learning can be demonstrated in this in vitro preparation, and many of our current ideas regarding the neuronal mechanisms that underlie learning have been obtained from these studies (Kandel and Schwartz, 1982; Hawkins et al., 1983; Mpitsos and Lukowiak, 1985). The hope is that, because of its relative simplicity (compared to the mammalian brain), the neuronal mechanisms of learning can be more fully worked out and then be applied to more complex and complicated organisms, including ourselves. That this assumption is correct remains to be determined; but it appears that mechanisms similar to those that exist in the molluscan nervous systems exist also in the more complicated nervous systems of vertebrates.

5. The Gill Withdrawal Reflex

A tactile stimulus applied to the siphon or gill of this in vitro preparation evokes a defensive gill withdrawal reflex (GWR). This reflex is mediated by the integrated activity of neurons in the abdominal ganglion (the CNS) and in the peripheral nervous system (PNS) between the siphon and gill (Lukowiak, 1979). In addition to the stimulus-evoked GWR, the preparation also exhibits

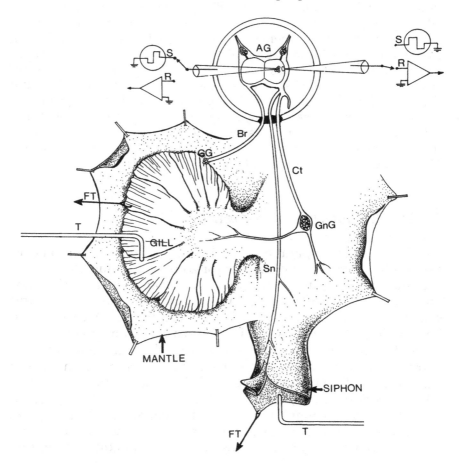

spontaneous gill respiratory movements (SGMs). The SGMs are mediated by an interneuron network (INT II) in the abdominal ganglion and are characterized by their constant amplitude and periodicity (Byrne and Koester, 1978; Pinsker, 1982). These contractions serve to help move freshly oxygenated blood through the gill to the heart and then to the rest of the animal. In contrast to the relative constant amplitude of the SGMs, the GWR amplitude is dependent on many parameters, including stimulus strength, the previous history of stimulation, and the preparation's behavioral state.

6. Behavioral State

Behavioral state can be simply defined as level of arousal, mood, or how an animal will respond to a particular stimulus. In everyday terms that we can all understand, it is the equivalent of determining if your Dean or Department Chairman is in a good mood before asking for more laboratory space or a raise in pay. Obviously, on certain days it would be better not to make such a request, whereas on other days the request will be granted.

←————————————————————————————————————

Fig. 9. Schematic diagram of the in vitro siphon, mantle, gill, and abdominal ganglion preparation of *Aplysia californica*. A tactile stimulus applied by a mechanical tapper (T) to either the siphon or gill evokes a defensive gill withdrawal reflex (GWR). This reflex undergoes both nonassociative and associative learning. The amplitude of the GWR and siphon withdrawal response is measured by a force transducer (FT) attached to the gill and siphon by fine surgical suture. The abdominal ganglion (AG) is placed in its own leak-proof perfusion chamber so that different artificial seawaters (ASW) can be superfused over only the abdominal ganglion. The abdominal ganglion innervates the peripheral tissues via the branchial (Br), ctenidial (Ct), and siphon (Sn) nerves. Intracellular recordings can be made from identifiable sensory and motor neurons in the ganglion. Also, current can be passed into the cells via a bridge circuit in the electrometer to either depolarize or hyperpolarize the neurons. The preparation is maintained in ASW at 15–18°C and normally remains viable for 12–18 h. Two peripherally located ganglia are shown—the gill ganglion (GG) located on the branchial nerve and the genital nerve ganglion (GnG) located on the ctenidial nerve. It is not entirely clear what roles are played by either ganglion.

There are three behavioral states in the in vitro preparation that appear to be correlated with behaviors in the intact, freely moving animal. *Aplysia* that are food-satiated or engaged in sexual activity just prior to dissection are suppressed. Normal states are those in which *Aplysia* are not food satiated nor engaged in sexual activity immediately prior to dissection. As of yet, we have not been able to correlate any easily observed behaviors with the facilitated state. However, now that an ethogram of *Aplysia* has been constructed (Leonard and Lukowiak, 1986), we should be able to more easily determine the behaviors that result in the facilitated state.

In the in vitro siphon, mantle, gill, and abdominal ganglion preparation, it has been possible to operationally define three behavioral states based primarily on the amplitude of the GWR evoked by a standard tactile siphon stimulus compared to the amplitude of the SGMs. The three states have been termed (1) suppressed; (2) normal; and (3) facilitated (Ruben et al., 1981).

In the suppressed behavioral state, the GWR amplitude evoked by a 1-g tactile stimulus applied to the siphon is small, less than 30% of the SGM amplitude. In addition, the reflex habituates rapidly. The normal behavioral state is characterized by a GWR amplitude of greater than 35% of the SGM, and the rate of habituation does not occur extremely rapidly. Finally, in the facilitated state, the amplitude of the GWR is similar to that of the SGM and the reflex does not habituate readily.

The behavioral state of the preparation is also reflected at the neuronal level. In the suppressed state, fewer action potentials (APs) are evoked in gill motor neurons L7 and LDG1 by the siphon stimulus compared to normal state preparation. In addition, the rate of decrement of the excitatory input to L7 or LDG1 evoked by the siphon stimulus decrements rapidly with repeated stimulation. In facilitated state preparations, the activity evoked by the stimulus is robust and does not readily decrement.

7. Habituation and Sensitization

Habituation is probably the simplest and most ubiquitous form of learning (nonassociative); the animal essentially learns to not respond to a nonthreatening stimulus that is repeatedly present. Thus, for instance, *Aplysia*, which live in the intertidal area, become habituated to the gentle action of the surf and soon do not respond by withdrawing their gill to each small wave. The in vitro *Aplysia* preparation behaves very similarly to the intact animal in regard to both associative and nonassociative learning (Mptisos and Lukowiak, 1985). In the in vitro *Aplysia* preparation, the amplitude of the GWR decrements exponentially with repeated siphon stimulation. Sensitization is a more complex example of nonassociative learning in which an animal learns to increase its response to a stimulus after it has received a strong, noxious stimulus. For example, in the in vitro preparation, the GWR evoked by the siphon stimulus is significantly larger following a gill pinch. The advantage offered by this preparation is that one can simultaneously study the neuronal changes that mediate the behavioral responses. At the neuronal level, habituation is caused in part by a decrease in transmitter release at the sensorimotor neuron synapse. This is caused in part by a decrease in the Ca^{2+} influx into the terminals of the sensory neurons. In sensitization, the amount of transmitter released at the sensory motor neuron synapse is increased in part because of an increased Ca^{2+} influx.

8. Pharmacology of Behavioral State

Where do peptides fit into the scheme and how does one begin to determine how and where they exert their effect? For example, do food satiation and sexual activity cause the release of peptides that then, by acting on neuronal elements in the CNS and PNS, bring about the observed behavioral state? We have begun our investigations by examining the effects of endogenous neural peptides on GWR amplitude and associated neuronal activity in an attempt to determine if any of the peptides could mimic the observed behavioral states as previously demonstrated for *Helisoma*. A major advantage of the in vitro preparation is that it is easy to superfuse only the abdominal ganglion with peptide-containing artificial sea water (ASW), while recording intracellularly from identified neurons and determining the peptide's effect at both the neuronal and behavioral level. Since the preparation remains viable for at least 12–18 h, it is possible to wash out the peptide and test for recovery. This is an important point; it is also possible to perfuse the peptide–ASW to only the gill and thus study the peripheral effects of the peptide.

Using such techniques, we have been able to provide evidence for the involvement of two endogenous peptides, AVT and SCP_B, in the mediation of the GWR in this preparation.

8.1. AVT

Using high-pressure liquid chromatography (HPLC) and radioimmunoassay (RIA) techniques, AVT was first shown to be present in the *Aplysia* nervous system (Moore et al., 1981). AVT has since been shown by immunohistochemical techniques to be present in the nervous systems of *Limax*, *Lymnaea*, and *Helisoma*.

When AVT-containing ASW (10^{-6} to $10^{-12}M$) was superfused over the abdominal ganglion of a normal state preparation, it produced a state similar to that observed in the suppressed behavioral state (Thornhill et al., 1981). Moreover, with washout of the peptide, the effect was reversed (Fig. 10). In other normal-state preparations, the GWR was initially habituated, and dishabituation (sensitization) was then brought about by the interposition of a novel stimulus (gill pinch). Following a 3-h rest, the preparation recovered completely from the effects of habituation. When AVT-containing ASW was then superfused over the ganglion, it not only caused suppression of GWR amplitude, but also increased the rate at which the reflex habituated. However, dishabituation of the reflex still occurred, indicating that AVT's suppressive effects could be overcome. Following washout of AVT and a further 3-h rest, the preparations returned to controlled levels (Thornhill et al., 1981). Thus the superfusion of AVT over the ganglion altered the behavioral state of the preparation. It behaved like those taken from food-satiated animals. It is important to note that AVT superfusion of the ganglion did not alter the amplitude or frequency of the SGMs. At the neuronal level, AVT brought about a reduction in the excitatory postsynaptic input (EPSP) to the gill motor neurons evoked by the siphon stimulus, but did not alter the amplitude of the synaptic potentials associated with the INT II network that drives the SGMs. Thus, AVT's effects were very selective, affecting only the inputs associated with the GWR.

9. AVT Effects on the Sensory Motor Neuron Synapse

How does AVT bring about its selective effects on the GWR, and does this give us any clue as to how behavioral state is medi-

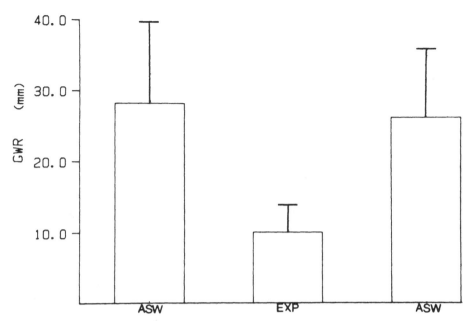

Fig. 10. The effect of arginine vasotocin (AVT) on the GWR evoked by siphon stimulation. The GWR was evoked by a 1-g siphon stimulus initially with ASW bathing the ganglion. Following a 1-h rest, AVT-containing ASW ($10^{-6}M$) was superfused over the ganglion for 5 min and the siphon stimulus was again applied (EXP). As can be seen, the GWR was significantly reduced. Following washout of the peptide and a 1-h rest, the GWR was again evoked (ASW) and the amplitude of the GWR was not significantly different from the initial control reflex ($n = 15$) (after Thornhill et al., 1981; Lukowiak et al., 1986).

ated? For these experiments, an LE sensory neuron (which innervate the siphon) and a gill motor neuron (normally L7) were simultaneously impaled (Fig. 11). When an AP is elicited in the LE neuron by the passage of depolarizing current through the recording electrode, a short-latency, monosynaptic EPSP is evoked in the gill motor neuron. If an AP is repeatedly elicited in the LE neuron at a frequency of 1/30 s, the amplitude of the EPSP decrements. This is known as low-frequency homosynaptic depression, and this depression, in part, is responsible for behavioral habituation (Fig. 12). When AVT-containing ASW was superfused over the ganglion, it brought about a reduction in the amplitude of the elicited EPSP and caused an increase in the rate of low-frequency homosynaptic depression (Goldberg, 1983; Goldberg et al., 1986). With washout,

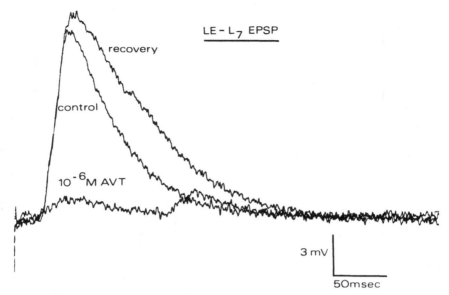

Fig. 11.   Intracellular recording obtained from gill motor neuron L7 and an LE sensory neuron (not shown). A depolarizing current pulse was applied to the LE neuron to elicit an action potential. The AP in the sensory neuron evoked a short latency, monosynaptic EPSP in L7 (control). When AVT ($10^{-6}M$) was superfused over the ganglion and an AP evoked in the sensory neuron, the EPSP recorded in L7 was markedly reduced. Following washout, the EPSP evoked by the sensory neuron AP recovered.

both the EPSP amplitude and the rate of low-frequency homosynaptic depression return to control levels. Although AVT brought about a reduction in the EPSP elicited by the LE neuron, it did not alter other synaptic inputs to the gill motor neuron. Thus, AVT did not appear to bring about its effects by altering the passive membrane properties of the motor neuron. When the input resistance and other passive membrane properties of the neuron were examined, they were found to be unaffected by AVT. Therefore, it appears that AVT brought about its effect by decreasing the amount of transmitter released by the sensory neuron.

AVT could affect transmitter release by decreasing the Ca^{2+} influx into the presynaptic nerve terminal during the AP. To test this possibility, the AP was broadened by the addition of 10 mM tetraethylammonium (TEA) to the ASW. TEA blocks the outward K^+ current, thus prolonging the AP (Fig. 13). This allowed the contribution of the inward calcium current to become more appar-

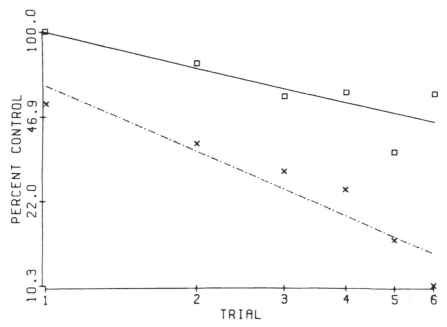

Fig. 12.   The effect of AVT on low frequency homosynaptic depression. Single LE APs were evoked at an ISI of 30 s to produce low-frequency homosynaptic depression at the sensorimotor neuron synapse. Control (ASW, □) and test (AVT $10^{-6}M$, x) runs were separated by 60 min. The rate of homosynaptic depression was measured as the slope of the least-squares fit of the log–log plot: EPSP amplitude vs. trials. In each experiment, the data were normalized relative to trial 1 of the control run. In ASW, the slope obtained was –0.44. When AVT was superfused over the abdominal ganglion, the rate of decrement was increased to –0.84. In addition to increasing the rate of synaptic depression by almost twofold, AVT also brought about suppression of the EPSP amplitude, as shown in the previous figure (Goldberg, 1983).

ent. In the presence of 10 mM TEA, AVT caused a reduction in AP duration. Upon washout of the AVT, the AP returned to its pre-AVT duration. Thus, it would appear that AVT affects Ca^{2+} influx into the sensory neuron and therefore causes less transmitter to be released. We do not yet know how AVT does this.

AVT could, itself, have a direct effect on the LE sensory neuron, or it could be exerting its effect via an interposed neuron and another transmitter. Preliminary, indirect evidence suggests that AVT activates interposed presynaptic inhibitory neuron. When the abdominal ganglion is bathed in a high-calcium, high-

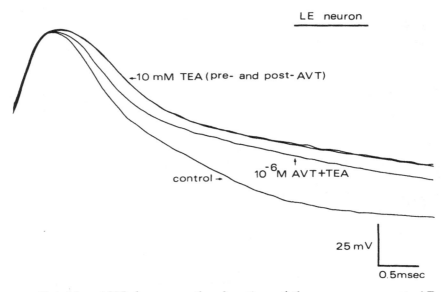

Fig. 13. AVT decreases the duration of the sensory neuron AP. Shown are the effects of 10 mM tetraethylammonium ion (TEA) and $10^{-6}M$ AVT on action potential duration of sensory neurons in the abdominal ganglion. In the control, the action potential was evoked in the LE sensory neuron in ASW. Twenty minutes later, while TEA (10 mM) was superfused over the ganglion, the action potential was again evoked and this caused an increase in the duration of the action potential half-amplitude from 1.61 to 2.25 ms. Twenty minutes later, $10^{-6}M$ AVT was added to the 10 mM TEA-containing ASW and the preparation was perfused at the same rate. The action potential evoked was reduced in its duration at half amplitude from 2.25 to 1.88 ms. Washout of the peptide for 15 min with 10 mM TEA-containing ASW caused an increase in the duration to 2.22 ms. Subsequent washout of TEA with ASW caused a recovery of AP duration to nearer control values (not shown). Similar data were obtained for nine other preparations (Colmers and Lukowiak, unpublished observations).

magnesium ASW (six times normal), AVT's suppressive effects are not apparent. This high divalent cation ASW increases the threshold of interneurons. These data are only suggestive, but they are consistent with a hypothesis proposed by Lukowiak and Peretz (1980). This hypothesis suggested that there are suppressive and facilitory control neurons in the abdominal ganglion, which both "gate" the input from sensory neurons onto the motor neurons and set the level of responsiveness of the PNS in the gill (Fig. 14). If this

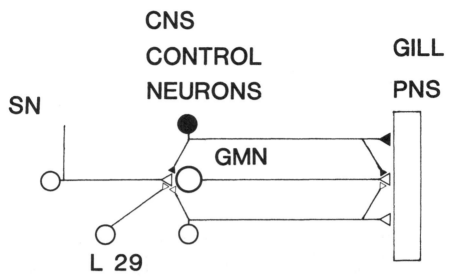

Fig. 14.   Schematic representation of a possible manner in which endogenous neuropeptides could affect the behavior of the gill withdrawal response. In this scheme, the peptides directly affect two CNS control neurons represented in the figure. AVT is hypothesized to activate the black filled-in cell. This cell or group of cells presynaptically gates the input to the gill motor neuron and at the same time brings about suppression of the ability of the motor neuron to elicit a gill movement. SCP_B, on the other hand, is hypothesized to work via the blank cell to facilitate the inputs to the gill motor neuron from the sensory neurons and to facilitate the ability of the gill motor neuron to elicit a gill withdrawal response. The manner in which the control neurons facilitate the ability of the gill motor neuron to elicit the gill withdrawal response is not yet known. It could be either by acting directly on the gill muscle or via neurons in the peripheral nervous system. Cell L29 is shown as an example of an identified neuron that has already been shown to affect the input from the sensory neuron to the gill motor neuron. Abbreviations: SN, sensory neuron; GMN, gill motor neuron) (Lukowiak, unpublished).

hypothesis is valid, then one would predict that AVT superfusion of the abdominal ganglion should affect the ability of a gill motor neuron to elicit a gill withdrawal response.

In a series of experiments (Fig. 15), L7 or LDG1 were impaled, and depolarizing current was passed into the neuron to evoke a train of APs. The APs cause a gill withdrawal response. If the same number of APs are evoked in the motor neuron once every 20 min, the amplitude of the elicited gill withdrawal response is constant.

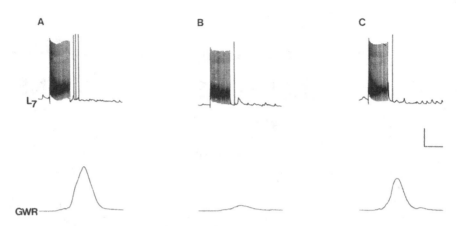

Fig. 15.   The effect of AVT superfusion over the abdominal ganglion on a gill motor neuron's ability to elicit a gill withdrawal response. (A) Individual data show the effect of AVT on the ability of motor neuron L7 to elicit a gill withdrawal response. L7 was depolarized to produce 19 APs, which resulted in the gill withdrawal response shown below. Twenty minutes later, 5 min in AVT, 18 APs evoked in L7 elicited a much smaller gill withdrawal response. Following washout and a 1-h rest, 17 APs elicited a much larger gill withdrawal response that came close to control values.

However, when AVT-containing ASW ($10^{-6}M$) was superfused over the ganglion and the same number of APs were evoked in the gill motor neuron, the amplitude of the elicited gill withdrawal response was significantly reduced. Following washout, the response amplitude returned to its pre-AVT level. Thus, it would appear that AVT may bring about its effects via the hypothesized central suppressive control neurons.

To make the story more complete, it will be necessary to find the AVT neurons using techniques described for the *Helisoma* preparation and demonstrate that activity in these neurons produces the effects seen with the exogenous application of AVT. To show that AVT is involved in the mediation of the suppressive state associated with food satiation, it will be necessary to show that these neurons are more activated in food-satiated animals than in control normal-state animals or that in a semi-intact preparation the presentation of food turns these neurons on.

Indirect evidence suggests that AVT is involved in the mediation of the suppressive state associated with food satiation. The blood taken from food-satiated animals and superfused over the ganglion of a control group preparation brings about suppression of the GWR and decreased the inputs of gill motor neurons (Lukowiak et al., 1986). It will need to be determined if the factor in the blood that brings about the suppression is indeed AVT or some other neuroactive agent.

10. Suppression of the GWR Caused by Sexual Activity

Suppression of the GWR and its concomitant neuronal activity is also associated with sexual activity (Lukowiak and Freedman, 1983). The suppression associated with sexual activity either as a male or female (*Aplysia* are hermaphroditic) is similar to the suppression associated with food satiation. However, AVT apparently does not mediate this behavioral state; it appears that an enkephalin-like peptide is involved. The superfusion of met-enkephalin over the abdominal ganglion, in much the same manner as AVT, brings about suppression of the GWR and its concomitant neuronal activity (Lukowiak et al., 1982). Naloxone, a relatively specific opiate antagonist, blocks met-enkephalin's ability to suppress the GWR but does not block AVT's effects. In preliminary experiments, naloxone did reverse a suppression associated with sexual activity, but not food satiation. Thus, it appears that a second peptidergic pathway is involved in the mediation of the suppressed behavioral state. Whether or not the pathways converge at the level of proposed central suppressive control neurons remains to be determined.

11. Facilitation of the GWR and Its Associated Neuronal Activity

Another endogenous peptide, SCP_B, has also been studied. SCP_B facilitates both the GWR and the excitatory synaptic input to the gill motor neuron.

Superfusion of the abdominal ganglion with SCP_B (10^{-6} to 10^{-9} M) facilitates the GWR evoked by siphon stimulation (Fig. 16A). SCP_B changes the behavioral state of the preparation from a normal state to one resembling the facilitated state. That is, following the superfusion of SCP_B over the ganglion, the GWR amplitude is similar to that of the SGM and the preparations do not readily

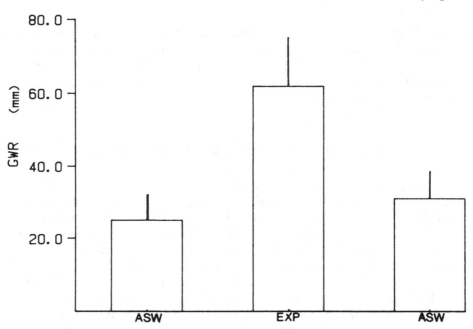

Fig. 16A.   The effect of small cardioactive peptide B (SCP$_B$) on the
GWR. Data were obtained as in Fig. 9, except that SCP$_B$ ($10^{-6}M$) was
superfused over the abdominal ganglion. As can be readily observed,
SCPB facilitated the GWR ($n = 8$) (Lukowiak, unpublished observations).

habituate. Following washout, the preparations return to their
normal behavioral state.

Superfusion of the ganglion with SCP$_B$ also selectively increases the amplitude of the EPSP at the sensory motor neuron
synapse elicited by an AP in the sensory neuron (Fig. 16B). Other
spontaneously occurring synaptic inputs to the motor neuron were
unaffected by SCP$_B$. Again, following washout, the facilitory effect
was reversed. Was this increase in EPSP amplitude caused by an
increase in AP duration? The duration of the sensory neuron's AP
is increased by SCP$_B$ in much the same manner as by 5-HT (Abrams
et al., 1985). Initially, using a low concentration of TEA (10 mM),
we could not see a broadening of the AP with SCP$_B$ (Lukowiak et
al., 1984). Using this same concentration of TEA, we did see a
decrease in the AP duration with AVT. However, more recent
experiments using a technique known as frequency-dependent
spike broadening (FDSB; Edstrom and Lukowiak, 1985) indicate
that SCP$_B$ does indeed bring about an increase in the duration of
the AP; however, this affect was not completely reversible.

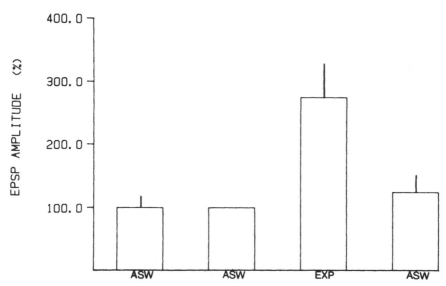

Fig. 16B. Normalized data showing the effect of SCP_B on the EPSP recorded in L7 evoked by an AP in an LE sensory neuron. Trials were separated by 20-min intervals. The amplitude of the EPSP in ASW was taken as 100%. Following a second control and a 15-min rest, SCP_B-containing ASW was superfused over the ganglion for 5 min. It can be seen that this significantly facilitated the EPSP (EXP). Following washout and a further 20-min rest, the EPSP amplitude was not significantly different from control ($n = 7$) (Lukowiak, unpublished observations).

Whether or not the increase seen in the EPSP amplitude is entirely caused by an increase in AP duration remains to be determined.

We then determined whether SCP_B could affect the ability of a motor neuron to elicit a gill withdrawal response. In much the same manner as AVT, SCP_B superfusion of the abdominal ganglion significantly altered the ability of a gill motor neuron to elicit a gill withdrawal response. However, SCP_B facilitated the response (Fig. 17). These results suggest that SCP_B may bring about its effects via the hypothesized central control neurons. However, SCP_B would activate the facilitatory control neurons, not the suppressive ones.

Again, it will be necessary to ascertain, using techniques described above for *Helisoma*, the neurons that contain SCP_B and to determine how and when they are activated and if their activation does indeed alter the behavioral state of the preparation.

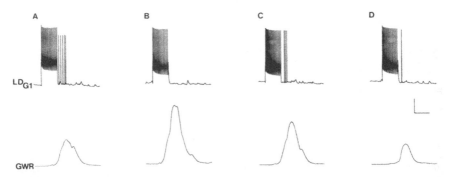

Fig. 17. The effect of superfusion of SCP$_B$ over the abdominal ganglion on the efficacy of LDG1 and L7 to elicit the gill withdrawal response. Data are similar to those presented in Fig. 14. However, in this case, SCP$_B$ instead of AVT was superfused over the ganglion. As can be seen, SCP$_B$ brought about a facilitation of the motor neurons' ability to elicit a gill withdrawal response. In (A), LDG1 was depolarized to produce 18 APs; this elicits a gill withdrawal response. When LDG1 was depolarized to produce 16 APs with SCP$_B$ bathing the ganglion (B), it produced a significantly larger response. Following a 1-h washout (C), 18 APs in LDG1 still brought about a larger gill withdrawal response than initially. Finally, 2 h later (D), LDG depolarization (17 APs) elicited a slightly smaller gill withdrawal response than the control.

12. Conclusion

The peptides we have briefly looked at as regards behavioral state in *Aplysia* also have potent effects in the periphery affecting the activity of the heart, gut, and gill reflex directly. It is beyond the scope of this review, however, to go into these effects.

The conclusion that we would like to leave you with is that a model system approach utilizing invertebrate preparations can be utilized to uncover the role and mechanisms of actions that peptides play in the mediation of adaptive and important homeostatic behaviors. A combined behavioral, electrophysiological, anatomical, and immunocytological approach can be utilized to directly answer some of the most important questions in neurobiology today. It may well be that as we progress and begin to uncover the mechanisms responsible for higher forms of learning, we may find that peptides play an important role.

References

Abrams T. W., Castellucci V. F., Camardo J. S., Kandel E. R., and Lloyd P. E. (1985) Two endogenous neuropeptides modulate the gill and siphon withdrawal reflex in *Aplysia* by presynaptic facilitation involving cAMP-dependent closure of a serotonin-sensitive potassium channel. *Proc. Natl. Acad. Sci. USA* **81**, 7956–7960.

Bahls F., Kater S. B., and Joyner R. W. (1980) Neuronal mechanisms for bilateral coordination of salivary gland activity in *Helisoma*. *J. Neurobiol.* **11**, 365–379.

Byrne J. H. and Koester J. (1978) Respiratory pumping: Neuronal control of a centrally commanded behavior in *Aplysia*. *Brain Res.* **143**, 87–105.

Coates C. J. and Bulloch A. G. M. (1985) Synaptic plasticity in the molluscan peripheral nervous system: Physiology and role for peptides. *J. Neurosci.* **5**, 2677–2684.

Edstrom J. and Lukowiak K. (1985) Frequency dependent action potential prolongation in *Aplysia* plural sensory neurons. *Neuroscience* **16**, 451–460.

Goldberg J. I. (1983) Transfer of habituation in *Aplysia californica:* Evidence for a heterosynaptic mechanism underlying habituation of the gill withdrawal reflex. Ph.D. Thesis, University of Calgary, Calgary, Canada.

Goldberg J. I., Colmers W. F., Edstrom J., and Lukowiak K. (1986) Suppression of sensory neuron synaptic transmission and narrowing of the sensory neuron potential by the peptide hormone, arginine vasotocin. *J. Exp. Biol.* (in press).

Gramsow B. and Kater S. B. (1977) Identified higher-order neurons controlling the feeding motor program of *Helisoma*. *J. Neurosci.* **2**, 1049–1063.

Hawkins R. D., Abrams T. W., Carew T. J., and Kandel E. R. (1983) A cellular mechanism of classical conditioning in *Aplysia:* Activity-dependent amplification of pre-synaptic facilitation. *Science* **219**, 400–405.

Hodgkin A. L. and Huxley A. F. (1952) A quantitative description of membrane current and its application to conduction and excitation in nerve. *J. Physiol.* (Lond.) **117**, 500–544.

Iversen L. L., Iversen S. D., and Snyder S. H. (eds.) (1983) *Handbook of Psychopharmacology. Neuropeptides* vol. 16, Plenum, New York.

Joosse J. and Geraerts W. P. M. (1983) Endocrinology, in *The Molluscan Physiology* part I, vol. 4, (Saleuddin A. S. M. and Wilbur K. M., eds.) Academic, New York.

Kaldany R. J., Nambu J. R., and Scheller R. H. (1985) Neuropeptides in identified *Aplysia* neurons. *Ann. Rev. Neurosci.* **8**, 431–455.

Kandel E. R. (1976) *Cellular Basis of Behavior* W. H. Freeman, San Francisco, California.

Kandel E. R. and Schwartz J. H. (1982) Molecular biology of learning: Modulation of transmitter release. *Science* **218,** 433–443.

Kaneko C. R., Merickel M., and Kater S. D. (1978) Centrally programmed feeding in *Helisoma:* Identification in characteristics of an electrically coupled premotor neuron network. *Brain Res.* **146,** 1–21.

Kater S. B. (1974) Feeding in *Helisoma trivolis:* The morphological and physiological bases of a fixed action pattern. *Am. Zool.* **14,** 1017–1036.

Kater S. D. and Rowell C. H. F. (1973) Integration of sensory and centrally programmed components in the generation of cyclical feeding activity of *Helisoma trivolvis. J. Neurophysiol.* **36,** 145–155.

Kater S. B., Murphy A. D., and Rued J. R. (1978) Control of the salivary glands of *Helisoma* by identified neurones. *J. Exp. Biol.* **72,** 91–106.

Koester J. and Byrne J. H. (1980) *Molluscan Nerve Cells: From Biophysics to Behavior.* Cold Spring Harbor Laboratory, Cold Spring Harbor, Long Island, New York.

Landis D. M. D. (1985) *Promise and Pitfalls in Immunocytochemistry* Elsevier, Amsterdam.

Leonard J. L. and Lukowiak K. (1986) The behavior of *Aplysia californica* Cooper (Gastropoda; Opisthobranchial: I. Ethogram). *Behavior* (in press).

Llinas R., Steinberg I. Z., and Walton K. (1981) Relationship between presynaptic calcium current and post-synaptic potential in squid giant synapse. *Biophys. J.* **33,** 323–351.

Lloyd P. E., Kupfermann I., and Weiss K. R. (1984) Evidence from parallel actions of a molluscan neuropeptide and serotonin in mediating arousal in *Aplysia. Proc. Natl. Acad. Sci. USA* **81,** 2934–2937.

Lloyd P. E., Mahon A. C., Kupfermann I., Cohen J. L., Scheller R. H., and Weiss K. R. (1985) Biochemical and immunocytochemical localization of molluscan small cardioactive peptides in the nervous system of *Aplysia californica. J. Neurosci.* **5,** 1851–1861.

Lukowiak K. (1979) The development of central nervous control of the gill withdrawal reflex evoked by siphon stimulation in *Aplysia. Can. J. Physiol. Pharmacaol* **57,** 987–997.

Lukowiak K. and Freedman L. (1983) The gill withdrawal reflex is suppressed in sexually active *Aplysia. Can. J. Physiol. Pharmacol.* **61,** 743–748.

Lukowiak K. and Peretz B. (1980) Control of gill reflex habituation and the rate of EPSP decrement of L7 by a common source in the CNS of *Aplysia. J. Neurobiol.* **11,** 425–433.

Lukowiak K., Thornhill J. A., and Edstrom J. (1982) Methionine enkephalin increases CNS suppressive control exerted over gill reflex behaviours and associated neural activity in *Aplysia californica. Reg. Peptides* **3**, 303–312.

Lukowiak K., Edstrom J. P., and Colmers W. F. (1984) Peptidergic (SCP$_B$ and FMRFamide) modulation of gill reflex behaviors and associated neuronal activity in *Aplysia. Soc. Neurosci. Abstr.* **14**, 151–158.

Lukowiak K., Goldberg J., Colmers W. F., and Edstrom J. P. (1986) Peptide Modulation of Neuronal Activity and Behavior in *Aplysia,* in *Comparative Aspects of Opioid and Related Neuropeptide Mechanisms* (Stefano G., ed.) CRC, Florida (in press).

Mackie G. O., Stell W. K., and Singla C. L. (1985) Distribution of nerve elements showing FMRFamide-like activity in hydromedusae. *Acta Zoologica* **66**, 199–210.

Moore G. J., Thornhill J. A., Gill V., Lederis K., and Lukowiak K. (1981) An arginine vasotocin-like neuropeptide is present in the nervous system of the marine mollusc *Aplysia californica. Brain Res.* **206**, 213–218.

Mptisos G. J. and Lukowiak K. (1985) Learning in Gastropod Molluscs, in *The Mollusca* vol. 8 *Neurobiology and Behavior* part I (Willows A. O. D., ed.) Academic, New York.

Murphy A. D. and Kater S. B. (1980) Sprouting and functional regeneration of an identified neuron in *Helisoma. Brain Res.* **186**, 251–272.

Murphy A. D., Hadley R. D., and Kater S. B. (1983) Axotomy-induced parallel increases in electrical and dye coupling between identified neurons of *Helisoma. J. Neurosci.* **3**, 1422–1429.

Murphy A. D., Barker D. L., Loring J. F., and Kater S. B. (1985a) Sprouting and functional regeneration of an identified serotonergic neuron following axotomy. *J. Neurobiol.* **16**, 137–151.

Murphy A. D., Lukowiak K., and Stell W. K. (1985b) Peptidergic modulation of patterned motor activity in identified neurons of *Helisoma. Proc. Natl. Acad. Sci. USA* **82**, 7140–7144.

Pinsker H. M. (1982) Integration of reflex activity and central pattern generation in intact *Aplysia. J. Physiol.* (Paris) **78**, 775–785.

Richmond J. E., Bulloch A. G. M., Murphy A. D., and Lukowiak K. (1985) Possible modulatory roles of arginine vasotocin on feeding motor output in *Helisoma. Soc. Neurosci. Abst.* **11**, 479.

Richmond J. E., Murphy A. D., Bulloch A. G. M., and Lukowiak K. (1986) Evidence for an excitatory effect of GABA on feeding patterned motor activity (PMA) of *Helisoma Trivolis. Soc. Neurosci. Abst.* **12**, 792.

Ruben P., Goldberg J., Edstrom J., Voshart K., and Lukowiak K. (1981) What the marine mollusc *Aplysia* can tell the neurologist about behavioral neurophysiology. *Can. J. Neurol. Sci.* **8**, 275–280.

Scheller R. H., Kaldany R. R., Kreiner T., Mahon A. C., Nambu J. R., Schaefer M., and Taussig R. (1984) Neuropeptides: Mediators of behavior in *Aplysia*. *Science* **225**, 1300–1308.

Thornhill J. A., Lukowiak K., Cooper K. E., Veale W. L., and Edstrom J. P. (1981) Arginine vasotocin, an endogenous neuropeptide of *Aplysia*, suppresses the gill withdrawal reflex and reduces the evoked synaptic input to central gill motor neurons. *J. Neurobiol.* **12**, 533–544.

Trimble D. L. and Barker D. L. (1984) Activation by dopamine of patterned motor output from the buccal ganglia of *Helisoma trivolis*. *J. Neurobiol.* **15**, 37–48.

Index